GEOTECHNICAL ENGINEERING

[SOIL MECHANICS]

GEOTECHNICAL ENGINEERING

[SOIL MECHANICS]

(For Students of B.E./B.Tech., Civil Engineering)

Prof. T.N. RAMAMURTHY

Department of Civil Engineering
R.V. College of Engineering
Bangalore – 560 059

Prof. T.G. SITHARAM

Department of Civil Engineering
Indian Institute of Science
Bangalore – 560 012

S. CHAND & COMPANY LTD.

(AN ISO 9001 : 2000 COMPANY)
RAM NAGAR, NEW DELHI-110055

S. CHAND & COMPANY LTD.

(An ISO 9001 : 2000 Company)

Head Office: 7361, RAM NAGAR, NEW DELHI - 110 055
Phone: 23672080-81-82, 9899107446, 9911310888
Fax: 91-11-23677446

Shop at: **schandgroup.com**; e-mail: **info@schandgroup.com**

Branches :

AHMEDABAD : 1st Floor, Heritage, Near Gujarat Vidhyapeeth, Ashram Road, **Ahmedabad** - 380 014, Ph: 27541965, 27542369, ahmedabad@schandgroup.com

BENGALURU : No. 6, Ahuja Chambers, 1st Cross, Kumara Krupa Road, **Bengaluru** - 560 001, Ph: 22268048, 22354008, bangalore@schandgroup.com

BHOPAL : Bajaj Tower, Plot No. 243, Lala Lajpat Rai Colony, Raisen Road, **Bhopal** - 462 011, Ph: 4274723. bhopal@schandgroup.com

CHANDIGARH : S.C.O. 2419-20, First Floor, Sector - 22-C (Near Aroma Hotel), **Chandigarh** -160 022, Ph: 2725443, 2725446, chandigarh@schandgroup.com

CHENNAI : 152, Anna Salai, **Chennai** - 600 002, Ph: 28460026, 28460027, chennai@schandgroup.com

COIMBATORE : Plot No. 5, Rajalakshmi Nagar, Peelamedu, **Coimbatore** -641 004, (M) 09444228242, coimbatore@schandgroup.com **(Marketing Office)**

CUTTACK : 1st Floor, Bhartia Tower, Badambadi, **Cuttack** - 753 009, Ph: 2332580; 2332581, cuttack@schandgroup.com

DEHRADUN : 1st Floor, 20, New Road, Near Dwarka Store, **Dehradun** - 248 001, Ph: 2711101, 2710861, dehradun@schandgroup.com

GUWAHATI : Pan Bazar, **Guwahati** - 781 001, Ph: 2738811, 2735640 guwahati@schandgroup.com

HYDERABAD : Padma Plaza, H.No. 3-4-630, Opp. Ratna College, Narayanaguda, **Hyderabad** - 500 029, Ph: 24651135, 24744815, hyderabad@schandgroup.com

JAIPUR : A-14, Janta Store Shopping Complex, University Marg, Bapu Nagar, **Jaipur** - 302 015, Ph: 2719126, jaipur@schandgroup.com

JALANDHAR : Mai Hiran Gate, **Jalandhar** - 144 008, Ph: 2401630, 5000630, jalandhar@schandgroup.com

JAMMU : 67/B, B-Block, Gandhi Nagar, **Jammu** - 180 004, (M) 09878651464 **(Marketing Office)**

KOCHI : Kachapilly Square, Mullassery Canal Road, Ernakulam, **Kochi** - 682 011, Ph: 2378207, cochin@schandgroup.com

KOLKATA : 285/J, Bipin Bihari Ganguli Street, **Kolkata** - 700 012, Ph: 22367459, 22373914, kolkata@schandgroup.com

LUCKNOW : Mahabeer Market, 25 Gwynne Road, Aminabad, **Lucknow** - 226 018, Ph: 2626801, 2284815, lucknow@schandgroup.com

MUMBAI : Blackie House, 103/5, Walchand Hirachand Marg, Opp. G.P.O., **Mumbai** - 400 001, Ph: 22690881, 22610885, mumbai@schandgroup.com

NAGPUR : Karnal Bag, Model Mill Chowk, Umrer Road, **Nagpur** - 440 032, Ph: 2723901, 2777666 nagpur@schandgroup.com

PATNA : 104, Citicentre Ashok, Govind Mitra Road, **Patna** - 800 004, Ph: 2300489, 2302100, patna@schandgroup.com

PUNE : 291/1, Ganesh Gayatri Complex, 1st Floor, Somwarpeth, Near Jain Mandir, **Pune** - 411 011, Ph: 64017298, pune@schandgroup.com **(Marketing Office)**

RAIPUR : Kailash Residency, Plot No. 4B, Bottle House Road, Shankar Nagar, **Raipur** - 492 007, Ph: 09981200834, raipur@schandgroup.com **(Marketing Office)**

RANCHI : Flat No. 104, Sri Draupadi Smriti Apartments, East of Jaipal Singh Stadium, Neel Ratan Street, Upper Bazar, **Ranchi** - 834 001, Ph: 2208761, ranchi@schandgroup.com **(Marketing Office)**

SILIGURI : 122, Raja Ram Mohan Roy Road, East Vivekanandapally, P.O., **Siliguri**-734001, Dist., Jalpaiguri, (W.B.) Ph. 0353-2520750 **(Marketing Office)**

VISAKHAPATNAM: Plot No. 7, 1st Floor, Allipuram Extension, Opp. Radhakrishna Towers, Seethammadhara North Extn., **Visakhapatnam** - 530 013, (M) 09347580841, visakhapatnam@schandgroup.com **(Marketing Office)**

First Edition 2005
Revised Editions 2008
Third Revised Edition 2010
Reprint 2011 (Twice)

ISBN : 81-219-2457-X   **Code** : 10A 301

PRINTED IN INDIA
*By Rajendra Ravindra Printers Pvt. Ltd., 7361, Ram Nagar, New Delhi -110 055
and published by S. Chand & Company Ltd., 7361, Ram Nagar, New Delhi -110 055.*

PREFACE TO THE THIRD REVISED EDITION

In this book, Geotechnical Engineering the topics related to soil mechanics have been covered. At the end of the book, a chapter on stability of slopes has been included as most of the universities cover this in the first course of Geotechnical Engineering. The contents of this volume are written at a basic level suitable for a first course in Geotechnical Engineering. This book highlights the basic principles of soil mechanics along with applications to many problems in Geotechnical Engineering. The material is covered in a very simple, clear and logical manner. A number of solved and exercise problems have been included in each chapter.

In this volume, contents have been presented in fourteen chapters. Chapter 1 provides a brief introduction along with history of soil mechanics. Basic terminology and interrelations along with index properties of soils are covered in Chapters 2 and 3 respectively. Chapters 4, 5 and 6 cover soil classification systems, soil formation & soil structure, and soil water respectively. Permeability and seepage analysis are covered in Chapters 7 and 8 respectively. Stress distribution in soil mass is presented in Chapter 9. Compaction, consolidation and shear strength of soils are covered in Chapters 10, 11 and 12 respectively. Chapter 13 deals with stability of slopes. At the last objective questions, additional problems and exercises are given.

During the preparation of this book many papers and books have been referred. The authors acknowledge all the individuals whose papers and books have been referred. We expect to receive suggestions from the readers both from faculty as well as students for the improvement of the book, which will be highly appreciated. Finally, we thank our family members for their patience and encouragement.

T.N. RAMAMURTHY

T.G. SITHARAM

CONTENTS

LIST OF SYMBOLS

a_c	air content	q	load per unit area
A_c	activity number	S_r	degree of saturation
a_v	coefficient of compressibility	S_n	Taylor stability number
C_m	meniscus correction	T_v	time factor
C_d	dispersing agent correction	u	pore pressure
C_T	temperature correction	$\bar{u}$	excess pore pressure
C_u	uniformity coefficient	V	volume of soil mass
C_c	compression index	V_a	volume of air
C_s	expansion index	V_w	volume of water
C_v	coefficient of consolidation	V_s	volume of soil solids
c	cohesion	V_v	volume of voids
D	particle size	W	weight of soil mass
D_{10}	effective size	W_w	weight of water
d	diameter of tube, size of voids	w	water content
D_f	depth factor	w_L	liquid limit
e	void ratio	w_p	plastic limit
e_w	water voids ratio	w_s	shrinkage limit
G	specific gravity of soil particles	Z	datum head
G_m	specific gravity of soil mass	γ_w	unit weight of water
h	total head	ρ_w	density of water
h_c	capillary rise	γ	bulk unit weight
h_w	pressure head	ρ	bulk density
I_d	density index	ρ_d	dry density
i	hydraulic gradient	γ_{sat}	dry unit weight
i_c	critical hydraulic gradient	γ_{sat}	saturated unit weight
i_e	exit gradient	γ_{sat}	saturated density
k	coefficient of permeability	γ'	submerged unit weight
m_v	coefficient of volume change	ρ'	submerged density
n	porosity	γ_s	unit weight of soil solids
n_a	percentage air voids	ρ_s	density of soil solids
N_f	number of flow channels	σ	total stress
N_d	number of potential drops	σ'	effective stress
p_s	seepage pressure	ϕ	angle of shearing resistance
		σ_d	deviator stress
		q_u	unconfined compressive strength

CHAPTER 1

INTRODUCTION

1.1 DEFINITIONS

a_v	coefficient of compressibility		p	load per unit area
A	activity number		S_r	degree of saturation
a_s			S_n	Taylor stability number
C_m	meniscus correction		I	flute factor
C_a	dispersing agent correction		u	pore pressure
C_t	temperature correction		$\bar{u}$	excess pore pressure
C_u	uniformity coefficient		V	volume of soil mass
C_c	compression index		V_a	volume of air
C_e	expansion index		V_w	volume of water
C_v	coefficient of consolidation		V_s	volume of soil solids
c	cohesion		V_v	volume of voids
D	particle size		W	weight of soil mass
D_{10}	effective size		W_w	weight of water
d	diameter of tube, size of voids		w	water content
D_f	depth factor		w_L	liquid limit
e	void ratio		w_P	plastic limit
e_w	water voids ratio		w_s	shrinkage limit
G	specific gravity of soil particles		Z	datum head
G_m	specific gravity of soil mass		γ_w	unit weight of water
h	total head		ρ_w	density of water
h_c	capillary rise		γ	bulk unit weight
h_w	pressure head		ρ	bulk density
I_d	density index		ρ_d	dry density
i	hydraulic gradient		γ_d	dry unit weight
i_c	critical hydraulic gradient		γ_{sat}	saturated unit weight
i_e	exit gradient		ρ_{sat}	saturated density
k	coefficient of permeability		γ'	submerged unit weight
m_v	coefficient of volume change		ρ'	submerged density
n	porosity		γ_s	unit weight of soil solids
n_a	percentage air voids		ρ_s	density of soil solids
N_f	number of flow channels		σ	total stress
N_d	number of potential drops		σ'	effective stress
p_s	seepage pressure		ϕ	angle of shearing resistance
			σ_d	deviator stress
			q_u	unconfined compressive strength

CHAPTER 1

INTRODUCTION

1.1 Definition of Soil

The term 'Soil', which has originated from the Latin word 'Solum' has different meanings to different professional groups. To an agriculturist, it means the loose material lying on earth's surface, formed by disintegration of rocks, with an admixture of organic matter, which can support plant life. To the geologist, it means the disintegrated rock material overlying the parent rock. To the civil engineer, it means all the inorganic material on the earth's surface, produced by weathering of rocks, being either 'residual' or 'transported'. It may or may not contain an admixture of organic matter. Both soil and rock consist of mineral grains. But the bond between the mineral grains in soil is relatively weak compared to the strong bond between mineral grains in rock. Terzaghi based on this distinction defined soil as a natural aggregate of mineral grains which can be separated by such gentle mechanical means as agitation in water.

1.2 Soil Mechanics, Rock Mechanics and Geotechnical Engineering

Soil mechanics as the name indicates deals with the mechanics of soils. It has developed as a branch of mechanics involving the application of both mechanics of solids and mechanics of fluids to problems involving soils. Of late due attention is being paid to the particulate nature of a soil medium, while studying its behaviour.

Rock mechanics has been developed necessitated by situations where heavy loads from structures on ground have to be transferred to rock below and also in the case of underground structures in rock.

The name 'soil engineering' was coined to denote that branch of engineering which deals with the application of principles of soil mechanics to engineering problems involving soils, such as in design and construction of foundations of structures, earthdams, highway embankments, etc.

Earlier to this, the name 'foundation engineering' was being used instead of soil engineering, even though foundation engineering should be restricted to problems dealing with foundations of structures.

The name in current usage is 'geotechnical engineering'. It has a much wider scope and refers in total to all engineering problems involving soil and/or rock as foundation material and construction material. Geotechnical engineering involves the application of principles of soil mechanics, rock mechanics and engineering geology to engineering problems involving soils and rocks.

Geotechnical engineering practice being quite vast, only a brief list of the various areas of applications is presented below:

1. Shallow foundations
2. Deep foundations
3. Machine foundations
4. Underground and earth retaining structures
5. Embankments & cuts for highways, railways and canals
6. Earth and rockfill dams

1.3   Brief History of Development of Soil Mechanics

The knowledge of the use of soil for construction of dwellings and roads dates back to prehistoric times. Soil was treated as foundation material under structures and construction material for embankments based on knowledge gained through trial and error approach.

Even as early as 2000 BC caissons of timber and stone are known to have been used in construction of structures on soft ground. The famous hanging gardens of Babylon consisted of terraces supported by huge retaining walls, which would have required some knowledge of earth pressure, though empirical.

The Romans are credited to have built large public buildings and other utilities such as harbours, aqueducts, bridges, network of roads and sewer lines wherein empirical knowledge of soil behaviour is known to have been applied as evidenced by writings of Roman engineer Vitruvious in first century BC. In India also, during the medieval period, some literature on construction techniques were published. Notable among them were the works of Mansar and Visvakarma. To tackle compressible soils Mansar had recommended the compaction of soil by cows and oxen and dewatering of foundations.

Many structures constructed on clays during the medieval period are reported to have undergone large settlements, which indicate lack of knowledge of consolidation during those days. A glaring example, for this, is the Leaning Tower of Pisa constructed between 1174 and 1350 AD.

In India the Taj Mahal constructed between 1632 and 1650 A.D. has its various portions supported on masonry cylindrical wells. In the later part of the 17th century French engineers engaged in large scale construction activity involving improvement of highways and building of canals, besides bridges and large buildings. The credit for the first major contribution to the scientific study of soil as engineering material goes to Coulomb (1776) who gave his wedge theory of earth pressure. Later Poncelet (1788-1867) gave the graphical procedure for finding earth pressure based on Coulomb's theory. In 1856 the famous Darcy's law and Stoke's law were presented. Rankine in 1857 presented his theory of earth pressure. Boussinesq analysis for determining stresses inside soil mass due to surface point loads was presented in 1885. In 1911, Atterberg defined the four states of consistency in the case of clay soils. Resal (1910) and Bell (1915) extended Rankine's analysis of earth pressure to cohesive soils. In 1920, Prandtl published his theory of plastic equilibrium based on which various bearing capacity theories were formulated later. Pavlovsky is said to have given solutions to problems of seepage below hydraulic structures in 1922-23. The work being in Russian it remained unknown to the world at that time. In 1923, Terzaghi published his theory of consolidation. In 1925 Terzaghi coined the name 'Soil Mechanics' when he published his first book 'Erdbaumechanik' in German. In recognition of his immense contribution to the development of Soil Mechanics, Terzaghi is called the 'Father of Soil Mechanics'. Following Terzaghi's work many notable contributions have been made to the development of this field, like those of R.R. Proctor and A. Casagrande.

Several countries in the world have formed national bodies to monitor research and development in the field of Geotechnical Engineering. In India we have the Indian Geotechnical Society affiliated to the International Society of Soil Mechanics and Geotechnical Engineering. So far fourteen International Conferences have been held under the auspices of the international body: at Harvard, Massachusetts (USA) 1936, Rotterdam (Netherlands) 1948, Zurich (Switzerland) 1953, London (U.K.) 1957, Paris (France) 1961, Montreal (Canada) 1965, Mexico City (Mexico) 1969, Moscow (USSR) 1973, Tokyo (Japan) 1977, Stockholm (Sweden) 1981, San Francisco (USA) 1985, Rio de Janeiro (Brazil) 1989, New Delhi (India) 1993, Hamburg (Germany) 1997, Istanbul (Turkey) 2001. These have helped in consolidating the efforts of research and practice from different parts of the world.

1.4   SI Units

SI Units are being used all over the world and to help the student the following units in SI system are summarised.

Unit of Force: The unit of force is Newton. It is the force which when acting on a mass of 1 kg produces an acceleration of 1 m/sec^2.

$$1 \text{ kilo-Newton (1kN)} = 10^3 \text{ N}$$

$$1 \text{ mega-Newton (1MN)} = 10^6 \text{ N}$$

Unit of Pressure or Stress: The unit of pressure or stress is N/m^2 which is known as Pascal (Pa)

$$1 \, Pa = 1 \text{ N/m}^2$$

$$1 \, kPa = 1000 \, Pa = 1 \text{ kN/m}^2$$

Unit of Energy: The unit of energy is Joule (J) which is defined as the work done when a force of 1 N moves through a distance of 1 m.

$$1 \, J = 1 \, N - m$$

Unit of Power: The unit of power is 1 Joule per second, which is known as Watt.

$$1 \text{ Watt } = 1 \text{ Joule/sec}$$

Unit of Temperature: The unit of temperature is Kelvin (K)

Conversion from MKS to SI

$$1 \, Kgf = 9.81 \text{ N}$$

$$1 \, Kgf\text{-}m = 9.81 \text{ J}$$

$$1 \, Kgf\text{-m/sec} = 9.81 \text{ W}$$

$$^\circ C + 273^\circ = \,^\circ K$$

EXERCISE – 1

1.1   Define the term 'soil' from an engineering point of view.

1.2   What is the role of a geotechnical engineer in civil engineering practice?

CHAPTER 2

BASIC TERMINOLOGY AND INTERRELATIONS

2.1 Soil mass

A soil mass is an aggregation of soil particles forming a porous structure. The soil particles in a soil mass are often referred to as soil solids. The pores, often referred to as voids, may be filled with air (or some other gas), water (or some other liquid) or both.

2.2 Soil mass as a three phase system

A partially saturated soil mass has its voids filled partially with water and the remaining with air. Thus the three constituents of a partially saturated soil mass are soil solids, water and air. Fig. 2.1 illustrates a partially saturated soil mass as a three phase system. It is only an idealised representation as the three constituents can never be shown to exist separately in practice.

2.3 Soil mass as a two phase system

A dry soil mass has its voids filled with air only. A fully saturated soil mass has its voids filled with water only.

Fig. 2.2 (a) and (b) are idealised representations of a dry soil mass and a fully saturated soil mass respectively, as two phase systems.

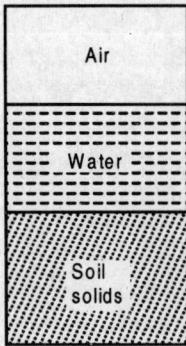

Fig. 2.1. Three phase system

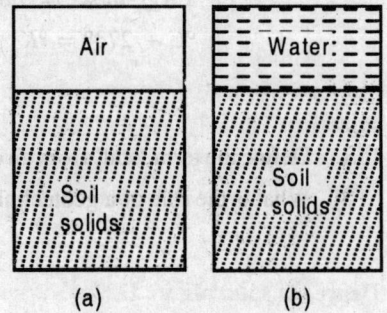

(a)   (b)

Fig. 2.2. Two phase systems

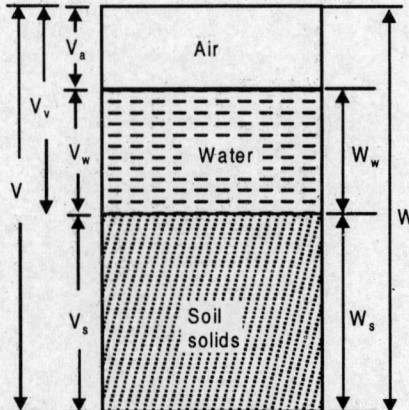

Fig. 2.3. Notations

4

2.4 Notations

The fundamental notations used in further discussions are illustrated in Fig. 2.3.

The explanation of the symbols used are as follows:

V_a = volume of air

V_w = volume of water

V_s = volume of soil solids

V_v = volume of voids

V = total volume of soil mass

W_w = weight of water

W_s = weight of soil solids

W = total weight of soil mass

No symbol is shown for weight of air which is considered negligible compared to that of soil solids.

From Fig. 2.3 it is clear that

$$V_v = V_a + V_w$$
$$V = V_v + V_s$$
$$W = W_w + W_s$$

2.5 Void ratio, Porosity, Degree of saturation, Air content and Percentage air voids

Void ratio: Void ratio, e of a soil mass is defined as the ratio of volume of voids, V_v to the volume of soil solids, V_s. It is expressed as a number in decimal form.

$$e = \frac{V_v}{V_s}$$

Although the void ratio can theoretically range from zero to infinity, the common range is between 0.5 and 0.9 for coarse-grained soils and between 0.7 and 1.5 for fine-grained soils.

Porosity : Porosity, n is defined as the ratio of volume of voids, V_v to the total volume of soil mass, V. It is usually expressed as a percentage and is also referred to as percentage voids. It falls in the range $0 < n < 100\%$.

$$n = \frac{V_v}{V} \times 100\%$$

Degree of saturation : Degree of saturation, S_r is defined as the ratio of volume of water, V_w present in a soil mass to the volume of voids, V_v. It is usually expressed as a percentage and is also referred to as percent saturation.

$$S_r = \frac{V_w}{V_v} \times 100\%$$

The degree of saturation ranges from 0 for dry soil mass to 100% for fully saturated soil mass.

Air content : Air content, a_c is defined as the ratio of volume of air, V_a present in soil mass to the volume of voids, V_v. It is expressed as a number in decimal form.

$$a_c = \frac{V_a}{V_v}$$

Further,

$$a_c = \frac{V_a}{V_v} = \frac{V_v - V_w}{V_v} = 1 - \frac{V_w}{V_v} = 1 - S_r$$

$$a_c = 1 - S_r$$

Percentage air voids : Percentage air voids, n_a is defined as the ratio of volume of air, V_a present in soil mass to the total volume of soil mass, V expressed as a percentage.

$$n_a = \frac{V_a}{V} \times 100\%$$

2.6   Water content

Water content, w of a soil mass is defined as the ratio of weight of water, W_w present in soil mass to the weight of soil solids, W_s. It is usually expressed as a percentage. It is also referred to as moisture content

$$w = \frac{W_w}{W_s} \times 100\%$$

The water content theoretically falls in the range 0 to ∞. However, in sands it is found to vary from about 10% to 30% and in clays from about 5% to possibly over 300%, the larger values being associated with very fine-grained clays when they are very loosely deposited.

2.7 Unit weights and densities

Unit weight of water : Unit weight of water, γ_w is the ratio of weight of a given volume of water, W_w to the volume of water, V_w at a stated temperature.

$$\gamma_w = \frac{W_w}{V_w}$$

Also density of water, $$\rho_w = \frac{M_w}{V_w}$$

ρ_w or γ_w depends on temperature and its value is 1 gm/cc or 9.81 kN/m³ at 4° C. The specific gravity of distilled water decreases from 0.999992 at 5° C to 0.99024 at 45° C. For practical purposes it may be taken as 1.

Bulk unit weight : Bulk unit weight of soil mass, γ is the ratio of weight of soil mass, W to the volume of soil mass, V

$$\gamma = \frac{W}{V}$$

In the case of saturated soil mass, γ is also referred to as moist unit weight.

Bulk density : Bulk density of soil mass, ρ is the mass per unit volume of soil mass.

$$\rho = \frac{M}{V}$$

Dry unit weight : Dry unit weight of soil mass, γ_d is the ratio of weight of soil solids, W_s to the volume of soil mass, V.

$$\gamma_d = \frac{W_s}{V}$$

Dry density : Dry density of soil mass is the ratio of mass of soil solids, M_s to the volume of soil mass, V.

$$\rho_d = \frac{M_s}{V}$$

Saturated unit weight : Saturated unit weight of soil mass is the ratio of weight of fully saturated soil mass, W_{sat} to the total volume of soil mass, V.

$$\gamma_{sat} = \frac{W_{sat}}{V}$$

Saturated density : Saturated density of soil mass, ρ_{sat} is the ratio of mass of fully saturated soil mass, M_{sat} to the total volume of soil mass, V.

$$\rho_{sat} = \frac{M_{sat}}{V}$$

For fully saturated soil mass, $\gamma = \gamma_{sat}$ or $\rho = \rho_{sat}$

Submerged unit weight : Submerged unit weight of soil mass, γ' is the ratio of submerged weight of soil mass, W_{sub} to the total volume of soil mass, V.

$$\gamma' = \frac{W_{sub}}{V}$$

Submerged unit weight is also known as bouyant unit weight or effective unit weight.

Submerged density : Submerged density of soil mass, ρ' is the ratio of submerged mass of soil mass, M_{sub} to the total volume of soil mass, V.

$$\rho' = \frac{M_{sub}}{V}$$

We can show that $\gamma' = \gamma_{sat} - \gamma_w$ or $\rho' = \rho_{sat} - \rho_w$

Unit weight of soil solids : Unit weight of soil solids, γ_s is the ratio of weight of soil solids, W_s to the volume of soil solids, V_s in a given soil mass.

$$\gamma_s = \frac{W_s}{V_s}$$

Density of soil solids : Density of soil solids is the mass of soil solids per unit volume of soil solids, in a given soil mass.

$$\rho_s = \frac{M_s}{V_s}$$

Note: To convert density to unit weight the following relation is convenient to remember.

$$1 \, gm/cm^3 \Rightarrow 9.81 kN/m^3$$

2.8 Specific gravity

Specific gravity of soil particles : Specific gravity of soil particles is defined as the ratio of the weight of a given volume of soil particles to the weight of an equivalent volume of pure water at a stated temperature. It is denoted by G_s or more usually G. It can also be defined as the ratio of unit weight of soil particles to the unit weight of pure water at a stated temperature.

$$G = \frac{\gamma_s}{\gamma_w} \, or \, \frac{\rho_s}{\rho_w}$$

Table 2.1 gives typical values of specific gravity of soil particles for selected soil types.

The specific gravities of different mineral constituents vary rather widely but that of the majority of soil particles range between 2.6 and 2.8.

Table 2.1: Typical values of specific gravity of soil particles

Soil type	Value of G
Sand	2.64 – 2.67
Silt	2.68 – 2.70
Clay	2.70 – 2.80
Soil containing mica or iron	2.85 – 2.90
Organic soils	1.26 – 2.20

Specific gravity of soil mass : Specific gravity of soil mass is the ratio of bulk unit weight of soil mass to the unit weight of pure water at a stated temperature. It is denoted by G_m.

$$G_m = \frac{\gamma}{\gamma_w} \, or \, \frac{\rho}{\rho_w}$$

2.9 Interrelations

1.   Relation between e and n

We have $e = \dfrac{V_v}{V_s}$. If we represent V_s by unity then

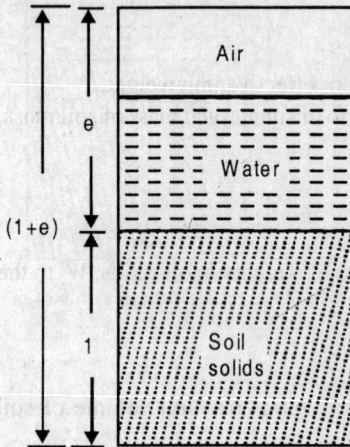

Fig. 2.4. Soil element in terms of e

$V_v = e$ and

$V = (1 + e)$ as illustrated in Fig. 2.4.

Referring to Fig. 2.4, Porosity,

$$n = \frac{V_v}{V} = \frac{e}{1 + e}$$

$$n = \frac{e}{1 + e} \qquad \qquad ....(2.1)$$

Relation (2.1) can also be derived for (i) dry soil mass and (ii) fully saturated soil mass as shown in the following discussion.

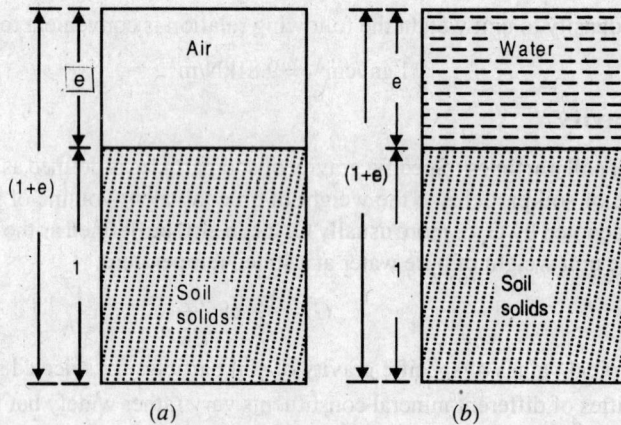

(a) (b)

Fig. 2.5. Soil element in terms of e

Referring to Fig 2.5 (a) and (b)

Porosity, $n = \dfrac{V_v}{V} = \dfrac{e}{1 + e}$

Further, taking reciprocals on both sides of Eq. 2.1, we obtain

$$\frac{1}{n} = \frac{1 + e}{e} = \frac{1}{e} + 1$$

$$\frac{1}{e} = \frac{1}{n} - 1 = \frac{1 - n}{n}$$

$$e = \frac{n}{1 - n} \qquad \qquad ...(2.2)$$

Alternatively, we have $n = \dfrac{V_v}{V}$

If we represent V by unity, then $V_v = n$ and $V_s = (1 - n)$ as illustrated in Fig 2.6.

Referring to Fig. 2.6,

$$\text{Void ratio } e = \frac{V_v}{V_s} = \frac{n}{1 - n}$$

2. Relation between e, S_r, w and G.

The symbol, e_w used for volume of water in Fig. 2.7 is referred to as water-voids ratio.

$$S_r = \frac{V_w}{V_v} = \frac{e_w}{e}$$

$\therefore$

$$e_w = e.\,S_r \qquad\qquad ....(2.3)$$

Referring to Fig. 2.7,

$$w = \frac{W_w}{W_s} = \frac{V_w\,\gamma_w}{V_s\,\gamma_s} = \frac{e_w.\gamma_w}{1.\gamma_s}$$

Substituting $e_w = e.\,S_r$ and $\gamma_s = G\gamma_w$,

we obtain

$$w = \frac{e\,S_r.\gamma_w}{G\gamma_w} = \frac{e\,S_r}{G}$$

$$e S_r = wG \qquad ....(2.4)$$

3. Relation between γ, γ_d and w

We have

$$w = \frac{W_w}{W_s}$$

Adding 1 to both sides,

$$1 + w = \frac{W_w}{W_s} + 1 = \frac{W_w + W_s}{W_s} = \frac{W}{W_s}$$

$\therefore$

$$W_s = \frac{W}{1 + w} \qquad\qquad ......(2.5)$$

Dividing both sides by V,

$$\frac{W_s}{V} = \frac{W/V}{1 + w}$$

$$\gamma_d = \frac{\gamma}{1 + w} \qquad\qquad .....(2.6)$$

4. Expressions for γ, γ_d, γ_{sat} and γ'

Referring to Fig. 2.7, we have $\quad \gamma = \dfrac{W}{V} = \dfrac{W_s + W_w}{V} = \dfrac{V_s\gamma_s + V_w.\gamma_w}{V}$

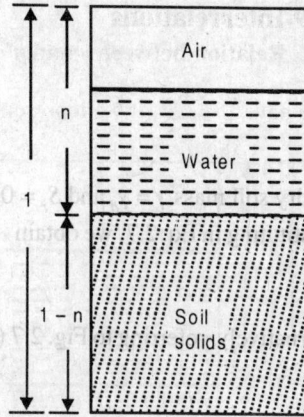

Fig. 2.6. Soil element in terms of n.

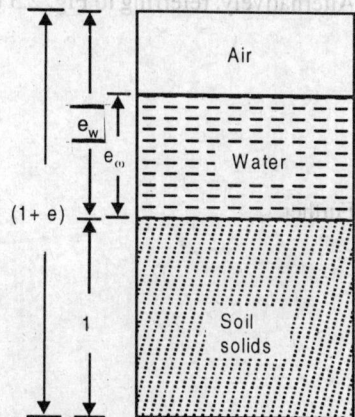

Fig. 2.7. Soil element in terms of e.

$$= \frac{1.\gamma_s + e_w\gamma_w}{1+e} = \frac{G\gamma_w + e\bar{S}_r\,\gamma_w}{1+e}$$

$$\gamma = \frac{(G + eS_r)\gamma_w}{1+e} \qquad\qquad ....(2.7)$$

For dry soil mass $\gamma = \gamma_d$ and $S_r = 0$

Substituting in Eq. 2.7, we obtain

$$\gamma_d = \frac{G\gamma_w}{1+e} \qquad\qquad ....(2.8)$$

Alternatively, referring to Fig. 2.7 (for three phase system) or Fig. 2.5 (for two phase system) we have

$$\gamma_d = \frac{W_s}{V} = \frac{V_s\gamma_s}{V} = \frac{1.\gamma_s}{1+e} = \frac{G\gamma_w}{1+e}$$

For fully saturated soil mass $\gamma = \gamma_{sat}$ and $S_r = 1$. Substituting in Eq. 2.7, we obtain

$$\gamma_{sat} = \frac{(G + e)\gamma_w}{1+e} \qquad\qquad ....(2.9)$$

Alternatively, referring to Fig. 2.5 (b), we have

$$\gamma_{sat} = \frac{W_{sat}}{V} = \frac{W_s + W_w}{V} = \frac{V_s\gamma_s + V_w\gamma_w}{V} = \frac{1.\gamma_s + e.\gamma_w}{1+e}$$

$$= \frac{G\gamma_w + e.\gamma_w}{1+e} = \frac{(G + e)\gamma_w}{1+e}$$

Further,

$$\gamma' = \gamma_{sat} - \gamma_w = \frac{(G + e)\gamma_w}{1+e} - \gamma_w$$

$$= \frac{(G + e)\gamma_w - \gamma_w(1+e)}{1+e} = \frac{G\gamma_w - \gamma_w}{1+e}$$

$$\gamma' = \frac{(G - 1)\gamma_w}{1+e} \qquad\qquad ....(2.10)$$

5. Expression for e in terms of G, γ_w and γ_d

We have $\qquad\qquad \gamma_d = \dfrac{G\gamma_w}{1+e} \qquad\qquad$ (Refer derivation of Eq. 2.8)

$$1 + e = \frac{G\gamma_w}{\gamma_d}$$

$$e = \frac{G\gamma_w}{\gamma_d} - 1 \qquad\qquad ....(2.11)$$

6. Expression for γ_d in terms of n_a, G, γ_w and w

We have $\qquad\qquad V = V_a + V_w + V_s$

Dividing by V,

$$1 = \frac{V_a}{V} + \frac{V_w}{V} + \frac{V_s}{V}$$

$$= n_a + \frac{W_w}{\gamma_w V} + \frac{W_s}{\gamma_s V}$$

$$(1 - n_a) = \frac{\omega W_s}{\gamma_w V} + \frac{W_s}{G\gamma_w V} = \frac{w.\gamma_d}{\gamma_w} + \frac{\gamma_d}{G\gamma_w}$$

$$= \frac{\gamma_d}{\gamma_w}\left(w + \frac{1}{G}\right) = \frac{\gamma_d}{\gamma_w}\left(\frac{wG+1}{G}\right)$$

$$\therefore \qquad \gamma_d = \frac{(1 - n_a)G\gamma_w}{1 + wG} \qquad \qquad ...(2.12)$$

Example 2.1

A soil sample in its undisturbed state was found to have volume of 105 cm³ and mass of 201g. After oven drying the mass got reduced to 168g. Compute (i) water content, (ii) void ratio, (iii) porosity (iv) degree of saturation and (v) air content. Take $G = 2.7$.

Solution:

Volume of soil mass, $V = 105$ cm³

Mass of soil mass, $M = 201$ g

Mass of dry soil mass, $M_d = 168$ g

Specific gravity of soil particles, $G = 2.7$

(i) Water content, $\qquad w = \dfrac{M_w}{M_s} = \dfrac{M - M_d}{M_d} = \dfrac{201 - 168}{168} = 0.196 = 19.6\%$

(ii) Dry density, $\qquad \gamma_d = \dfrac{M_s}{V} = \dfrac{M_d}{V} = \dfrac{168}{105} = 1.6 g / cm^3$

Void ratio, $\qquad e = \dfrac{G\gamma_\omega}{\gamma_d} - 1 = \dfrac{2.7(1)}{1.6} - 1 = 0.69$

Example 2.2

For a soil sample the specific gravity of soil mass is 1.7 and specific gravity of soil particles is 2.7. Determine the void ratio (i) assuming the soil sample is dry and (ii) the sample has a water content of 12 percent.

Solution: $G_m = 1.7 \qquad G = 2.7$

(i) For dry soil mass $\qquad \gamma = \gamma_d$

$$G_m = \frac{\gamma}{\gamma_w} = \frac{\gamma_d}{\gamma_w}$$

$$\gamma_d = G_m \gamma_w$$

Void ratio, $\qquad e = \dfrac{G\gamma_w}{\gamma_d} - 1 = \dfrac{G\gamma_w}{G_m\gamma_w} - 1 = \dfrac{G}{G_m} - 1$

$$= \frac{2.7}{1.7} - 1 = 0.59$$

(ii) When $w = 12\% = 0.12$

$$G_m = \frac{\gamma}{\gamma_w} \ or \ \gamma = G_m\gamma_w = (1.7)(9.81) = 16.68 \, kN / m^3$$

$$\gamma_d = \frac{\gamma}{1 + w} = \frac{16.68}{1 + 0.12} = 14.89 \, kN / m^3$$

$$e = \frac{G\gamma_w}{\gamma_d} - 1 = \frac{(2.7)(9.81)}{14.89} - 1 = 0.78$$

Example 2.3

Calculate the bulk unit weight of soil mass and specific gravity of soil particles for (*i*) a soil sample composed of only quartz and (*ii*) a soil sample composed of 65% quartz, 23% mica and 12% iron oxide if the average value of specific gravity of soil particles is 2.65 for quartz, 3.0 for mica and 3.8 for iron oxide. Assume that both soils are fully saturated and have void ratio of 0.6.

Solution: $S_r = 1$ $e = 0.6$

For quartz $G_{qa} = 2.65$

For mica $G_{mi} = 3.0$

For iron oxide $G_{io} = 3.8$

(a) Soil sample composed of only quartz

For the soil sample, $G = G_{qa} = 2.65$

Bulk unit weight, $\gamma = \gamma_{sat} = \dfrac{(G + e)\gamma_w}{1 + e} = \dfrac{(2.65 + 0.6)(9.81)}{1 + 0.6} = 19.93\,kN/m^3$

(b) Soil sample composed of 65% quartz, 23% mica and 12% iron oxide

$$\text{For the soil sample, } G = \frac{\gamma_s}{\gamma_w} = \frac{W_s}{V_s \gamma_w}$$

Weight of quartz particles, $W_{qa} = 0.65\,W_s$

Weight of mica particles, $W_{mi} = 0.23\,W_s$

Weight of iron oxide particles, $W_{io} = 0.12\,W_s$

Volume of quartz particles, $V_{qa} = \dfrac{W_{qa}}{\gamma_{qa}} = \dfrac{0.65\,W_s}{(2.65)\gamma_w} = 0.245\,\dfrac{W_s}{\gamma_w}$

Volume of mica particles, $V_{mi} = \dfrac{W_{mi}}{\gamma_{mi}} = \dfrac{0.23 W_s}{(3.0)\gamma_w} = 0.077\,\dfrac{W_s}{\gamma_w}$

Volume of iron oxide particles, $V_{to} = \dfrac{W_{io}}{\gamma_{io}} = \dfrac{0.12 W_s}{(3.8)\gamma_w} = 0.031\,\dfrac{W_s}{\gamma_w}$

$$V_s = V_{qa} + V_{mi} + V_{io} = (0.245 + 0.077 + 0.031)\,\frac{W_s}{\gamma_w} = 0.353\,\frac{W_s}{\gamma_w}$$

$$G = \frac{W_s}{V_s \gamma_w} = \frac{W_s}{0.353\dfrac{W_s}{\gamma_w}\gamma_w} = \frac{1}{0.353} = 2.83$$

$$\gamma = \gamma_{sat} = \frac{(G + e)\gamma_w}{1 + e} = \frac{(2.83 + 0.6)(9.81)}{1 + 0.6} = 21.03\,kN/m^3$$

Example 2.4

A soil sample assumed to consist of spherical grains all of same diameter will have maximum void ratio when the grains are arranged in a cubical array. Find void ratio and dry unit weight. Take unit weight of grains as 20 kN/m³.

Solution: We consider a unit cube packed with the spherical grains of diameter d.

Number of spherical grains in the container $= \dfrac{1}{d} \times \dfrac{1}{d} \times \dfrac{1}{d} = \dfrac{1}{d^3}$

Volume of each spherical grain $= \dfrac{\pi d^3}{6}$

Volume of soil solids, $\quad V_s = \dfrac{1}{d^3} \times \dfrac{\pi d^3}{6} = \dfrac{\pi}{6}$

Total volume of cube, $\quad V = 1 \times 1 \times 1 = 1\, m^3$

Volume of voids, $\quad V_v = V - V_s = 1 - \dfrac{\pi}{6} = \dfrac{6 - \pi}{6}$

Void ratio $\quad e = \dfrac{V_v}{V_s} = \dfrac{(6 - \pi)(6)}{6(\pi)} = \dfrac{6 - \pi}{\pi} = 0.91$

Dry unit weight $\quad \gamma_d = \dfrac{W_s}{V} = \dfrac{V_s \gamma_s}{V} = \dfrac{\pi}{6}(20) = 10.47\, kN/m^3$

Example 2.5

1000 m³ of earthfill is to be constructed. How many cubic metres of soil is to be excavated from borrow pit in which the void ratio is 0.95, if the void ratio of earthfill is to be 0.7?

Solution: Volume of earth fill, $V_1 = 1000\, m^3$

Void ratio of earth fill, $e_1 = 0.7$

Void ratio in borrow pit, $e_2 = 0.95$

Let volume of soil to be excavated from borrow pit = V_2

We have $\quad e = \dfrac{V_v}{V_s}$

Adding 1 to both sides, $\quad 1 + e = \dfrac{V_v}{V_s} + 1 = \dfrac{V_v + V_s}{V_s} = \dfrac{V}{V_s}$

For soil in earth fill, $\quad 1 + e_1 = \dfrac{V_1}{V_s}$ ...(i)

For soil to be excavated from borrow pit $\quad 1 + e_2 = \dfrac{V_2}{V_s}$ ...(ii)

Dividing (ii) by (i), $\dfrac{V_2}{V_1} = \dfrac{1 + e_2}{1 + e_1}$ ($\because$ V_s is same for earthfill and soil excavated from borrow pit)

$$V_2 = \left(\dfrac{1 + e_2}{1 + e_1} \right) V_1 = \left(\dfrac{1 + 0.95}{1 + 0.7} \right)(1000) = 1147\, m^3$$

Example 2.6

A sample of clay with a weight of 6.7 N was coated with paraffin wax. The combined weight of clay and wax was found to be 6.78 N. The volume of the wax coated sample was found, by immersion in water, to be $350 \times 10^3\, mm^3$. The sample was then broken open and moisture content was found to be 17%. If the specific gravities of soil particles and wax are 2.67 and 0.89, determine the bulk unit weight, void ratio and degree of saturation of the soil sample.

Solution:

Weight of clay sample,	$W = 6.7\, N$
Weight of soil sample + wax,	$W_1 = 6.78\, N$
Volume of wax coated soil sample,	$V_1 = 350 \times 10^3\, mm^3$
Water content of soil sample,	$w = 17\%$

Specific gravity of soil solids, $G = 2.67$

Specific gravity of paraffin wax, $G_p = 0.89$

We have

$$V_p = \frac{W_p}{\gamma_p} = \frac{W_p}{G_p \gamma_w} = \frac{6.78 - 6.7}{(0.89)(9.81 \times 10^{-6})} = 9.16 \times 10^3 \ mm^3$$

Volume of soil sample,

$$V = V_1 - V_p = (350 - 9.16) \times 10^3 = 340.84 \times 10^3 \ mm^3$$

Bulk unit weight,

$$\gamma = \frac{W}{V} = \frac{6.7}{340.84 \times 10^3} = 19.66 \times 10^{-6} \ N/mm^3$$

Dry unit weight

$$\gamma_d = \frac{\gamma}{1 + w} = \frac{19.66 \times 10^{-6}}{1 + 0.17} = 16.80 \times 10^{-6} \ N/mm^3$$

Void ratio,

$$e = \frac{G\gamma_w}{\gamma_d} - 1 = \frac{2.67 \times 9.81 \times 10^{-6}}{16.8 \times 10^{-6}} - 1 = 0.56$$

Degree of saturation,

$$S_r = \frac{wG}{e} = \frac{(0.17)(2.67)}{0.56} = 81.05\%$$

Example 2.7

In an earthen embankment under construction the bulk unit weight is 16.5 kN/m³ at water content of 11%. If the water content is to be raised to 15%, compute the quantity of water required to be added per cubic meter of soil? Assume no change in the void ratio.

Solution :

$$\gamma_1 = 16.5 \ kN/m^3$$
$$w_1 = 11\%$$
$$w_2 = 15\%$$
$$V = 1m^3$$

$$(\gamma_d)_1 = \frac{\gamma_1}{1 + w_1} = \frac{16.5}{1 + 0.11} = 14.86 \ kN/m^3$$

$$W_s = (\gamma_d)_1 . V = (14.86)(1) = 14.86 \ kN$$

$$w_1 = \frac{Ww_1}{W_s}$$

$$Ww_1 = w_1 . W_s = (0.11)(14.86) = 1.635 \ kN$$

$$Vw_1 = \frac{Ww_1}{\gamma_w} = \frac{1.635}{9.81} = 0.167 \ m^3 = 167 \ litres$$

Similarly,

$$Ww_2 = w_2 . W_s = (0.15)(14.86) = 2.229 \ kN$$

$$Vw_2 = \frac{Ww_2}{\gamma_w} = \frac{2.229}{9.81} = 0.227 \ m^3 = 227 \ litres$$

Answer: Required quantity of water to be added per cubic meter of soil

$$= 227 - 167 = 60 \ litres.$$

EXERCISE-2

2.1. Derive relation between void ratio and porosity for (i) dry soil mass and (ii) fully saturated soil mass.

2.2. A soil mass has a bulk unit weight of 20kN/m³ with water content of 19%, Compute dry unit weight, void ratio, porosity, degree of saturation and submerged unit weight. Take G = 2.7.

2.3. A fully saturated soil sample has a water content of 27% and bulk unit weight of 19.5kN/m³. Compute (i) dry unit weight and void ratio of soil mass, (ii) specific gravity of soil solids and (iii) bulk unit weight of soil sample when the degree of saturation is reduced to 80% without change in the void ratio.

2.4. How many cubic meters of earth fill can be constructed at a void ratio of 0.67 from 190000 m³ of borrow material that has a void ratio of 1.1?

2.5. With the usual notations show that

(a) $\gamma = \gamma_d + S_r (\gamma_{sat} - \gamma_d)$

(b) $S_r = \dfrac{w}{\dfrac{\gamma_w}{\gamma}(1+w) - \dfrac{1}{G}}$

2.6. A sampling tube of 38 mm internal diameter was used to extract a sample of cohesive soil from a test pit. The length of the extracted sample was 102 mm and it had a mass of 220 gm and water content of 18%. Find void ratio, porosity, degree of saturation and percentage air voids Take G =2.7

2.7. A soil sample with porosity of 38% has degree of saturation of 50%. Taking specific gravity of soil solids as 2.67, compute dry unit weight, saturated unit weight, submerged unit weight and bulk unit weight.

2.8. A sample of sand has a porosity of 40%. Taking G = 2.7 find (i) dry unit weight, (ii) saturated unit weight (iii) submerged unit weight and (iv) bulk unit weight when the degree of saturation is 60%.

2.9. A soil has been compacted in an embankment, at a bulk unit weight of 21.5 kN/m³ and water content of 12%. Taking G = 2.65, calculate dry density, void ratio, degree of saturation and air content.

2.10. Distinguish between

(a) Water content and air content

(b) Per cent voids and percentage air voids

(c) Bulk density and bulk unit weight

(d) Saturated unit weight and submerged unit weight.

(e) Specific gravity of soil particles and specific gravity of soil mass.

CHAPTER 3

INDEX PROPERTIES OF SOILS

3.1 Introduction

The various properties of soils are grouped under two heads: index properties and engineering properties. The three engineering properties of soils are permeability, compressibility and shear strength. Index properties of soils are those soil properties which are mainly used in the identification and classification of soils and help the geotechnical engineer in predicting the suitability of soils as foundation/construction material. Following is the list of index properties:

1. Specific gravity of soil particles
2. Particle size distribution
3. Consistency limits and indices
4. Density index

The determination of water content and field density have been included in this chapter as they are required in computations relating to other properties.

3.2 Specific Gravity

Specific gravity of soil solids is useful in several computations. For example, it is used to compute void ratio, degree of saturation, different unit weights etc. It finds application in computations in seepage and consolidation problems. Hence it should be determined with sufficient accuracy.

The specific gravity of soil solids is determined by using a 50 ml/100 ml density bottle or a 500 ml pycnometer. While the density bottle method is more accurate and is suitable for all types of soils, the pycnometer method is used in the case of coarse grained soils only. Both the methods involve the same sequence of observations. The mass M_1 of the clean, dry bottle is found. Suitable quantity of oven-dried soil sample, cooled in a desiccator is put in the bottle and the mass M_2 of the bottle with soil is found. Distilled water is then added to the soil inside bottle until the bottle is full, care being taken to see that entrapped air is fully expelled (either by applying vacuum or by gentle heating and shaking). The mass M_3 of the bottle with soil and water is found. The bottle is then emptied of its contents, cleaned and filled with distilled water only. The outer surface of the bottle is wiped dry and the mass M_4 of the bottle with water is found. The specific gravity of soil solids is computed as :

$$G = \frac{(M_2 - M_1)}{(M_2 - M_1) - (M_3 - M_4)}$$

Derivation: If we remove mass of water whose volume is equivalent to that of soil solids from mass M_4 and add mass of soil solids, M_d we get M_3.

$$\text{Mass of equivalent volume of water} = \left(\frac{M_s}{\gamma_s}\right)\gamma_w = \left(\frac{M_d}{G\gamma_w}\right)\gamma_w = \frac{M_d}{G}$$

$$M_4 - \frac{M_d}{G} + M_d = M_3$$

$$\therefore \qquad G = \frac{M_d}{M_d - (M_3 - M_4)} \qquad \text{where } M_d = (M_2 - M_1)$$

The expression for G can also be derived as follows :

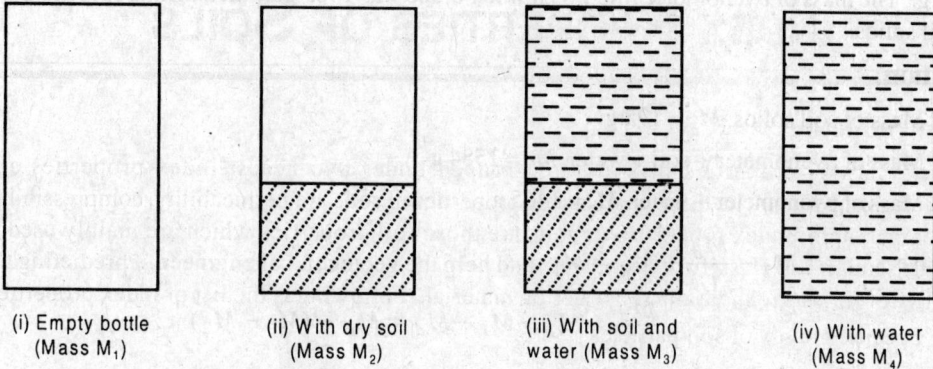

(i) Empty bottle (Mass M_1)

(ii) With dry soil (Mass M_2)

(iii) With soil and water (Mass M_3)

(iv) With water (Mass M_4)

Fig. 3.1. Specific gravity determination

Referring to Fig 3.1, from (i) and (ii)

Mass of soil solids, $M_d = (M_2 - M_1)$

Mass of water in $(iii) = (M_3 - M_2)$

Mass of water in $(iv) = (M_4 - M_1)$

Hence, mass of water equivalent in volume to that of soil solids = mass of water in (iv) – mass of water in $(iii) = (M_4 - M_1) - (M_3 - M_2)$

$$G = \frac{Mass \text{ of soil solids}}{Mass \text{ of equivalent volume of water}}$$

$$= \frac{(M_2 - M_1)}{(M_4 - M_1) - (M_3 - M_2)}$$

$$= \frac{(M_2 - M_1)}{(M_2 - M_1) - (M_3 - M_4)}$$

In the case of pycnometer method the accuracy of mass measurements is usually 0.1 g to 1 g, whereas in the case of density bottle method it is 0.01 g to 0.001 g. In the case of clays, kerosene which has better wetting capacity is preferred to distilled water in the density bottle method. The specific gravity of soil solids is computed as :

$$G = \frac{M_d G_k}{M_d - (M_3 - M_4)}$$

where $\qquad\qquad M_d = (M_2 - M_1)$

G_k = specific gravity of kerosene at test temperature, $T_t°C$

Conventionally, the specific gravity is reported at temperature of 27°C. For this purpose the masses M_3 and M_4 are found by filling the bottle with water (or kerosene) at 27°C using constant temperature bath. Alternatively, the test temperature $T_t°C$ is noted and specific gravity of soil solids at 27°C is computed as :

$$G_{27°C} = G_{Tt°C} \times \frac{\text{Specific gravity of distilled water at } T_t°C}{\text{Specific gravity of distilled water at } 27°C}$$

Example 3.1

An oven-dried soil sample having mass of 195 g was put inside a pycnometer which was then completely filled with distilled water. The mass of pycnometer with soil and water was found to be 1584 g. The mass of pycnometer filled with water alone was 1465 g. Calculate the specific gravity of soil solids.

Solution:

Mass of soil solids, $M_d = 195$ g

Mass of pycnometer + soil + water, $M_1 = 1584$ g

Mass of pycnometer + water, $M_2 = 1465$ g

We have
$$M_2 - \frac{M_d}{G} + M_d = M_1$$

$\therefore$
$$\frac{M_d}{G} = M_2 - M_1 + M_d = M_d - (M_1 - M_2)$$

$\therefore$
$$G = \frac{M_d}{M_d - (M_1 - M_2)} = \frac{195}{195 - (1584 - 1465)}$$

$$= 2.56$$

Example 3.2

The specific gravity of soil solids for a given soil sample was determined by density bottle method using kerosene. Following observations were recorded. Compute the specific gravity of soil solids at test temperature which was maintained at 27^oC. Also report the value at 4^0C. Take specific gravity of kerosene at 27^oC as 0.773.

Mass of density bottle, $M_1 = 61.45$ g

Mass of bottle + soil, $M_2 = 82.24$ g

Mass of bottle + soil + kerosene, $M_3 = 261.12$ g

Mass of bottle + kerosene, $M_4 = 246.49$ g

Solution:

Mass of soil solids,
$$M_d = M_2 - M_1 = 82.24 - 61.45$$
$$= 20.79 \text{ g}$$

We have
$$M_4 - \frac{M_d}{G} + M_d = M_3$$

$\therefore$
$$G = \frac{M_d . G_k}{M_d - (M_3 - M_4)} = \frac{20.79 \times 0.773}{20.79 - (261.12 - 246.49)} = 2.61$$

$$G_{27^oC} = 2.61$$

If the value of G has to be reported at 4^0C we have

$$G_{4^oC} = G_{27^oC} \times \frac{\text{Sp. gr. of water at } 27^o \text{ C}}{\text{Sp. gr. of water at } 4^o \text{ C}}$$

$$= 2.61 \times \frac{0.9965}{1.0000} = 2.60$$

Example 3.3 If in problem of Ex. 3.1, while finding mass of pycnometer with soil and water, 2 cm^3 of air got entrapped, will the computed value of G be higher or lower than the correct value and what will be the percentage error?

Using the data given in Example 3.1,

$$G = \frac{M_d}{M_d - (M_1 - M_2)} = \frac{195}{195 - (1584 - 1465)} = 2.57$$

If some air got entrapped while finding M_1, then the value of M_1 will be less than that when water replaces the entrapped air. Since M_1 occurs with negative sign in the denominator, the denominator will increase and hence computed value of G will be less than the correct value.

Corrected value of G is obtained as

$$G = \frac{195}{195 - (1586 - 1465)} = 2.63$$

Percentage error $= \dfrac{2.63 - 2.57}{2.63} \times 100 = 2.28\%$

3.3 Water Content

The water content of a soil sample can be determined by the following methods:

(*i*)   *Oven-drying method*

(*ii*) *Pycnometer method*

(*iii*) *Rapid methods*

(*i*) *Oven-drying method*

This is the most accurate method and is recommended as the standard method for determining water content in the laboratory. A cup with tight fitting lid, of non-corrodible material, is used. The mass, M_1, of cup with lid is found. Suitable quantity of wet soil sample whose water content is to be determined is put inside the cup and the lid replaced. The mass, M_2, of the cup with soil and lid is found. The lid is removed and the cup with soil is kept inside a thermostatically controlled oven and the soil is dried for 24 hours at 105 to 110° C. The cup is then taken out of the oven, the lid replaced and cooled. The mass, M_3, of the cup with dry soil and lid is found. The water content is calculated as shown below.

Mass of soil solids, $M_s = (M_3 - M_1)$

Mass of water, $M_w = (M_2 - M_3)$

Water content, $w = \dfrac{M_w}{M_s} = \dfrac{M_2 - M_3}{M_3 - M_1} \times 100\%.$

It may be noted that sandy soils require about 4 hours and clays about 15 hours of drying. However, to standardize the procedure 24 hours of drying is recommended. In most of the soils, there is chance of breaking the soil structure when heated beyond 110°C. Hence the temperature range of 105-110°C is recommended. But in the case of soils containing organic matter or calcareous compounds a lower temperature 60-80°C will have to be used and the soil dried for a longer period.

(*ii*) *Pycnometer method*

When the specific gravity of soil solids, G, is known this method can be used for quick determination of water content. The mass, M_1, of pycnometer with cone fitted to it is found. The cone is removed and suitable quantity of soil sample is put inside pycnometer. The cone is refitted and the mass, M_2, of the pycnometer with soil is found. Water is added to the soil inside the pycnometer until the excess water oozes out of the hole in the cone. The outer surface of the pycnometer is wiped dry and the mass, M_3, of pycnometer with soil and water is found. The pycnometer is emptied of its contents, cleaned and filled with water only. The outer surface is wiped dry and the mass, M_4, of pycnometer with water is found. In the last two steps care is taken to remove entrapped air from water inside the pycnometer. The water content is calculated as shown below.

If from mass M_4 we deduct mass of water whose volume is equivalent to volume of soil solids and add mass of soil solids, we get mass M_3.

Mass of water equivalent in volume to that of soil solids

$$= \left(\frac{M_s}{\gamma_s}\right)\gamma_w = \left(\frac{M_s}{G\gamma_w}\right)\gamma_w = \frac{M_s}{G}$$

$$\therefore \qquad M_4 - \frac{M_s}{G} + M_s = M_3$$

$$M_s\left(1 - \frac{1}{G}\right) = M_3 - M_4$$

$$M_s\left(\frac{G-1}{G}\right) = M_3 - M_4$$

$$M_s = \frac{G}{G-1}(M_3 - M_4)$$

$$M_w = (M_2 - M_1) - M_s$$

Water Content, $$w = \frac{M_w}{M_s} = \frac{(M_2 - M_1) - M_s}{M_s}$$

$$= \frac{(M_2 - M_1)}{M_s} - 1$$

$$w = \left[\frac{M_2 - M_1}{M_3 - M_4}\left(\frac{G-1}{G}\right) - 1\right] \times 100\%$$

This method is not convenient for use in the case of fine grained soils as it becomes difficult to find M_3 accurately.

(iii) *Rapid Methods*: Infra-red lamp and torsion balance method, calcium carbide method and Proctor needle method are used for rapid determination of water content. The student is advised to refer to a laboratory manual which gives full details about the principle, apparatus and procedure for these methods.

Example 3.4

Following observations were made while determining water content of a soil sample by pycnometer method.

Weight of wet soil sample, $W = 2.308$ N

Weight of pycnometer + soil + water, $W_1 = 30.930$ N

Weight of pycnometer + water, $W_2 = 29.661$ N

Calculate water content of soil sample, if $G = 2.69$.

Solution:

We have $W_2 - \dfrac{W_s}{G} + W_s = W_1$, where W_s = weight of soil solids

$$W_s\left(1 - \frac{1}{G}\right) = W_1 - W_2$$

$$W_s = (W_1 - W_2)\frac{G}{G-1}$$

$$= (30.930 - 29.661)\frac{2.69}{(2.69-1)}$$

$$= 2.02\,N$$

Weight of water present in soil sample,

$$W_w = W - W_s = 2.308 - 2.02 = 0.288\,N$$

$$\therefore \qquad w = \frac{W_w}{W_s} = \frac{0.288}{2.02} = 0.1426$$

$$= 14.26\%$$

Example 3.5

In determining water content of a soil sample by pycnometer method, following observations were made

Mass of pycnometer, $M_1 = 803$ g

Mass of pycnometer + wet sample, $M_2 = 1165$ g

Mass of pycnometer + soil + water, $M_3 = 2008$ g

Mass of pycnometer + water, $M_4 = 1802$ g

Calculate water content of soil sample, if $G = 2.65$

We have

$$M_4 - \frac{M_s}{G} + M_s = M_3 \quad \text{where } M_s = \text{mass of soil solids}$$

$$M_s\left(1 - \frac{1}{G}\right) = M_3 - M_4$$

$$M_s = (M_3 - M_4)\left(\frac{G}{G-1}\right)$$

$$M_w = (M - M_s)$$

where total mass of soil sample, $\qquad M = M_2 - M_1$

$$w = \frac{M_w}{M_s} = \frac{(M_2 - M_1) - M_s}{M_s}$$

$$= \frac{(M_2 - M_1)}{M_s} - 1 = \left(\frac{M_2 - M_1}{M_3 - M_4}\right)\left(\frac{G-1}{G}\right) - 1$$

$$= \left(\frac{1165 - 803}{2008 - 1802}\right)\left(\frac{1.65}{2.65}\right) - 1 = 9.41\%$$

3.4 Particle Size Distribution

The particle size distribution for a given soil sample is quantitatively expressed by giving the percentage by mass of different soil components such as silt, clay etc. based on a particle size classification chart. To enable the student to understand the discussion that follows, the I.S. particle size classification chart is given in the following page.

Clay (size)	Silt (size)	Sand			Gravel	
		Fine	Medium	Coarse	Fine	Coarse

2μ   75μ   425μ   2 mm   4.75 mm   20 mm   80 mm

The particle size distribution is determined by conducting grain size analysis, also known as mechanical analysis, which consists of two parts; sieve analysis for gravel and sand (coarse grained fraction) and sedimentation analysis or wet analysis for silt and clay (fine grained fraction).

3.4.1 Sieve Analysis

In the Indian standard (IS:460-1962), the sieves are designated by the size of the aperture in mm, as indicated in Appendix I.

The soil sample to be analysed is thoroughly dried with the lumps being pulverized by means of wooden mallet or mortar with rubber pestle. Suitable quantity of representative soil sample depending on soil type is taken and is sieved through IS 4.75 mm sieve, The soil retained on 4.75 mm sieve (+ 4.75 mm fraction) is subjected to coarse sieve analysis which consists of sieving the soil through a nest of sieves which may consist of 40 mm, 20 mm and 10 mm IS sieves. The soil passing through 4.75 mm sieve (-4.75 mm fraction) is subjected to fine sieve analysis which consists of sieving the soil through a nest of sieves which may consist of 2.00 mm, 1.00 mm, 600 μ, 300 μ, 212 μ, 150 μ, and 75 μ sieves. The results of sieve analysis can be tabulated as shown in Table 3.1. (1μ = 0.001 mm).

Table 3.1 – Typical table for sieve analysis

Sieve Designation (IS)	Sieve Size D mm	Mass of soil retained (gm)	Percentage mass retained	Cumulative percentage retained	Percent finer N
4.75 mm	4.75				
2.00 mm	2.00				
1.00 mm	1.00				
600 μ	0.600				
425 μ	0.425				
300 μ	0.300				
212 μ	0.212				
150 μ	0.150				
75 μ	0.075				
Pan					

The particle size distribution curve is obtained by plotting percent finer N as ordinate on natural scale against particle size D mm as abscissa on logarithmic scale.

3.4.2 Sedimentation Analysis

Sedimentation analysis or wet mechanical analysis is conducted on soil fraction finer than 75micron, which is kept in suspension in a liquid medium (usually water).

Sedimentation analysis is based on Stoke's law according to which the terminal velocity of sinking of a spherical particle in a still liquid medium is dependent upon size of particle, density of material of particle, density of liquid and viscosity of liquid. The following equation has been derived using Stoke's law :

$$v = \frac{2}{9} r^2 \frac{(\gamma_s - \gamma_1)}{\eta}$$ 3.4 (i)

where

v = terminal velocity of sinking of particle in cm/sec.

r = radius of particle in cm

γ_s = density of particle in gm/cm^3

γ_1 = density of liquid in gm/cm^3

η = Coefficient of viscosity of liquid in gm-sec/cm^2

When a particle is placed on the surface of a still liquid medium, it first sinks with increasing velocity due to action of gravity but after sometime attains a constant velocity which is referred to as terminal velocity of sinking of particle.

Use of Stoke's law in sedimentation analysis:

Let a soil particle of size D mm fall through height, H_e cms, in time interval t minutes, when suspended in still water. Then velocity of fall v is given by

$$v = \frac{H_e}{60t} \text{ cm/sec}$$ 3.4(ii)

From Eq 3.4(i)

$$v = \frac{2}{9} \left(\frac{D}{20} \right)^2 \frac{(\gamma_s - \gamma_w)}{\eta}$$

$$= \frac{D^2}{1800} \frac{(G\gamma_w - \gamma_w)}{\eta}$$

i.e.

$$v = \frac{D^2}{1800} \frac{(G-1)}{\eta} \gamma_w$$ 3.4(iii)

Comparing equations 3.4 (ii) and 3.4 (iii), we get

$$\frac{D^2}{1800} \frac{(G-1)\gamma_w}{\eta} = \frac{H_e}{60t}$$

$$D = \sqrt{\frac{30\eta}{(G-1)\gamma_w} \cdot \frac{H_e}{t}}$$

or

$$D = M \sqrt{\frac{H_e}{t}}$$ 3.4(iv)

where factor

$$M = \sqrt{\frac{30\eta}{(G-1)\gamma_w}}$$

Factor M varies with temperature. Equation 3.4 (iv) is used to compute size D mm of largest particle still in suspension at depth H_e cms at end of elapsed time interval t minutes.

Brief procedure of sedimentation analysis:

About 50 gms of dry soil sample passing through 75 micron sieve is weighed accurately and is taken in a porcelain dish. 50 cc of dispersing agent solution is added to it. Some quantity of distilled water is also added to form a soil slurry which is gently stirred using a glass rod. The contents of the dish are then transferred to the cup of a high-speed stirrer, care being taken to see that no particles of soil are left behind in the dish. For this purpose plastic wash bottle with nozzle which produces jet of water when pressed is used. The soil-water mixture in the cup is stirred for 7 to 10 minutes to ensure thorough dispersion of soil particles. The contents of the cup are then transferred to a/1000 cc sedimentation jar. Again using jet of water care is taken to see that all soil particles are transferred to the jar. Required quantity of distilled water is added to the soil suspension in the jar to make up 1000 cc. The mouth of the jar is closed tightly with the palm of hand and the jar is inverted several times to ensure uniform distribution of soil particles in the soil suspension. The jar is then kept on a level surface and a stopwatch started simultaneously. At end of different elapsed time intervals, t minutes, (usually 1min, 2 min, 4 min, 8 min, 15 min, 30 min, 1hour, 2 hours, 4 hours, 8 hours, 16 hours, 1 day ....) the size D mm of the largest particle still in suspension at depth H_e cms is determined either by pipette method or hydrometer method and using equation 3.4 (iv). The corresponding percent finer, N, is computed using the following equation:

$$N = \frac{W_D}{W_d / V} \times 100\% \qquad\qquad 3.4 \ (v)$$

where W_D = Weight of soil particles finer than size D mm, per ml, still in suspension at depth H_e cms at elapsed time interval t minutes

W_d = Weight of soil particles taken for sedimentation analysis

V = Volume of soil suspension.

Pipette Method

In this method 10 ml of soil suspension is sucked from a depth, H_e, equal to 10 cms, at different elapsed time intervals, t minutes, using sampling pipette. The weight, W_D, of soil particles per ml of soil suspension is obtained by drying the sucked soil suspension and dividing the weight of dry soil recovered by 10. The size D mm of largest particle still in suspension at depth H_e cms at elapsed time interval, t minutes, is computed using equation 3.4(iv) and the corresponding percent finer, N, using equation 3.4(v).

Hydrometer Method

This method first requires preparation of calibration chart which is used to find effective depth, H_e, corresponding to any reading, R_h, of hydrometer. A hydrometer used in sedimentation analysis is usually graduated from 0.995 to 1.030, as shown in Fig.3.2 As indicated in this figure, the effective depth H_e corresponding to a reading , R_h, on the stem is given by

$$H_e = h + \frac{H}{2} \qquad\qquad \text{(without immersion correction)}$$

$$H_e = h + \frac{H}{2} - \frac{V_H}{2Aj} \qquad\qquad \text{(with immersion correction)}$$

where h = distance from reading R_h on stem to the neck, in cms

H = height of bulb, in cms

V_H = Volume of hydrometer, in cm^3

Aj = area of cross section of jar used for sedimentation analysis, in cm^2.

The distances h from different readings R_h on stem to the neck and the height of bulb H are measured with the help of a graduated scale. The volume of hydrometer, V_H is determined by finding the volume of water displaced by hydrometer, when it is immersed in water taken in a graduated jar. After computing effective depth, H_e, corresponding to different hydrometer reading R_h, the calibration chart is obtained by plotting effective depth, H_e as ordinate against reading of hydrometer R_h as abscissa.

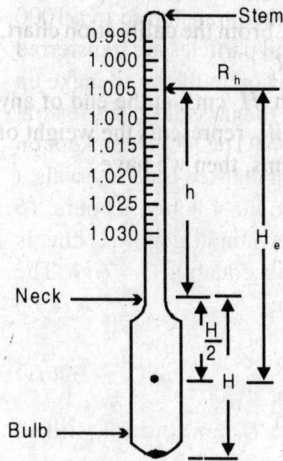

Fig. 3.2 (a)	**Fig. 3.2 (b)**
Hydrometer	Calibration chart

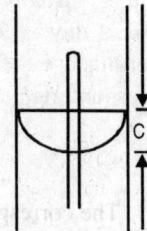

Fig. 3.2 (c)
Meniscus correction

Corrections to be applied to the observed hydrometer reading:

The following corrections are to be applied to the observed hydrometer readings: (*i*) meniscus correction, C_m, (*ii*) dispersing agent correction, C_d and (*iii*) temperature correction, C_T.

As the soil suspension will be opaque it will not be possible to read the hydrometer against the bottom of meniscus and hence one has to read against the top of meniscus. This necessitates the application of meniscus correction, C_m which is always additive. The value of C_m is determined by reading the hydrometer against both top and bottom of meniscus when the hydrometer is floating in clear water, and taking the difference between the two readings.

The addition of dispersing agent solution necessitates the application of dispersing agent correction, C_d that is always negative. The value of C_d is determined by reading the hydrometer first when it is floating in clear water and then when it is floating in water to which dispersing agent solution is added in the same proportion as used in the test and taking the difference between the two readings.

The temperature of soil suspension may vary during the test and be different from the temperature at which the hydrometer is calibrated. This necessitates the application of temperature correction, C_T, which will be positive or negative depending on whether the temperature of soil suspension is above or below the calibration temperature of hydrometer. The value of C_T can be obtained from the temperature correction chart prepared for the particular hydrometer.

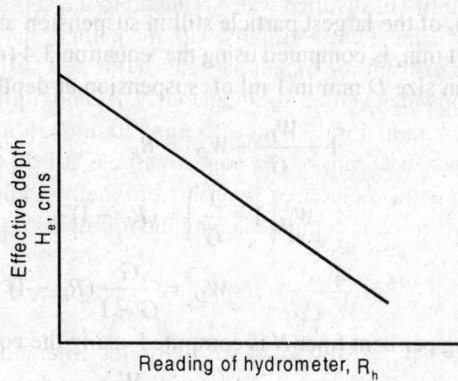

The composite correction C is given by
$$C = C_m - C_d \pm C_T$$
The corrected hydrometer reading R_h is then computed as

$$R_h = R_h' + C \quad (C \text{ may be positive or negative}) \qquad 3.4\,(vi)$$

where, R_h' = observed hydrometer reading

Use of hydrometer in sedimentation analysis

The hydrometer is inserted into the soil suspension prepared in 1000 ml sedimentation jar at the end of different elapsed time intervals, *t* min, usually 1, 2, 4, 8, 15, 30, 60 minutes, 2, 4, 8, 16 hrs, 1 day, etc...... and the reading of hydrometer R'_h noted. The temperature of soil suspension is also noted each time. The corrected hydrometer reading R_h is computed. From the calibration chart, the effective depth, H_e, is found corresponding to hydrometer reading R_h.

The size, D mm, of the largest particle still in suspension at depth, H_e cms, at the end of any elapsed time interval, t min, is computed using the equation 3.4 (iv). If W_D represents the weight of soil particles finer than size D mm in 1 ml of suspension at depth H_e cms, then we have

$$1 - \frac{W_D}{G} + W_D = R_h$$

$$W_D\left(1 - \frac{1}{G}\right) = (R_h - 1)$$

$$W_D = \frac{G}{G-1}(R_h - 1)$$

The corresponding per cent finer N is computed using the equation

$$N = \frac{W_D}{W_d / V} \times 100\%$$

where W_d = Weight of dry soil taken for sedimentation analysis
           V = Volume of soil suspension.

Limitations in the use of Stoke's Law in Sedimentation analysis

1. Stoke's law is applicable for spherical particles only. The fine clay particles are not spherical in shape. While applying Stoke's law, the concept of equivalent diameter is used. The equivalent diameter of a soil particle is defined as the diameter of an imaginary sphere which has the same specific gravity as the soil particle and settles with the same terminal velocity as that of the soil particle.

2. It is assumed that every particle settles independently without interference from other particles as well as from the sides of the jar. To minimize error due to this assumption, it is recommended that not more than 50 gms of soil particles be taken in 1000 ml of soil suspension.

3. The soil particles in the soil suspension may have different values of specific gravity. But in the computations, an average value of G is used.

4. The lower limit of particle size for validity of Stoke's Law is 0.0002 mm. However, the upper limit for the same is 0.2 mm. For particles of size less than 0.0002 mm, Brownian movement affects their settlement and in the case of particles larger than 0.2 mm, turbulence affects the settlement.

Use of particle size distribution curve

As already stated a particle size distribution curve is obtained by plotting percent finer, N as ordinate on natural scale against particle size, D mm as abscissa on logarithmic scale and is used to find the particle size distribution. Based on the particle size distribution soils may be described as well graded, poorly graded or uniformly graded. In a well-graded soil there is good representation of all particle sizes between the maximum and minimum sizes. In a poorly graded soil there will be an excess or deficiency of one or more particle sizes. A soil is said to be uniformly graded if most of the particles are nearly of the same size.

In the case of coarse grained soils three particle sizes D_{10}, D_{30}, and D_{60} are obtained from particle size distribution curve. They represent particle sizes corresponding to $N = 10\%$, 30% and 60% respectively.

D_{10} is referred to as effective size. 10% by mass of the soil sample will be finer than size D_{10}. A measure of particle size range is given by coefficient of uniformity C_u (also known as uniformity coefficient) which is defined as

$$C_u = \frac{D_{60}}{D_{10}} \qquad\qquad 3.4(vii)$$

The shape of the particle size curve can be represented by coefficient of curvature C_c, defined as:

$$C_c = \frac{(D_{30})^2}{D_{10}\cdot D_{60}} \qquad\qquad 3.4(viii)$$

C_u will be nearly unity for a uniformly graded soil. Generally for a well-graded soil C_c lies between 1 and 3 and in addition C_u will be greater than 4 for gravels and greater than 6 for sands.

Example 3.6

50 gm of oven-dried soil passing 75 μ test sieve was taken in a hydrometer analysis. The corrected hydrometer reading in 1000 ml soil suspension at 2 min elapsed time interval was 25. The effective depth corresponding to R_h= 25 is H_e= 12.1 cm. Taking G = 2.7 and η = 1 centipoise, calculate the coordinates of the point on the grain size distribution curve.

Solution:

Mass of soil taken for hydrometer analysis, M_d= 50g

Volume of soil suspension, $\qquad V = 1000$ cc

Elapsed time interval, $\qquad\qquad t = 2$ min

Corrected hydrometer reading, $\quad R_h$= 1.025

Effective depth, $H_e = 12.1$ cm

Specific gravity of soil particle., $\;G = 2.7$

Coefficient of viscosity of water, $\;η = 0.01$ poise

$$= \frac{0.01}{981} \text{ gm-sec/cm}^2$$

Particle size, $\qquad\qquad D = M\sqrt{\frac{H_e}{t}} = \sqrt{\left(\frac{30η}{G-1}\right)\left(\frac{H_e}{t}\right)}$

$$= \sqrt{\frac{30 \times 0.01}{9.81(2.7-1)} \times \frac{12.1}{2}} = 0.033 \text{ mm}$$

Per cent finer, $\qquad\qquad N = \frac{M_D}{M_d/V} \times 100\%$

where $\qquad\qquad M_D = \left(\frac{G}{G-1}\right)(R_h - 1)$

$$= \left(\frac{2.7}{1.7}\right)(1.025 - 1) = 0.04 \text{ gm}$$

$$N = \frac{0.04}{50/1000} \times 100\% = 80\%$$

Example 3.7

Particles of five different sizes are mixed in the proportions shown in the following table and enough water is added to make 1000 cc of suspension.

Particle Size	Mass
(mm)	(gm)
0.050	6
0.020	20
0.010	15
0.005	5
0.001	4

It is ensured that the suspension is thoroughly mixed so as to have a uniform distribution of the particles. Taking $G = 2.7$ and $\eta = 0.01$ poise, find

(i) Size of largest particle still in suspension at depth 6 cm, 5 minutes after start of sedimentation.

(ii) The specific gravity of suspension at depth 6 cms, 5 minutes after start of sedimentation.

(iii) The time interval in which all the particles will have settled below 6 cm.

Solution:

(i) Factor
$$M = \sqrt{\frac{30\eta}{(G-1)\gamma_w}} = \sqrt{\frac{30 \times 0.01}{9.81 \times 1.7 \times 1}} = 0.0134$$

$$D = M\sqrt{\frac{H_e}{t}} = 0.0134\sqrt{\frac{6}{5}} = 0.015 \text{ mm}$$

Size of largest particle still in suspension at depth 6 cm, 5 minutes after commencement of sedimentation = 0.010 mm

(ii) We have
$$W_D = \frac{15 + 5 + 4}{1000} = 0.024 \text{ gm}$$

Specific gravity of suspension at depth 6 cm, 5 min after start of sedimentation

$$1 - \frac{W_D}{G} + W_D = 1 - \frac{0.024}{2.7} + 0.024 = 1.015$$

(iii)
$$D = M\sqrt{\frac{H_e}{t}}$$

$$D^2 = M^2 \frac{H_e}{t}$$

$\therefore$
$$t = \left(\frac{M}{D}\right)^2 H_e$$

Time interval in which all the particles will have settled below 6 cm is given by

$$t = \left(\frac{0.0134}{0.001}\right)^2 6$$

$$= 1077.4 \text{ min} = 17 \text{ hrs } 57.4 \text{ min}$$

3.5 Consistency of Soils

The term consistency refers to the relative ease with which a soil mass can be deformed and is used to describe the degree of firmness of fine-grained soils for which consistency relates to a large extent to water content. The four states of consistency suggested by Atterberg are indicated in Fig. 3.3.

In the solid state there will be no change in volume of soil mass accompanying change in water content. In the remaining three states increase in water content is accompanied by increase in volume of soil mass and decrease of water content by reduction in volume of soil mass. In the liquid state the soil mass behaves like a liquid possessing very less shear strength. In the plastic state the soil mass can be deformed without cracking. In the semi-solid state the soil mass cannot be deformed without cracking.

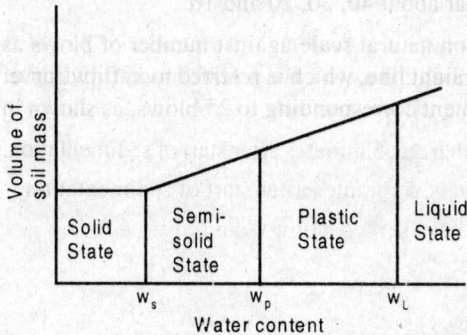

Fig. 3.3. States of consistency

The water contents, which arbitrarily define the boundary between the four states of consistency, are referred to as consistency limits or Atterberg limits.

The three consistency limits are liquid limit, plastic limit and shrinkage limit.

Liquid limit is denoted by w_L and is the boundary between plastic and liquid states of consistency. It is the minimum water content at which the soil mass still flows like a liquid.

Liquid limit is defined as the water content at which a groove, cut with a standard grooving tool, in soil pat taken in the cup of a standard liquid limit device closes for a distance of 13 mm when the cup is imparted 25 blows.

Plastic limit is denoted by w_p and is the boundary between semi-solid and plastic states of consistency. It is the minimum water content at which the soil mass can still be deformed without cracking.

Plastic limit is defined as the water content at which the soil mass can be rolled into a thread of 3 mm diameter and the thread first shows signs of cracking.

Shrinkage limit is denoted by w_s and is the boundary between solid and semi-solid states of consistency. It is defined as the maximum water content at which there is no reduction in volume of soil mass accompanying reduction in water content.

3.5.1 Determination of Liquid Limit

The liquid limit is determined in the laboratory using the standard liquid limit apparatus designed by A. Casagrande. The apparatus consists of a brass cup, which can be raised and lowered to fall on a micarta base of specified hardness by means of a cam operated by a handle. The height of fall of the cup can be adjusted with the help of adjusting screws.

To cut standard groove in the soil pat taken in the cup of the liquid limit apparatus, a grooving tool is used. There are two types of grooving tool: (*i*) Casagrande (*BS*) tool and (*ii*) ASTM Tool. The Casagrande tool should be preferred as it provides control over depth of groove. The size of groove will be 2 mm wide at the bottom, 11.0 mm wide at the top and 8 mm in depth. The ASTM tool cuts a groove 2 mm wide at the bottom 13.6 mm at the top and 10mm in depth. In the case of sandy soils, ASTM tool is to be used instead of Casagrande tool as the latter has the tendency to tear the sides of the groove.

About 100 gm of the soil sample passing through 425 micron IS sieve is taken in a porcelain dish. Some quantity of distilled water is added to it and thoroughly mixed to form a soil paste of uniform colour. The height of fall of cup of the liquid limit device is adjusted to be 1 cm. A portion of the soil paste in the porcelain dish is placed in the cup of liquid limit device and levelled by means of a spatula. Using standard grooving tool a groove is cut in the soil pat formed in the cup. The cup is given blows by manual operation of handle or by electrically operated motorized system, the rotation of handle being at the rate of 2 revolutions per second. The number of blows required to close the groove for a distance of 13 mm is noted down. Some quantity of soil at the place where the groove has closed is taken for water content determination. The above steps are repeated to get atleast 4 concurrent sets of number of blows and water content. It is convenient to increase the water content in successive steps and obtain blow counts near about 40, 30, 20 and 10.

The water content values are plotted as ordinate on natural scale against number of blows as abscissa on logarithmic scale to obtain a best fitting straight line, which is referred to as flow curve. From this plot the liquid limit is obtained as water content corresponding to 25 blows, as shown in Fig. 3.4.

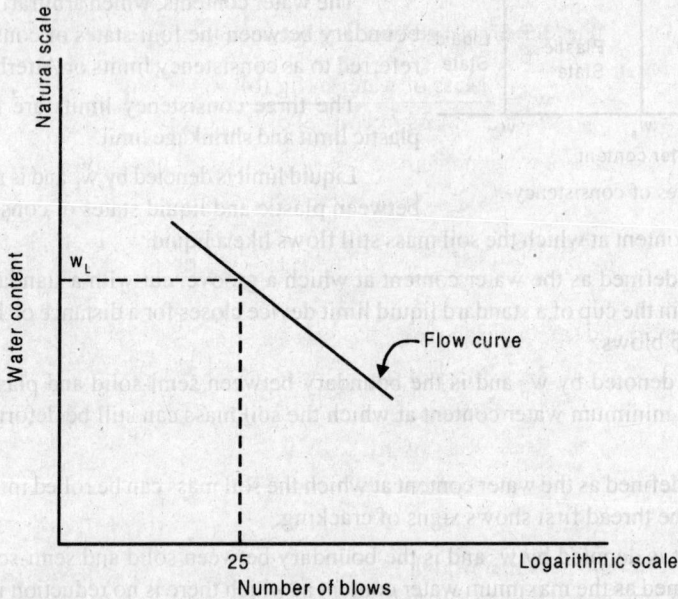

Figure 3.4. Liquid limit determination

3.5.2 Determination of Plastic Limit

About 30 gms of soil sample passing through 425 micron IS sieve is taken and some quantity of distilled water is added and thoroughly mixed to form soil paste which can be rolled into a ball between palms of hands. A small portion of the ball is then rolled on a smooth plate into a thread of 3 mm diameter, and the thread is looked for signs of cracking. If no cracks are seen, the thread is picked up and again rolled into a ball between palms, to reduce water content. The ball is then rolled on smooth plate into a thread of 3 mm diameter. The steps are repeated until a 3mm diameter thread first shows signs of cracking. A portion of the thread is taken for water content determination which gives the plastic limit.

3.5.3 Determination of Shrinkage Limit

Derivation of expression:

To derive expression for shrinkage limit we consider the two-phase diagrams shown in Fig 3.5 which correspond to the three states of soil pat during the experimental procedure.

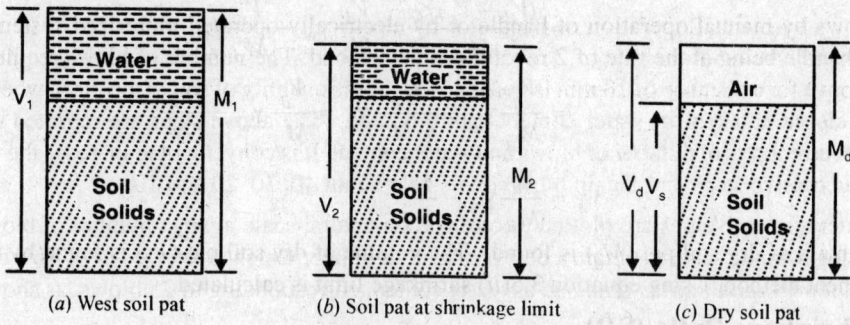

Fig. 3.5. Shrinkage limit determination

Shrinkage limit, $\qquad w_s = \dfrac{\text{Mass of water in fig (b)}}{\text{Mass of soil solids}}$

Mass of water in fig (b) = Mass of water in fig (a) minus loss in mass of water from fig (a) to fig (b)

$$= (M_1 - M_d) - (V_1 - V_2)\gamma_w \text{ where } V_2 = V_d$$

$\therefore \qquad w_s = \dfrac{(M_1 - M_d) - (V_1 - V_d)\gamma_w}{M_d}$ ....3.5(i)

As is clear from Figs (a) and (c), in the above expression for w_s,

$\qquad M_1$ = Mass of wet soil pat

$\qquad V_1$ = Volume of wet soil pat

$\qquad M_d$ = mass of dry soil pat

$\qquad V_d$ = Volume of dry soil pat

Experimental procedure:

About 50 gms of soil sample passing through IS 425 micron sieve is taken in a porcelain dish, distilled water added to it, and mixed thoroughly to form a soil paste of slightly flowing consistency. The shrinkage dish (non-corrodible cup of 45 mm dia and 15 mm ht) is weighed after coating inner side of the cup with a thin layer of grease or oil. The shrinkage cup is filled with the soil paste in three layers, the cup being gently tapped on a cushioned surface after filling with each layer to ensure expulsion of air bubbles. The surface of soil is levelled and outer side of cup is cleaned. The mass of shrinkage cup with wet soil pat is found and this is deducted from mass of shrinkage cup to get mass of wet soil pat (M_1). The wet soil pat is allowed to dry in air for sometime, then kept in a thermostatically controlled oven and dried for 24 hours at 105-110°C. After oven drying the mass of dry soil pat (M_d) is found. The volume of dry soil pat (V_d) is found by mercury displacement method. The volume of wet soil pat (V_1) is equal to the volume of shrinkage dish which is found by filling it with mercury and finding the mass of mercury required to fill if after removing the convex portion at the top by pressing with a flat plate. The volume is obtained by dividing this mass by density of mercury. The shrinkage limit is calculated using the Eq. 3.5 (i).

Alternative method:

The shrinkage limit can be determined using dry soil pat if the value of G is known.

We have

Shrinkage limit, $\qquad w_s = \dfrac{\text{Mass of water at shrinkage limit}}{\text{Mass of soil solids}}$

$$= \frac{(V_d - V_s)\gamma_w}{M_d} = \frac{V_d}{M_d}\gamma_w - \frac{V_s}{M_d}\gamma_w$$

$$= \frac{V_d}{M_d}\gamma_w - \frac{M_d}{\gamma_s M_d}\gamma_w = \frac{V_d}{M_d}\gamma_w - \frac{\gamma_w}{G\gamma_w}$$

$$\therefore \qquad w_s = \frac{V_d}{M_d}\gamma_w - \frac{1}{G} \qquad\qquad\qquad 3.5(ii)$$

The mass of dry soil pat (M_d) is found. The volume of dry soil pat (V_d) is found by mercury displacement method. Using equation 3.5(ii) shrinkage limit is calculated.

3.5.4 Shrinkage Ratio (SR)

When a wet soil mass with its water content above shrinkage limit is dried to a water content greater than or equal to shrinkage limit, then whatever reduction in volume of soil mass takes place will be equal to the volume of water evaporated. Shrinkage ratio (SR) is defined as the ratio of reduction in volume of soil mass expressed as a percentage of its dry volume to the corresponding reduction in water content.

$$SR = \frac{\dfrac{V_1 - V_2}{V_d} \times 100}{w_1 - w_2} \qquad\qquad\qquad 3.5(iii)$$

where V_1 = Volume of soil mass at water content w_1
            V_2 = Volume of soil mass at water content w_2
            V_d = Volume of dry soil mass
    w_1, w_2 = water contents (expressed as percentage)

If the soil is dried to shrinkage limit, then $w_2 = w_s$ and $V_2 = V_d$. Substituting in Eq. 3.5(iii), we get

$$SR = \frac{\dfrac{V_1 - V_d}{V_d} \times 100}{w_1 - w_s} \qquad\qquad\qquad 3.5(iv)$$

Also, we have

$$w_1 - w_2 = \frac{W_{w_1}}{W_d} - \frac{W_{w_2}}{W_d} = \frac{(W_{w_1} - W_{w_2})}{W_d} = \frac{(V_1 - V_2)\gamma_w}{W_d}$$

Then $$SR = \frac{\dfrac{(V_1 - V_2)}{V_d}}{w_1 - w_2} = \frac{\left(\dfrac{V_1 - V_2}{V_d}\right)}{(V_1 - V_2)\gamma_w / W_d} = \frac{W_d}{V_d \gamma_w} = \frac{(\gamma)\,\text{dry state}}{\gamma_w}$$

Thus shrinkage ratio is equal to the mass specific gravity of soil in the dry state.

3.5.5 Volumetric Shrinkage (VS)

Volumetric shrinkage (or volumetric change) is defined as the reduction in volume of soil mass expressed as a percentage of its dry volume when the soil mass is dried from a water content above shrinkage limit to shrinkage limit.

$$VS = \frac{V_1 - V_d}{V_d} \times 100\% \qquad\qquad\qquad 3.5(v)$$

where V_1 = Volume of soil mass at any water content $w_1 > w_s$.
            V_d = Volume of dry soil mass

Also, we can write

$$SR = \frac{\frac{V_1 - V_d}{V_d} \times 100}{w_1 - w_s} = \frac{VS}{w_1 - w_s}$$

$$\therefore \qquad VS = SR\,(w_1 - w_s) \qquad\qquad 3.5(vi)$$

Example 3.8

The mass and volume of a saturated clay specimen were 29.8g and 17.7cm^3 respectively. On oven drying the mass got reduced to 19g and the volume to 8.9cm^3. Calculate shrinkage limit, shrinkage ratio and volumetric shrinkage. Also compute G of soil.

Solution:

Mass of wet soil specimen, $M = 29.8$ g

Volume of wet soil specimen, $V = 17.7$ cm^3

Mass of dry soil specimen, $M_d = 19.0$ g

Volume of dry soil specimen, $V_d = 8.9$ cm^3

Shrinkage limit,
$$w_s = \frac{\text{Mass of water at shrinkage limit}}{\text{Mass of soil solids}}$$

$$= \frac{(M - M_d) - (V - V_d)\gamma_w}{M_d}$$

$$= \frac{(29.8-19.0)-(17.7-8.9)(1)}{19.0} = 0.1053$$

$$= 10.53\%$$

Shrinkage ratio,
$$SR = \frac{M_d}{V_d \cdot \gamma_w} = \frac{19.0}{8.9(1)} = 2.13$$

Volumetric shrinkage,
$$V_s = \frac{V - V_d}{V_d} \times 100\%$$

$$= \frac{17.7 - 8.9}{8.9} = 98.8\%$$

We have
$$w_s = \frac{V_d}{M_d}\gamma_w - \frac{1}{G}$$

$$\frac{1}{G} = \frac{V_d}{M_d}\gamma_w - w_s = \frac{8.9}{19.0}(1) - 0.1053$$

$$= 0.3631$$

$$G = \frac{1}{0.3631} = 2.75$$

Example 3.9

A fully saturated sample of clay was found to have mass specific gravity of 1.91 and water content of 29%. The soil sample was oven-dried and its mass specific gravity was found to reduce to 1.83. Calculate the shrinkage limit of soil.

Solution

For saturated soil sample,

$$e = wG \qquad = 0.29\,G$$

$$G_m = \frac{\gamma}{\gamma_w} = \left[\frac{(G + e)\gamma_w}{1 + e} \right] \frac{1}{\gamma_w}$$

$$1.91 = \frac{G + 0.29G}{1 + 0.29G}$$

$$1.91 + 0.554\,G = G + 0.29G = 1.29\,G$$

$$G = \frac{1.91}{0.736} = 2.59$$

Shrinkage limit,

$$w_s = \frac{V_d}{W_d} \gamma_w - \frac{1}{G}$$

$$= \frac{\gamma_w}{(\gamma)_{dry\ state}} - \frac{1}{G}$$

$$= \frac{1}{(G_m)_{dry\ state}} - \frac{1}{G} = \frac{1}{1.83} - \frac{1}{2.59} = 16\%$$

Example 3.10

The liquid limit and shrinkage limit of a soil sample are 49% and 16% respectively. If the volume of a specimen of this soil decreases, on drying, from 37.2 cm³ at liquid limit to 22.4 cm³ at shrinkage limit, compute the specific gravity of soil particles.

Solution

Liquid limit, $w_L = 49\%$

Shrinkage limit, $w_S = 16\%$

Volume of soil specimen at liquid limit, $V_L = 37.2$ cm³

Volume of soil specimen at shrinkage limit, $V_d = 22.4$ cm³

we have

Shrinkage ratio,

$$SR = \frac{\left(\dfrac{V_L - V_d}{V_d} \right) \times 100}{w_L - w_s} = \frac{\left(\dfrac{37.2 - 22.4}{22.4} \right) \times 100}{49 - 16}$$

$$= 2$$

Also

$$SR = \frac{W_d}{V_d \gamma_w}$$

∴

$$\frac{W_d}{V_d} = (SR)\gamma_w = 2\gamma_w$$

$$w_s = \frac{V_d}{W_d} \gamma_w - \frac{1}{G} = \frac{\gamma_w}{2\gamma_w} - \frac{1}{G}$$

$$0.16 = \frac{1}{2} - \frac{1}{G}$$

$$\frac{1}{G} = \frac{1}{2} - 0.16 = 0.34$$

$$G = \frac{1}{0.34} = 2.94$$

Example 3.11

The liquid limit and plastic limit of a soil are 34% and 26% respectively. When the soil is dried from its state at liquid limit to dry state the reduction in volume was found to be 35% of its volume at liquid limit. The corresponding volume reduction from the state of plastic limit to dry state was 25% of its volume at plastic limit. Calculate (i) shrinkage limit and (ii) shrinkage ratio

Solution: $\quad w_L = 34\% \quad w_p = 26\%$

(i) Let volume at liquid limit be denoted by V_L and that at plastic limit by V_p.

Then $V_d = V_L - 0.35\ V_L = 0.65\ V_L$

Also $V_d = V_p - 0.25\ V_p = 0.75\ V_p$

$\therefore \qquad V_p = \dfrac{0.65}{0.75} V_L = 0.87 V_L$

Referring to Fig.3.6, $\quad w_s = w_L - AB$

From similar triangles, ABE and CDE,

$$\frac{AB}{CD} = \frac{BE}{DE}$$

$\therefore \qquad AB = \dfrac{BE}{DE} . CD$

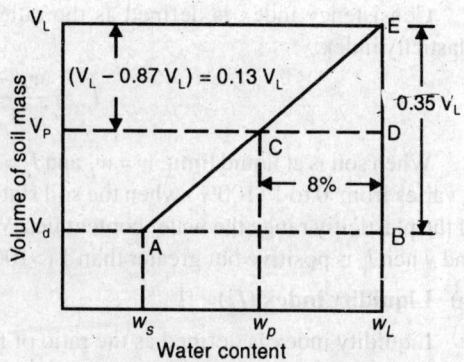

Fig. 3.6: For Example 3.11

$$= \frac{0.35 V_L}{0.13 V_L} \times 0.08 = 0.215$$

$\therefore \qquad w_s = 0.340 - 0.215 = 12.5\%$

(ii) Shrinkage ratio, $\quad SR = \dfrac{\left(\dfrac{V_L - V_P}{V_d}\right) \times 100}{w_L - w_p} = \dfrac{(V_L - 0.87 V_L) \times 100}{0.65 V_L (34 - 26)}$

$$= \frac{0.13 \times 100}{0.65 \times 8} = 2.5$$

3.5.6 Atterberg Indices

Following is the list of Atterberg indices:

1. Plasticity index
2. Flow index
3. Toughness index
4. Consistency index
5. Liquidity index

(i) Plasticity index

Plasticity index is defined as liquid limit minus plastic limit.

$$I_p = w_L - w_p \qquad\qquad\qquad 3.5(vii)$$

(ii) Flow index

Flow index is the slope of flow curve obtained by plotting water content as ordinate on natural scale against number of blows as abscissa on logarithmic scale.

$$I_F = \frac{w_1 - w_2}{\log_{10} \dfrac{N_2}{N_1}}$$

3.5(viii)

where w_1 = water content corresponding to number of blows, N_1
w_2 = water content corresponding to number of blows, N_2

(iii) Toughness index

Toughness index is defined as the ratio of plasticity index to flow index.

$$I_T = \frac{I_p}{I_F}$$

3.5(ix)

(iv) Consistency index (I_c)

Consistency index is defined as the ratio of liquid limit minus natural water content to the plasticity index.

$$I_c = \frac{w_L - w}{I_p}$$

3.5(x)

When soil is at liquid limit, $w = w_L$ and $I_c = 0$. When soil is at plastic limit, $w = w_p$ and $I_c = 1$. Thus I_c varies from 0 to 1 (100%) when the soil state is in the plasticity range and indicates in which part of the plasticitiy range the water content lies. When I_c is negative the soil mass will be in liquid state and when I_c is positive but greater than 1 (>100%), the soil mass will be in semi-solid or solid state.

(v) Liquidity index (I_L)

Liquidity index is defined as the ratio of natural water content minus plastic limit to plasticity index.

$$I_L = \frac{w - w_p}{I_p}$$

3.5(xi)

When the soil mass is at plastic limit, $w = w_p$ and $I_L = 0$. When the soil mass is at liquid limit $w = w_L$ and $I_L = 1$. Thus when the soil state is in the plasticity range I_L will lie between 0 and 1 (100%) and indicates in which part of the plasticity range the water content of soil mass lies. When $I_L > 1$ the soil mass will be in liquid state and when I_L is negative, it will be in semi-solid or solid state.

Brief discussion on Atterberg limits and indices

Liquid limit and plasticity index are used to classify fine-grained soils. For inorganic clays liquid limit values are usually never greater than 100%. But values greater than 100% are possible in the case of highly organic clays and clays of volcanic origin. In the case of bentonite, liquid limit is found in the range of 400% to 600%. Based on plasticity index, soils have been classified by Atterberg as indicated below :

Plasticity index (%)	Plasticity
0	Non-plastic
<7	Low plastic
7–17	Medium plastic
>17	Highly plastic

The shear strength of clay at plastic limit is a measure of its toughness. Toughness of two clays having the same plasticity index are inversely proportional to the flow index. The value of toughness index generally lies between 0 and 3 for most clays.

Clay soils are characterised by their ability to shrink on drying and swell on wetting. The degree of shrinkage is given by reduction in volume of soil mass expressed as a percentage of its initial volume. On the basis of degree of shrinkage, Schedig (1948) classified soils as indicated below.

Degree of shrinkage	Quality of soil
(%)	(Resistance to shrinkage)
< 5	Good
5–10	Medium
10–15	Poor
> 15	Very poor

According to Skempton the change in the volume of a clay soil during shrinking or swelling is a function of plasticity index and the quantity of clay-size particles present in the soil. Skempton studied the relation between plasticity index and soil fraction less than 2μ, defined a quantity called Activity Number and classified clay soils as indicated below.

Activity Number, $$A_c = \frac{I_p}{\text{Percent finer than } 2\mu}$$

A_c	Soil type
< 0.75	Inactive
0.75–1.40	Normal
> 1.40	Active

3.6 Determination of Field Density

The in-situ bulk density of a soil deposit in the field is commonly determined by (i) core-cutter method in the case of cohesive soils and (ii) sand replacement method in the case of cohesionless soils.

3.6.1 Core Cutter Method

A core-cutter is a steel cylinder open at both ends with one end sharpened to form the cutting edge. The usual dimensions are 10 cm internal diameter and height about 12 to 15 cm. The internal diameter, height and mass of core-cutter are noted. The place where the field density is to be determined is cleared of shrubs, if any, levelled and the core-cutter is placed vertically on the ground surface. Dolly (a steel ring about 2.5 cm height) is placed on top of the core cutter and is gently driven into the ground by blows of a rammer, until the top of the dolly is nearly flush with the ground surface. Sufficient soil is excavated from around the core cutter to enable a person to put his hands and lift the core-cutter with soil inside off the ground. The core cutter with soil inside is brought to the laboratory, the ends are trimmed, levelled and weighed. The soil is removed from the core cutter and samples are taken from top, middle and bottom portions for water content determination. The average of three determinations gives the in-situ water content. Using the experimental data computations are done as follows:

Internal volume of core cutter $= V_1$

Mass of core cutter $= M_1$

Mass of core cutter + soil $= M_2$

Field bulk density, $\gamma = \dfrac{M}{V} = \dfrac{(M_2 - M_1)}{V}$

Field water content $= w$

Field dry density, $\gamma_d = \dfrac{\gamma}{1 + w}$

3.6.2 Sand Replacement Method

The apparatus consists of sand pouring cylinder, calibrating cylinder and tray with circular hole. Sand passing through IS 600 μ sieve and retained on IS 300 μ sieve is taken in the sand-pouring cylinder. The method consists in determining the masses and volumes outlined below.

Part A Calibration

1. Mass of sand pouring cylinder + sand before pouring into calibrating cylinder $= M_1$
2. Mass of sand pouring cylinder + sand after pouring into calibrating cylinder $= M_2$
3. Mass of sand filling calibrating cylinder and cone portion, $M_3 = (M_1 - M_2)$
4. Mass of sand filling calibrating cylinder (after removing carefully the heap of sand above the top of cylinder) $= M_4$
5. Internal volume of calibrating cylinder $= V_1$
6. Density of sand filling calibrating cylinder, $\gamma_1 = \dfrac{M_4}{V_1}$
7. Mass of sand filling cone portion, $M_5 = (M_3 - M_4)$

Part B Field density determination

8. Mass of sand pouring cylinder + sand before pouring into pit $= M_6$
9. Mass of sand pouring cylinder + sand after pouring into pit $= M_7$
10. Mass of sand filling pit and cone portion, $M_8 = M_6 - M_7$
11. Mass of sand filling pit, $M_9 = (M_8 - M_5)$
12. Volume of pit $V = \dfrac{M_9}{\gamma_1}$
13. Mass of soil excavated from pit $= M$
14. Field bulk density $\gamma = \dfrac{M}{V}$

For step 4 the sand from the cone portion above the top of cylinder is carefully removed layer by layer without applying any downward pressure, using a straight edge. In order to ensure perfect simulation the mass in step 8 is taken equal to mass in step 1. In the field the spot where the field density is to be determined is cleared of shrubs, if any, levelled and the tray with hole is placed over it. A pit equal in diameter to the hole in the tray (and equal to that of calibrating cylinder) is dug upto a depth equal to that of calibrating cylinder. The excavated soil is carefully collected in the tray. The sand-pouring cylinder is placed centrally over the hole and shutter opened. After ensuring that the pit and cone portion are full and there is no further movement of sand in the sand pouring cylinder the shutter is closed and the sand pouring cylinder brought to the laboratory for weighing. The excavated soil collected in the tray is weighed and samples taken for water content determination in order to determine field dry density.

3.7 Density Index

Density Index (I_D), in the earlier days referred to as Relative Density (D_R), is the ratio of difference between maximum void ratio and natural void ratio to the difference between maximum void ratio and minimum void ratio.

$$I_D = \frac{e_{max} - e}{e_{max} - e_{min}}$$

3.6 (i)

where,

e_{max} = void ratio in loosest state

e_{min} = void ratio in densest state

e = natural void ratio

When $e = e_{max}, I_D = 0$

When $e = e_{min}, I_D = 1$ or 100%

I_D lies between 0 and 100%. Density index is a measure of the state of packing in the case of cohesionless soils and is usually expressed in percentage. Depending on value of I_D, states of packing of coarse-grained soils are described as follows.

Density index (%)	Description
< 35	Loose
35 to 65	Medium dense
65-85	Dense
> 85	Very dense

I_D can be conveniently expressed in terms of dry densities and the derivation is presented below.

$$e = \frac{G\gamma_w}{\gamma_d} - 1$$

$$e_{max} = \frac{G\gamma_w}{(\gamma_d)_{min}} - 1$$

$$e_{min} = \frac{G\gamma_w}{(\gamma_d)_{max}} - 1$$

$$I_D = \frac{e_{max} - e}{e_{max} - e_{min}}$$

$$= \frac{\left\{\dfrac{G\gamma_w}{(\gamma_d)_{min}} - 1\right\} - \left\{\dfrac{G\gamma_w}{\gamma_d} - 1\right\}}{\left\{\dfrac{G\gamma_w}{(\gamma_d)_{min}} - 1\right\} - \left\{\dfrac{G\gamma_w}{(\gamma_d)_{max}} - 1\right\}} = \frac{\dfrac{1}{(\gamma_d)_{min}} - \dfrac{1}{\gamma_d}}{\dfrac{1}{(\gamma_d)_{min}} - \dfrac{1}{(\gamma_d)_{max}}}$$

$$= \frac{(\gamma_d) - (\gamma_d)_{min}}{(\gamma_d)_{max} - (\gamma_d)_{min}} \times \frac{(\gamma_d)_{min}(\gamma_d)_{max}}{(\gamma_d)_{min}\gamma_d}$$

$$\therefore \qquad I_D = \frac{(\gamma_d) - (\gamma_d)_{min}}{(\gamma_d)_{max} - (\gamma_d)_{min}} \times \frac{(\gamma_d)_{max}}{\gamma_d}$$

3.6(ii)

Example 3.12

The natural dry density of a soil deposit was found to be 17.5 kN/m³. A sample of the soil was brought to the laboratory and the minimum and maximum dry densities were found as 16.0 kN/m³ and 19.0 kN/m³ respectively. Calculate the density index for the soil deposit.

Solution:

$$\gamma_d = 17.5 \text{ kN/m}^3$$

$$(\gamma_d)_{min} = 16.0 \text{ kN/m}^3$$

$$(\gamma_d)_{max} = 19.0 \text{ kN/m}^3$$

$$I_D = \frac{17.5 - 16}{19 - 16} \times \frac{19}{17.5} = 0.543 = 54.3\%$$

EXERCISE–3

3.1 5000 N of soil sample was taken for sieve analysis. The weight of soil retained on each sieve is as follows:

IS sieve	Weight of soil retained (N)
4.75 mm	12
2.36 mm	18
1.00 mm	950
425 μ	933
150 μ	1338
75 μ	895

Plot the grain size distribution curve. Determine C_c and C_u.

3.2 A soil sample has 97% of the particles (by weight) finer than 1 mm, 58% finer than 0.1 mm, 25% finer than 0.01 mm and 12% finer than 0.001 mm. Draw the grain size distribution curve and determine C_u and C_c.

3.3 A soil sample consisting of particles ranging in size from 0.01 mm to 0.5 mm is put on the surface of still water of 5 m depth. Compute (i) the time taken for the first particle and (ii) the time taken for all the particles to have settled at the bottom of tank. Take $G = 2.7$ and $\eta = 10^{-6}$ kN-sec/m².

Note: The student is advised to note the following change while working a problem on sedimentation analysis in SI units. From Stoke's law we have

$$v = \frac{2}{9} r^2 \frac{\gamma_s - \gamma_w}{\eta}$$

If we substitute r in m, γ_s and γ_w in kN/m³, and η in kN-sec/m², we will get v in m/sec.
If a particle of size D mm sinks through a height H_e m in time t min, then

$$v = \frac{H_e}{60t} m/\sec$$

Therefore, we can write

$$\frac{H_e}{60t} = \frac{2}{9}\left(\frac{D}{2000}\right)^2 \frac{(G-1)\gamma_w}{\eta}$$

Rearranging, we get

$$D = M\sqrt{\frac{H_e}{t}}$$

where $$M = 100 \sqrt{\frac{30\eta}{(G-1)\gamma_w}}$$

3.4 During a sedimentation analysis, the corrected hydrometer reading in a 1000 ml uniform soil suspension is 1.030 at the commencement of sedimentation. After 30 minutes, the corrected hydrometer reading is observed to be 1.015 and the corresponding effective depth is 10 cm. If $G = 2.7$ and viscosity of water is 1 centipoise, find

(i) the total mass of soil particles taken

(ii) the size of largest particle still in suspension at depth 10 cm at end of elapsed time interval 30 min. and the percent finer than this size.

3.5 The oven-dry weight of a clay pat was 0.12 N. Its volume was determined by immersion in mercury and the weight of displaced mercury was 0.81 N. Find the shrinkage limit of clay assuming $G = 2.65$

3.6 The following results refer to a liquid limit test:

Number of blows: 33   23   18   11

Water content (%): 41.5   49.5   51.5   55.6

The plastic limit is 23.5%. Determine the plasticity index and toughness index for the soil.

3.7 The following properties were determined for two soils A and B:

	A	B
Water content	37%	25%
Liquid limit	61%	35%
Plastic limit	25%	20%
Specific gravity of soil solids	2.72	2.68
Degree of saturation	100%	100%

Which of these soils:

(i) contains more clay particles,

(ii) has a greater saturated unit weight

(iii) has a greater dry unit weight

and (iv) has a greater void ratio?

Your answers should be supported by computations.

3.8 A field test gave the following results:

Mass of core-cutter + soil	= 3200 g
Mass of core-cutter	= 1500 g
Internal volume of core-cutter	= 1000 cc
Water content	= 12%
Specific gravity of soil solids	= 2.67

Calculate (i) In-place bulk density, (ii) dry density, (iii) degree of saturation and (iv) saturated density.

3.9   The results of a field density test conducted using sand replacement method are as follows:

     (*i*)   Mass of excavated soil = 923 g

     (*ii*)   Mass of sand pouring cylinder + sand, before pouring into pit = 5330 g

     (*iii*)   Mass of sand pouring cylinder + sand, after pouring into pit = 4152 g

     Earlier, the mass of sand filling the calibrating cylinder of 1000 ml capacity was found to be 1540 g. The mass of sand filling the cone portion of sand pouring cylinder was found to be 430 g. If the in-situ water content is 9%, calculate the in-situ dry density.

3.10   The in-situ bulk density of a sandy stratum is 1.9 g/cc and it has a water content of 8%. For determining the density index, dried sand from the stratum was first filled loosely in a 300 cc mould and then vibrated to give maximum density. The loose dry weight in the mould was 478 g and the dense dry weight at maximum compaction was 572 g. Calculate the density index of the stratum. Take $G = 2.7$.

CHAPTER 4

SOIL CLASSIFICATION SYSTEMS

4.1 Purpose of Soil Classification

The purpose of soil classification is to arrange various types of soils into specific groups based on physical properties and engineering behaviour of soils with the objective of finding the suitability of soils for different engineering applications, such as in the construction of earth dams, highways, and foundations of buildings, etc.

For different areas of applications and with the need for simplicity and acceptable terminology, several soil classification systems have been developed over the years, three of which are listed below.

1. Highway Research Board classification system
2. Unified Soil Classification system
3. Indian Standard soil classification system

4.2 Particle Size Classification Systems

In these systems, soils are arranged according to particle size ranges only, without consideration of other characteristics.

Three such systems, which have been widely used, are

(a) U.S Bureau of Soil and Public Roads Administration (PRA) classification system.

(b) M.I.T classification system (proposed by Prof. Gilboy of Massachussets Institute of Technology)

(c) Indian Standard particle size classification system (based on M.I.T system)

(a) U.S Bureau of Soils and PRA classification system

(b) M.I.T. Classification System

43

Clay (size)	Silt (size)	F	M	C	F	C	Cobble	Boulder
			Sand			Gravel		

Particle sizes (mm): 0.002mm | 0.075 | 0.425 | 2.00 | 4.75 | 20 | 80 | 300 mm

(c)　Indian Standard Classification (IS: 1498–1970)

Fig. 4.1. Particle size classification charts

Silts and clays should be distinguished, based on plasticity characteristics. But in these charts they are being distinguished, based only on particle size, without consideration of plasticity characteristics and to emphasize this, term 'size', is written along with 'clay' and 'silt'.

The textural classification chart of U.S. Public Roads Administration has been developed to classify composite soils. Based on particle sizes, the soil components are identified as sand (0.05 – 2 mm), silt (0.005 – 0.05 mm) and clay (< 0.005 mm). Knowing the percentages of sand, silt and clay, lines are drawn as suggested by the key accompanying the triangular chart. The region in which the point of intersection of the three lines lies gives the type of soil. It should be noted that the chart ignores soil fraction of size greater than 2.00 mm.

The term 'loam' used in this chart denotes a mixture of sand, silt and clay in varying proportions. It is primarily an agricultural term and has been adopted by highway engineers who deal with surface soil layers.

Example 4.1. A soil sample is found to consist of 30% sand, 30% silt and 40% clay. Classify the soil using textural classification chart of U.S.P.R.A.

Solution: The textural classification chart of U.S.P.R.A is shown in Fig 4.2. The three lines drawn

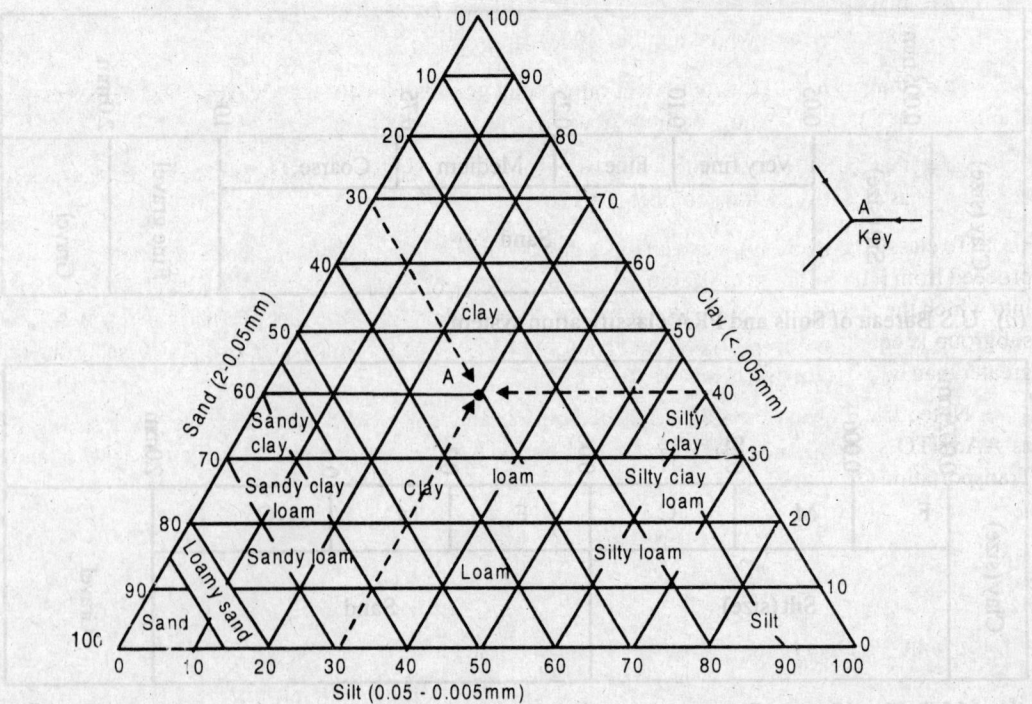

Fig. 4.2. Textural classification chart (*Adopted from U.S. Public Roads Administration*)

from respective points on the three sides, as guided by the key, intersect at point A, which lies in the region marked 'clay' Hence the soil can be classified as 'clay'.

4.3 Highway Research Board (HRB) Classification System

The Highway Research Board classification system, also known as Revised Public Roads Administration classification system, is used to find the suitability of a soil, as subgrade material in pavement construction. This classification system is based on both particle size ranges and plasticity characteristics. Soils are divided into 7 primary groups designated as A-1, A-2,.........A-7, as shown in Table 4.1.

Group A-1, is divided into two subgroups A-1-a and A-1-b and group A-2 into four subgroups, A-2-4, A-2-5, A-2-6, and A-2-7. A characteristic group index is used to describe the performance of a soil as subgrade material.

Group index is *not* used to place a soil in a particular group; it is actually a means of rating the value of a soil as a subgrade material within its own group. The higher the value of the group index, the poorer is the quality of the material.

The group index of a soil depends upon

     (i)   amount of material passing the 75-micron sieve,

     (ii)   liquid limit

and   (iii)   plastic limit

     Group index is given by the following equation:

     Group index $= 0.2\,a + 0.005\,ac + 0.01\,bd$ ...4.3(*i*)

where

    $a =$ that portion of percentage passing 75 micron sieve greater than 35 and not exceeding 75 expressed as whole number (0 to 40)

    $b =$ that portion of percentage passing 75 micron sieve greater than 15 and not exceeding 55 expressed as a whole number (0 to 40)

    $c =$ that portion of the numerical liquid limit greater than 40 and not exceeding 60 expressed as a positive whole number (0 to 20) and

    $d =$ that portion of the numerical plasticity index greater than 10 and not exceeding 30 expressed as a positive whole number (0 to 20).

To classify a given soil, sieve analysis data, liquid limit and plasticity index are obtained and we proceed from left to right in the Table 4.1 and by process of elimination find the first group from left into which the test data will fit. This gives the correct classification. The plasticity index of A-7-5 subgroup is equal to or less than liquid limit minus 30. The plasticity index of A-7-6 subgroup is greater than liquid limit minus 30.

Note : The PRA system was introduced in 1928 and revised in 1945 as HRB system. It is known as AASHTO system since 1978 after adoption by American Association of State Highway and Transportation Officials.

Table 4.1. HRB Classification of Soils and Soil-Aggregate Mixtures (also known as AASHTO system)

General Description	Granular material (35% or less passing 75 micron sieve)							Silt clay materials (more than 35% passing 75 micron sieve)				
Group classification	A-1		A-3	A-2				A-4	A-5	A-6	A-7	
	A-1-a	A-1-b		A-2-4	A-2-5	A-2-6	A-2-7				A-7-5	A-7-6
Sieve analysis, percent passing												
2.0 mm sieve*	50 max											
425 micron sieve	30 max	50 max	51 max									
75 micron sieve	15 max	25 max	10 max	35 max	35 max	35 max	35 max	36 min	36 min	36 min	36 min	36 min
Characteristics of fraction passing 425 micron sieve												
Liquid Limit				40 max	41 min	40 max	41 min	40 max	41 min	40 max	41 min	41 min
Plasticity Index	6 max	6 max	NP	10 max	10 max	11 min	11 min	10 max	10 max	11 min	11 min	11 min
Group Index	Zero					4 max		8 max	12 max	16 max	20 max	20 max
Usual type of significant constituent materials	Stone fragments, gravel and sand		Fine Sand	Silty or clayey gravel and sand				Silty soils		Clayey soils		
General rating as sub-grade	Excellent to good							Fair to poor				

* IS sieve sizes equivalent to ASTM sieves No. 10, 40, 200

Table 4.2 Extension of HRB Classification for Indian Black Cotton Soils

Group	A-7	A-7a	A-7b	A-7c
Per cent passing 75 micron sieve	upto 75	55–95	80–95	85–100
Characteristics of fraction passing 425 micron sieve				
Liquid limit	below 55	below 65	below 65	above 65
Plasticity index	below 25	below 42	below 42	above 42
Group index	20 max	20–30	30–40	40–50
Usual type of significant constituent material	Clayey soil			
General rating as sub-grade	Fair to Poor	Poor		

Classification of Black Cotton Soils of India

The division of A-7 group, on the basis of the demarcation line ($I_p = w_L - 30$) into A-7-5 and A-7-6 sub-group does not appear to divide the black cotton soils of India into two distinct groups, having maximum value of group index as 20 only. Based on investigations carried out at the Central Road Research Institute, New Delhi (1953), a classification of black cotton soil into narrow sub-groups (Table 4.2) has been suggested extending the maximum value of group index from 20 to 50. Accordingly the factors appearing in Eq. 4.3(i) for the calculation of group index have the following maximum values: $a = 65, b = 65, c = 45$ and $d = 34$. A majority of black cotton soils of India have been found to be well classified by this method for use in highways after treatment.

Example. 4.2. The properties of a subgrade soil are found as

(i) per cent finer than 0.075mm = 55%

(ii) liquid limit = 50%

(iii) plastic limit = 40%

Classify the soil according to HRB classification system (*revised PRA system or AASHTO system*)

Solution: From Table 4.1, we see that as more than 35% of soil passes through 0.075 mm sieve, it is fine grained.

Plasticity Index = $(50 - 40)\% = 10\%$

With $w_L = 50\%$ and $I_p = 10\%$ the soil belongs to the group A-5.

Group Index = $0.2\,a + 0.005\,ac + 0.01\,bd$

where, $a = 55 - 35 = 20$

$b = 55 - 15 = 40$

$c = 50 - 40 = 10$

$d = 10 - 10 = 0$

∴ Group index = $0.2\,(20) + 0.005\,(20)\,(10) + 0.001\,(40)\,(0)$

$= 5.0$

Soil classification is A-5.

4.4   Unified Soil Classification System

The Unified Soil Classification system is based on the Airfield Classification system that was developed by A Casagrande. The system is based on both grain size and plasticity characteristics of soils. The Unified Soil Classification (USC) system was adopted jointly by the Corps of Engineers, U.S. Army and U.S. Bureau of Reclamation during 1950s.

In this system soils are broadly divided into three divisions:

    1. Coarse-grained soils - if more than 50% by weight is retained on No. 200 ASTM sieve

    2. Fine-grained soils - if more than 50% by weight passes through No. 200 ASTM sieve.

    3. Organic soils

The soil components are assigned group symbols as indicated below:

Coarse-grained soils:

Gravel: G   Sand: S

Fine grained soils:

Silt: M   Clay: C   Organic soil: O

Table 4.3 Unified Soil Classification System

Major Divisions			Group Symbols	Typical Names
Coarse-Grained Soils More than 50% retained on No. 200 sieve*	Gravels 50% or more of coarse fraction retained on No. 4 sieve	Clean gravels –200 fraction <5%	GW	Well-graded gravels and gravel–sand mixtures, little or no fines
			GP	Poorly graded gravels and gravel-sand mixtures, little or no fines
		Gravels with fines –200 > 12% fraction	GM	Silty gravels, gravel–sand–silt mixtures
			GC	Clayey gravels, gravel–sand–clay mixtures
	Sands More than 50% of coarse fraction passes No. 4 sieve	Clean sands –200 < 5% fraction	SW	Well-graded sands and gravelly sands, little or no fines
			SP	Poorly graded sands and gravelly sands, little or no fines
		Sands with fines –200 > 12% fraction	SM	Silty sands, sand–silt mixtures
			SC	Clayey sands, sand–clay mixtures
Fine-grained soils 50% or more passes No. 200 sieve*	Silts and Clays Liquid limit 50% or less		ML	Inorganic silts, very fine sands, rock flour, silty or clayey fine sands
			CL	Inorganic clays of low to medium plasticity, gravelly clays, sandy clays, silty clays, lean clays

Contd.

	OL	Organic silts and organic silty clays of low plasticity
Silts and Clays Liquid limit greater than 50%	MH	Inorganic silts, micaceous or diatomaceous fine sands or silts, plastic silts
	CH	Inorganic clays of high plasticity, fat clays
	OH	Organic clays of medium to high plasticity
Highly Organic Soils	PT	Peat, muck, and other highly organic soils

* No. 200 sieve is of aperture size 0.074 mm.
No. 4 sieve is of aperture size 4.76 mm.

Table 4.3 gives the details of Unified Soil Classification system. The original Casagrande plasticity chart used for classifying fine-grained soils is given in Fig 4.3

The symbol M for silt is derived from the Swedish word 'Mo' for silt.

Example 4.3. Classify the soil with composition indicated in Example 4.2, using USC system.

Solution: Since more than 50% of soil passes through 0.074 mm sieve the soil is fine grained.

Plasticity Index = $(50 - 40)\% = 10\%$

From Fig 4.3, for $w_L = 40\%$ and $I_P = 10\%$ the soil can be classified as ML or OL

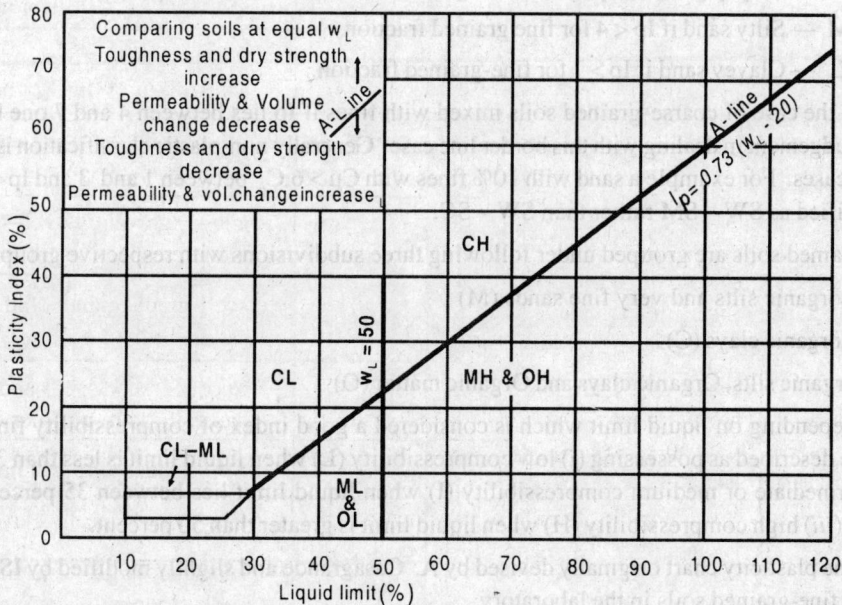

Fig. 4.3. Plasticity chart (Unified Soil Classification)

4.5 Indian Standard Soil Classification System

Indian Standard Soil Classification System (IS 1498–1970 Classification and identification of soils for general engineering purposes) is essentially based on Unified Soil Classification system and the salient features are given in the following discussion.

In this system soils are broadly divided into three divisions :

1. Coarse-grained soils - if more than 50 per cent by mass is retained on 75 micron IS sieve.

2. Fine grained soils - if more than 50 percent by mass passes through 75 micron IS sieve.

3. Highly organic soils and other miscellaneous soil materials: These soils contain large percentages of fibrous organic matter, such as peat and particles of decomposed vegetation. In addition, certain soils containing shells, concretions, cinders and other non-soil materials in sufficient quantities are also grouped in this division.

Coarse-grained soils are grouped as gravels and sands with group symbols G and S respectively.

Gravels (G) - if more than 50 percent by mass of the coarse-grained fraction (+75 micron) is retained on 4.75 mm IS sieve.

Sands (S) - if more than 50 percent by mass of the coarse-grained fraction passes through 4.75 mm IS sieve.

Depending on gradation Gravels (G) and Sands (S) are further described using group symbols as indicated below.

GW — Well graded gravel for which Cu > 4 and Cc lies between 1 and 3

GP — Poorly graded gravel which does not meet all gradation requirements of GW

SW — Well graded sand for which Cu > 6 and Cc lies between 1 and 3.

SP — Poorly graded sand which does not meet all gradation requirements of SW

GM — Silty gravel if Ip < 4 for fine-grained fraction.

GC — Clayey gravel if Ip > 7 for fine-grained fraction.

SM — Silty sand if Ip < 4 for fine grained fraction

SC — Clayey sand if Ip > 7 for fine-grained fraction.

In the case of coarse-grained soils mixed with fines if Ip lies between 4 and 7 one has to use proper judgement in dealing with this border line case. Generally non-plastic classification is favoured in such cases. For example a sand with 10% fines with Cu > 6,C_c between 1 and 3 and Ip = 6 would be classified as SW – SM rather than SW – SC.

Fine -grained soils are grouped under following three subdivisions with respective group symbols:

Inorganic silts and very fine sands (M)

Inorganic clays (C)

Organic silts, Organic clays and Organic matter (O)

Depending on liquid limit which is considered a good index of compressibility fine grained soils are described as possessing (*i*) low compressibility (L) when liquid limit is less than 35 percent (*ii*) intermediate or medium compressibility (I) when liquid limit lies between 35 percent and 50 percent (*iii*) high compressibility (H) when liquid limit is greater than 50 percent.

The plasticity chart originally devised by A. Casagrande and slightly modified by IS is used to classify fine-grained soils in the laboratory.

The A-line having the equation:

$$Ip = 0.73 (w_L - 20)$$

and the two vertical lines at $w_L = 35$ and $w_L = 50$ divide the chart into six regions with group symbols marked as shown in Fig. 4.4. If the plotted position lies below A-line, the soil has to be checked for organic odour by slight heating. If no organic odour is smelt then only it should be classified as inorganic silt. In case of doubt, the soil should be oven-dried and its liquid limit determined again. In the case of organic soils there will be large reduction in liquid limit on drying (reduction generally > 25%).

Table 4.4 Basic Soil Components (IS Classification)

(IS : 1498—1970)

Soil	Soil Component	Symbol	Particle size range and description
Coarse-grained Components	Boulder	None	Round to angular, bulky hard, rock particle, average diameter more than 300 mm
	Cobble	None	Round to angular, bulky hard, rock particle, average diameter smaller than 300 mm but retained on 80 mm sieve
	Gravel	G	Rounded to angular, bulky hard, rock particle, passing 80 mm sieve but retained on 4.75 mm sieve *Coarse* : 80 mm to 20 mm sieve *Fine* : 20 mm to 4.75 mm sieve
	Sand	S	Rounded to angular bulky hard, rocky particle, passing 4.75 mm sieve retained on 75 micron sieve *Coarse* : 4.75 mm to 2.0 mm sieve *Medium* : 2.0 mm to 425 micron sieve *Fine* : 425 micron to 75 micron sieve
Fine-grained Components	Silt	M	Particles smaller than 75-micron sieve indentified by behaviour; that is slightly plastic or non-plastic regardless of moisture and exhibits little or no strength when air dried
	Clay	C	Particles smaller than 75-micron sieve identified by behaviour, that is, it can be made to exhibit plastic properties within a certain range of moisture and exhibits considerable strength when air dried
	Organic matter	O	Organic matter in various sizes and stages of decomposition.

Table 4.5 IS Soil Classification (IS : 1498—1970)
(Including Field Identification and Description)

Major Divisions (1)	(2)	Group Symbol (3)	Typical Names (4)	Field identification procedures (excluding particles larger than 80 mm and basing fractions on estimated weights) (5)	Information required for describing soils (6)
COARSE-GRAINED SOILS — More than half of material is larger than 75-micron sieve size. The 75-micron sieve size is about the smallest particle visible to the naked eye					
GRAVELS — More than half of coarse fraction is larger than 4.75 mm sieve size	Clean gravels (Little or no fines)	GW	Well-graded gravel, gravel-sand mixtures, little or no fines	Wide range in grain size and substantial amounts of all intermediate particle sizes	For undisturbed soil stratification, degree of compactness, cementation, moisture conditions and drainage characteristics.
	Clean gravels (Little or no fines)	GP	Poorly graded gravels or gravel-sand mixtures, little or no fines	Predominantly one size or a range of sizes with some intermediate sizes missing	Give typical name, indicate approximate percentages of sand, gravel, maximum size, angularity, surface condition and hardness of the coarse grains, local or geologic name and other pertinent descriptive information and symbol in parentheses.
	Gravel with fines (Appreciable amount of fines)	GM	Silty gravels, poorly graded gravel-sand-silt mixtures	Non-plastic fines or fines with low plasticity (for identification procedures see ML and MI on p.53)	
	Gravel with fines (Appreciable amount of fines)	GC	Clayey gravels, poorly graded gravel-sand-silt mixtures	Plastic fines (for Identification procedures see CL and CI on p.53)	Example : Silty sand gravelly, about 20% hard angular gravels 10 mm maximum size, rounded and subangular sand grains, about 15%.
SANDS — More than half of coarse fraction is smaller than 4.75 mm sieve size (For visual classification, the 5 mm size may be used as equivalent to the 4.75 mm sieve size)	Clean sands (Little or no fines)	SW	Well graded sands, gravelly sands, little or no fines	Wide range in grain size and substantial amounts of all intermediate particle sizes	
	Clean sands (Little or no fines)	SP	Poorly graded sands or, gravelly sands, little or no fines	Predominantly one size or a range of sizes with some intermediate sizes missing	
	Sands with fines (Appreciable amount of fines)	SM	Silty sands, poorly graded sand-silt mixtures	Non-plastic fines or fines with low plasticity (for identification procedures see MI and ML on p.53)	Non-plastic fines with low dry strength, well compacted and moist in place, alluvial sand. (SM).
	Sands with fines (Appreciable amount of fines)	SC	Clayey sands, poorly graded sand-clay mixtures	Plastic fines (for Identification procedures see CL and CI on p.53)	

Contd.

1	2	3	4	5 Field Identification procedures (on fraction smaller than 425 μ sieve size) Dry strength	Dilatancy	Toughness	6
FINE-GRAINED SOILS — More than half of material is smaller than 75-micron sieve size	SILTS AND CLAYS — With low compressibility: liquid limit is less than 35%	ML	Inorganic silts and very fine sands, rock flour, silty or clayey fine sand or clayey silts with none to low plasticity	None to low	Quick	None	For undisturbed soils, add information on structure, stratificaion, consistency in undisturbed and remoulded states, moisture and drainage conditions.
		CL	Inorganic clays, gravelly clays, sandy clays, silty clays, lean clays of low plasticity	Medium	None to very slow	Medium	
		OL	Organic silts of low plasticity	Low	Slow	Low	
	SILTS AND CLAYS — With medium compressibility: liquid limit is greater than 35% and less than 50%	MI	Inorganic silts, silty or clayey fine sand or clayey silts of medium plasticity	Low	Quick to slow	None	
		CI	Inorganic clays, gravelly clays, sandy clays, silty clays, lean clays of medium plasticity	Medium to high	None	Medium	Give typical names, indicate degree and character of plasticity, amount and maximum size of coarse grains, colour in wet condition, odour, if any, local or geologic name and other pertinent descriptive information, and symbol in parentheses. Example: Clayey silt, brown, slightly plastic, small percentage of fine sand, numerous vertical root holes, firm and dry in place; loess, (ML).
		OI	Organic silts and organic silty clays of medium plasticity	Low to medium	Slow	Low	
	SILTS AND CLAYS — With high compressibility: liquid limit is greater than 50%	MH	Inorganic silts of high compressibility, micaceous or diatomaceous fine sandy or silty soils plastic silts	Low to medium	Slow to none	Low to medium	
		CH	Inorganic clays of high plasticity, fat clays	High to very high	None	High	
		OH	Organic clays of medium to high plasticity	Medium to high	Slow to very slow	Low to medium	
Highly organic soil		Pt	Peat and other highly organic soils with very high compressibility	Readily identified by colour, odour, spongy feel and frequently by fibrous texture			

Note. Boundary classification, Soil possessing characteristics of two groups are designated by combination of group symbols.
For example : GW-GC, well graded gravel sand mixture with clay binder.

Table 4.6 IS Soil Classification : Laboratory Classification Criteria for Coarse Grained Soils (IS : 1498 - 1970)

Group Symbols	Laboratory Classification Criteria	
GW	C_U Greater than 4	Determine percentages of gravel and sand from grain sizecurve depending on percentage of fines (fraction smaller than No. 75 micron sieve size); coarse-grained soils are classified as follows:
	C_C Between 1 and 3	
GP	Not meeting all gradation requirements for GW	
GM	I_P less than 4	I_P between 4 and 7 are border line cases requiring use of dual symbols.
GC	I_P greater than 7	Less than 5% : GW, GP, SW, SP
SW	C_U Greater than 6 C_C Between 1 and 3	More than 12%: GM, GC, SM, SC
		5% to 12% : border line cases requiring use of dual symbols
SP	Not meeting all gradation requirements for SW	$C_U = \dfrac{D_{60}}{D_{10}}$ (uniformity coefficient)
SM	I_P less than 4	I_P between 4 and 7 are border line cases requiring use of dual symbols
SC	I_P greater than 7	$C_C = \dfrac{(D_{30})^2}{D_{10} \times D_{60}}$ (coefficient of curvature)

Eq. of A-line $I_P = 0.73 (w_L - 20)$

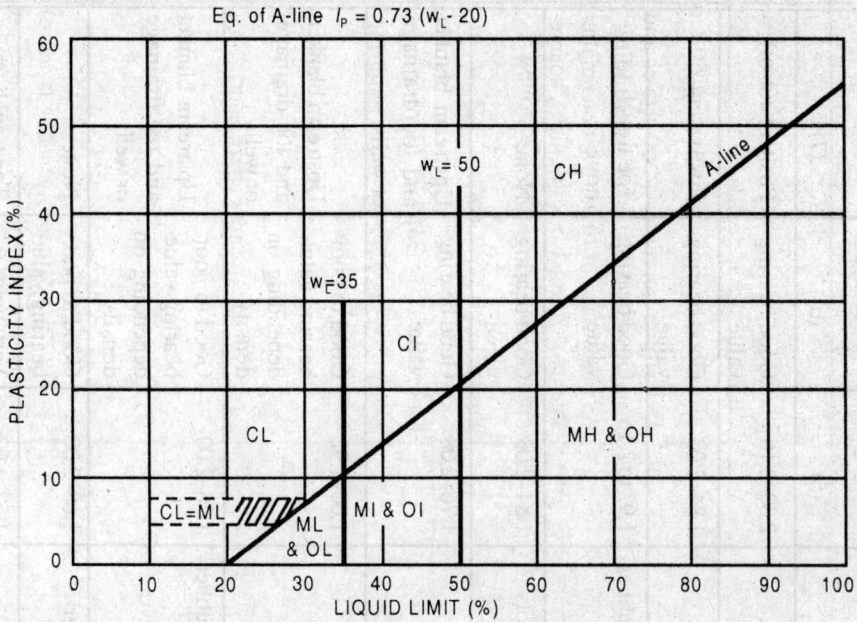

Fig. 4.4 Plasticity Chart (IS Soil Classification System).

Table 4.7 Characteristics Pertinent to Embankments and Foundations

Soil group (1)	Value for Embankment (2)	Permeability cm per sec. (3)	Compaction characteristics (4)	Unit dry density g/cm³ (5)	Value for foundation (6)	Requiremets for seepage control (7)
GW	Very stable, pervious shells of dikes and dams	$k > 10^{-2}$	Good, tractor, rubber-tyred, street-wheeled roller	2.00–2.16	Good bearing value	Positive cut-off
GP	Reasonably stable, pervious shells of dikes and dams	$k > 10^{-2}$	Good, tractor, rubber tyred, street-wheeled roller	1.84–2.00	Good bearing value	Postive cut-off
GM	Reasonably stable, not particularly suited to shells, but may be used for impervious cores or blankets	$k = 10^{-3}$ to 10^{-8}	Good, with close control, rubber tyred, sheepfoot roller	1.92 to 2.16	Good bearing value	Toe trench to none
GC	Fairly stable, may be used for impervious core	$k = 10^{-6}$ to 10^{-8}	Fair, rubber-tyred, sheepfoot roller	1.84–2.08	Good bearing value	None
SW	Very stable, pervious sections, slope protection required	$k > 10^{-3}$	Good, tractor	1.76–2.08	Good bearing value	Upstream blanket and toe drainage or wells
SP	Reasonably stable, may be used in dike section with flat slopes	$k > 10^{-3}$	Good, tractor	1.60–1.92	Good to poor bearing value depending on density	Upstream blanket and toe drainage or wells
SM	Fairly stable, not particularly suited to shells, may be used for impervious cores or dikes	$k = 10^{-3}$ to 10^{-6}	Good, with close control, rubber tyred, sheepfoot roller ·	1.76–2.00	Good to poor bearing value depending on density	Upstream blanket and toe drainage or wells
SC	Fairly stable, used for impervious core for flood control structures	$k = 10^{-6}$ to 10^{-8}	Fair, sheepfoot roller, rubber-tyred	1.68–2.00	Good to poor bearing value	None
ML, MI	Poor stability, may be used for embankments with proper control	$k = 10^{-3}$ to 10^{-6}	Good to poor, close control essential, rubber-tyred roller, sheepfoot roller	1.52–1.92	Very poor; susceptible to liquefaction	Toe trench to none

CONTINUATION OF TABLE **4.7** ON PAGE **56**

Soil group (1)	Value for Embankment (2)	Permeability cm per sec. (3)	Compaction characteristics (4)	Unit dry density g/cm³ (5)	Value for foundadtion (6)	Requirements for seepage control (7)
CL, CI	Stable impervious cores and blankets	$k = 10^{-6}$ to 10^{-3}	Fair to good, sheepfoot roller, rubber tyred	1.52–1.92	Good to poor bearing	None
OL, OI	Not suitable for embankments	$k = 10^{-4}$ to 10^{-6}	Fair to poor, sheepfoot roller	1.28–1.60	Fair to poor bearing, may have excessive settlements	None
MH	Poor stability, cores of hydraulic fill dam, not desirable in rolled fill construction	$k = 10^{-4}$ to 10^{-6}	Poor to very poor, sheepfoot roller	1.12–1.52	Poor bearing	None
CH	Fair stability with flat slopes, thin cores, blanket and dike sections	$k = 10^{-6}$ to 10^{-8}	Fair to poor, sheepfoot roller	1.20–1.68	Fair to poor bearing	None
OH	Not suitable for embankments	$k = 10^{-6}$ to 10^{-8}	poor to very poor, sheepfoot roller	1.04–1.60	Very poor bearing	None
Pt	Not used for construction		Compaction not practical		Remove from foundation	

Note

1. Values in columns 2 and 6 are for guidance only. Design should be based on test results.
2. In Column 4, the equipment listed will usually produce densities with a reasonable number of passes when misture conditions and thickness of lift are properly controlled.
3. Column 5, unit dry weights are for compacted soils at optimum water content for IS light compaction effort.

Table 4.8 Characteristics Pertinent to Roads and Airfields

Soil group	Value as subgrade when not subject to frost action	Value as subbase when not subject to frost action	Value as base when not subject to frost action	Potential frost action	Compressibility and Expansion	Drainage Characteristics	Compaction Equipment	Unit dry density gm/cm³
1	2	3	4	5	6	7	8	9
GW	Excellent	Excellent	Good	None to very slight	Almost none	Excellent	Crawler-type tractor, rubber-tyred roller, steel-wheeled roller	2.00–2.24
GP	Good to excellent	Good	Fair to good	None to very slight	Almost none	Excellent	Crawler-type tractor, rubber-tyred roller close control of moisture	1.76–2.24
GM d	Good to excellent	Good	Fair to good	Slight to medium	Very slight	Fair to poor	Rubber-tyred roller, sheepfoot roller, close control of moisture	2.00–2.32
GM u	Good	Fair	Poor to not suitable	Slight to medium	Slight	Poor to practically impervious	Rubber-tyred roller, sheepfoot roller	1.84–2.16
GC	Good	Fair	Poor to not suitable	Slight to medium	Slight	Poor to practically impervious	Rubber-tyred roller, sheepfoot roller	2.08–2.32
SW	Good	Fair to good	Poor	None to very slight	Almost none	Excellent	Crawler-type tractor, rubber-tyred roller	1.76–2.08
SP	Fair to good	Fair to good	Poor to not suitable	None to very slight	Almost none	Excellent	Crawler-type tractor, rubber-tyred roller	1.68–2.16
SM d	Fair to good	Fair to good	Poor	Slight to high	Very slight	Fair to poor	Rubber-tyred roller, sheepfoot roller, close control of moisture	1.92–2.16
SM u	Fair	Poor to fair	Not suitable	Slight to high	Slight to medium	Poor to practically impervious	Rubber-tyred roller, sheepfoot roller	1.60–2.08
SC	Poor to Fair	Poor	Not suitable	Slight to high	Slight to medium	Poor to practically impervious	Rubber-tyred roller, sheepfoot roller	1.60–2.16
ML, MI	Poor to Fair	Not suitable	Not suitbale	Medium to very high	slight to medium	Fair to poor	Rubber-tyred roller, sheepfoot roller, close control of moisture	1.44–2.08
CL, CI	Poor to Fair	Not suitable	Not suitable	Medium to high	Medium	Practically impervious	Rubber-tyred roller, sheepfoot roller	1.44–2.08

Continuation of Table 4.8 on page 58

1	2	3	4	5	6	7	8	9
OL, OI	Poor	Not suitable	Not suitable	Medium to high	Medium to high	Poor	Rubber-tyred roller, sheepfoot roller	1.44–1.68
MH	Poor	Not suitable	Not suitable	Medium to very high	High	Fair to poor	Sheepfoot roller, rubber-tyred roller	1.28–1.68
CH	Poor to very poor	Not suitable	Not suitable	Medium	High	Practically impervious	Sheepfoot roller, rubber-tyred roller	1.44–1.84
OH	Poor to very poor	Not suitable	Not suitable	Medium	High	Practically impervious	Sheepfoot roller, rubber-tyred roller	1.28–1.76
Pt	Not suitable	Not suitable	Not suitable	Slight	Very high	Fair to poor	Compaction not practical	—

Note

1. Column 1 : Division of GM and SM groups into Sub-division of "*d*" and "*u*" are for roads and airfields only. Sub-division is on basis of Atterberg limits suffix d (*i.e*, *GMd*) will be used when the liquid limit is 25 or less and the plasticity index is 5 or less ; the suffix "*u*" will be used otherwise.

2. In column 8, the equipment listed will usually produce the required densities with a reasonable number of passes when moisture condition and thickness of lift are properly controlled. In some instances, several types of equipment are listed because variable soil characteristics within which a given soil group may require different equipment. In some instances, a combination of two types may be necessary.

   (*a*) *Processed base materials and other angular materials*. Steel-wheeled and rubber-tyred rollers are recommended for hard, angular materials with limited fines or screening. Rubber-tyred equipment is recommended for softer materials subjected to degradation.

   (*b*) Finishing. Rubber-tyred equipment is recommended for rolling during final shaping operations for moist soils and processed materials.

   (*c*) *Equipment size*. The following sizes of equipment; are necessary to assure the high densities required for airfield construction.
   Crawler-type tractor-total weight in excess of 136 kN. Rubber-tyred equipment-wheel load in excess of 68 kN, wheel loads as high as 181 kN may be necessary to obtain the required densities for some materials (based on contact pressure of approximately 457 to 1050 kN/m²).
   Sheeps foot roller-unit pressure (on 38.7 to 77.4 cm²) to be in excess of 1750 kN/m² and unit pressure as high as 4600 kN/m² may be necessary to obtain the required densities for some materials. The areas of the feet should be at least 5 percent of the total peripheral area of the drum, using the diameter measured to the faces of the feet.
   Column 9 : Unit dry mass are for compacted soil at optimum water content for IS heavy compaction effort.

3. Column 9 : Unit dry mass are for compacted soil at optimum water content for IS heavy compaction effort.

Example 4.4. A soil sample is found to have the following properties. Classify the soil according to I.S. classification system.

Passing 75 μ sieve	= 10%
Passing 4.75 *mm* sieve	= 70%
Uniformity coefficient	= 8
Coefficient of curvature	= 2.8
Plasticity Index	= 4

Solution:

Since more than 50% is retained on 75 μ sieve the soil is coarse- grained.

Since more than 50% passes through 4.75 mm sieve , the soil is sandy (S).

Further, $C_u > 6$ and C_c lies between 1 and 3.

Therefore it is well graded. Since $I_p < 4$, it is silty.

∴ Soil classification is SW-SM.

EXERCISE – 4

4.1 What is the purpose of soil classification?

4.2 · Explain how soils are classified according to IS soil classification system.

4.3 Explain salient features of plasticity chart.

4.4 Test on subgrade soil gave the following data :

Percentage of soil passing IS 75 μ sieve = 65%

Liquid limit of soil = 48%

Plastic limit of soil = 34%

Classify the soil using HRB soil classification system.

CHAPTER 5

SOIL FORMATION AND SOIL STRUCTURE

5.1 Introduction

The vast and different soil deposits on the earth's surface are a result of one or more of the following four geological cycle of events continually taking place : (*i*) weathering, (*ii*) transportation, (*iii*) deposition and (*iv*) upheavel.

5.2 Weathering

Weathering of rocks is of two types, (*i*) mechanical or physical weathering and (*ii*) chemical weathering. In mechanical weathering, a rock gets simply broken down to particles of various sizes, without any change in chemical composition. The soil formed due to mechanical weathering will be cohesionless (sand and gravel).

The physical agencies causing mechanical weathering of rocks are (*i*) daily and seasonal temperature changes, (*ii*) flowing water, glaciers and wind, which produce impact and abrasive action on rock, (*iii*) splitting action of ice and (*iv*) growth of roots of plants in rock fissures and to a minor degree burrowing activities of small animals like earthworms. Rock and landslides can cause further fragmentation of rock pieces.

The chemical weathering of rocks also referred to as decomposition is the result of following reactions; hydration, oxidation, carbonation and leaching. Oxidation greatly affects rocks containing iron. An example of hydration is the chemical decomposition of mineral feldspar in granite to form kaolinite. Carbonation of rock material is caused by carbon dioxide in the presence of water. Limestones are very much affected by carbonation. Leaching is the process in which percolating water washes out water-soluble salts from the soil. Soil produced by chemical weathering of rocks will be cohesive (silt and clay).

As already introduced in previous chapter, based on particle-size ranges soils are classified as:

(*i*) Coarse-grained soils (sand and gravel)
(*ii*) Fine-grained soils (silt and clay)

Soils are also classified as residual soils and transported soils. Residual soils are those which have remained over the parent rock from which they have been formed. Transported soils are those which have been transported from the place of formation and redeposited elsewhere.

Residual soil deposits are relatively shallow in depth. They are characterized by a gradual transition from soil through partially weathered rock, fractured and fissured rock, to bedrock. Transported soils may be found as deposits of considerable depth. The homogeneity or heterogeneity of these deposits depends on the manner of their transportation and deposition.

5.3 Transportation and Deposition

Water, wind, ice and gravity are the natural agencies of transportation and deposition of soils. Alluvial, marine and lacustrine soils are the three types of water-transported soils. Alluvial soil or alluviums are the soils which have been transported and subsequently deposited by flowing water. An alluvial fan is formed when the velocity of a soil-laden stream suddenly decreases due to abrupt decreases in gradient. Flood-plains are formed on the sides of a stream due to overflowing of flood water. A delta is formed just before a stream reaches the standing water of the sea. Alluvial soil deposits are usually stratified because of fluctuations in velocity of flowing water. The average particle size of alluvial deposits decreases with increasing distance from the source of stream. The delta soils are soil deposits farthest from the source of a stream and usually consist of silt and clay. Marine deposits are formed when fine-grained soils are carried beyond deltas into the sea. Lacustrine soils are soils deposited at the bed of lakes.

Dune sand and loess are the two types of soils transported by wind and subsequently deposited. They are known as aeolin deposits. Dune sand can be found in arid regions and in some seashore where the wind blows consistently from the same direction. Loess is the wind transported fine-grained silt or silty clay. It has little or no stratification and is capable of standing with nearly vertical slopes.

Glacial soils are soils, which have been transported and re-deposited by glaciers. Glacial drift is the term used to refer to all the material picked up, mixed, transported and re-deposited by glaciers. Material transported only by ice and deposited underneath the ice sheet is referred to as glacial till. Boulder clay is the name given to an unstratified glacial till, when it contains a high clay content and particles ranging from boulders to clay sizes. Varved clay is the name given to the layers of clay deposited at the bottom of a glacial lake formed by blocked molten water. The yearly laminated deposit in a glacial lake is generally known as varve. The thickness of a varve may vary from less than 1mm to over 12mm and consists of fine-grained material such as clay, silt or rock flour.

Hard pan is a glacial deposit consisting of a layer of extremely dense soil formed when a glacial drift got cemented by products of chemical weathering and compressed subsequently by ice pressures.

When a glacier reached a line where it melted away, glacial till got deposited in the form of ridges at the outer edges of the ice sheets. This type of glacial deposit is called terminal moraine. Ground moraine is the name given to vertical deposition of glacial till due to rapid melting of vast sheet of ice.

Glacial stream resulting from melting of glaciers have carried and deposited glacial drift in stratified layers. This type of deposit is referred to as outwash plain.

It should be noted that except in the case of varves and outwash plains, glacial deposits in general consist of a heterogeneous mixture of rock fragments and soil of varying size and proportions and without any normal stratification. Colluvial soils are those transported by gravitational forces. Talus is a colluvial soil deposit consisting of rock fragments and soil collected at the foot of cliffs or steep slopes.

5.4 Upheavel

One can visualize tremendous earth upheavels as having taken place subjecting soil deposits and rock formations to enormous uplifting, tilting and folding. The cycle of weathering, transportation and deposition can then be expected to start again as soon as upland is formed, leading to further formation of soils.

5.5 Soil Structure

The term soil structure in general, refers to the arrangement or state of aggregation of particles in a soil mass. But deeper understanding of soil structure demands consideration of mineralogical composition, shape and orientation of soil particles; the nature and properties of soil water, and the forces of interaction between soil particles and soil water. The engineering behavior of soils is influenced by soil structure to varying degrees.

Types of Soil Structure

Following are the types of soil structure which have been recognized in various soil deposits.

1. Single grained structure - in the case of coarse-grained soil deposits.
2. Honeycomb structure - in the case of silt deposits

3. Flocculated structure and dispersed structure - in the case of clay deposits

4. Coarse - grained skeleton structure and cohesive matrix structure - in the case of composite soils.

1.   Single grained structure

This type of structure will be found in the case of coarse - grained soil deposits. When such soils settle out of suspension in water, the particles settle independently of each other. The major force causing their deposition is gravitational and the surface forces are too small to produce any effect. There will be particle-to-particle contact in the deposit. The void ratio attained depends on the relative size of grains.

2.   Honeycomb structure

This type of structure is associated with silt deposits. When silt particles settle out of suspension, in addition to gravitational forces the surface forces also play a significant role. When particles approach the lower region of suspension they will be attracted by particles already deposited as well as the neighbouring particles leading to the formation of arches. The combination of a number of arches leads to the honeycomb structure as shown in Fig.5.2. As the deposit has high void ratio, when disturbed as in pile driving, there will be large reduction in volume due to breakdown of structure.

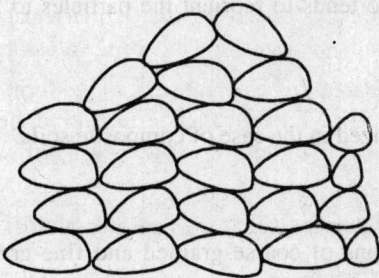

Fig. 5.1. Single grained structure

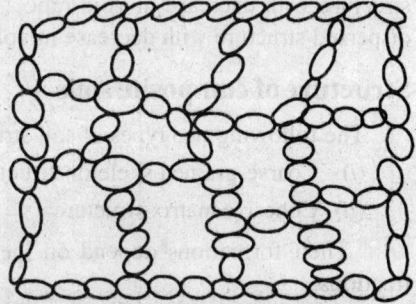

Fig. 5.2. Honeycomb structure

3.   Flocculated structure and Dispersed structure

These are the two types of structures found in clay deposits. In the case of flocculated structure there will be edge-to-edge and edge-to-face contact between the particles [Fig. 5.3(a)]. This type of formation is due to the net electrical forces between the adjacent particles at the time of deposition being attractive in nature. The concentration of dissolved minerals in water leads to formation of flocculated structure with very high void ratio as in the case of marine deposits.

(a) Flocculated structure   **(b) Dispersed structure**

Fig. 5.3. Soil structure in clay deposits

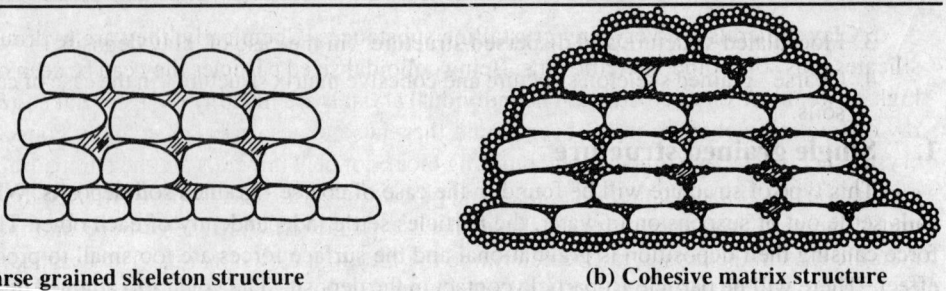

(a) Coarse grained skeleton structure   (b) Cohesive matrix structure

Fig. 5.4. Soil structure in composite soil deposits.

In the case of dispersed or oriented structure, the particles will have face to face contact [Fig. 5.3 (b)]. This type of formation is due to net electrical forces between adjacent soil particles at the time of deposition being repulsive in nature. This type of structure is common in fresh water deposits.

Clays with flocculated structure will have relatively high void ratio. Remoulding of such soils or application of pressure as in compaction leads to slippage of particles resulting in dispersed structure with decrease in void ratio. Consolidation also tends to reorient the particles to form dispersed structure with decrease in volume.

Structure of composite soils

The following two types of soil structures are recognized in the case of composite soils.

(*i*) Coarse-grained skeleton structure

(*ii*) Cohesive matrix structure

Their formations depend on the relative proportions of coarse-grained and fine-grained fractions.

The coarse-grained skeleton structure can be found in the case of composite soils in which the coarse-grained fraction is greater in proportion compared to fine-grained fraction. The coarse-grained particles form the skeleton with particle to particle contact and the voids between these particles will be occupied by the fine-grained particles [Fig.5.4 (a)].

The cohesive matrix structure can be found in composite soils in which the fine-grained fraction is more in proportion compared to coarse-grained fraction. In this case the coarse-grained particles will be embedded in fine-grained fraction and will be prevented from having particle-to-particle contact. This type of structure is relatively more compressible compared to the more stable coarse-grained skeleton structure.

5.6   Soil Particles in Soil Mass

Soil particles are composed of minerals which can be classified as primary minerals and secondary minerals. Primary minerals will be the same as in parent rock. Sand and gravel, which are products of mechanical weathering, are composed of primary minerals. They do not posses cohesion and plasticity characteristics. They are bulky particles and their behaviour is governed by gravitational forces.

Secondary minerals are found in silts and clays and are the results of chemical weathering. They can be further classified as non-clay minerals and clay minerals. Non-clay minerals are amorphous and impart little or no cohesion and plasticity characteristics to soil.

Clay minerals are very tiny crystalline substances. Chemically, they are hydrous alumino silicates plus some other metallic ions. Being colloidal sized particles they can be seen only with a high-power electron microscope. The individual crystals are like tiny plates or flakes and from X-ray diffraction technique it has been found that these plates or flakes actually consist of many crystal sheets. The two fundamental building blocks of clay minerals are the tetrahedral silica unit and octahedral gibbsite unit. Silica sheet and gibbsite sheet are each combinations of the respective units. The particular way in which these sheets are stacked, the type of bond and the different metallic ions in the crystal lattice determine the different clay minerals. The tetrahedral sheet is basically a combination of silica tetrahedral units. Each tetrahedral unit consists of four oxygen atoms at the corner surrounding a single silicon atom as shown in Fig 5.5 (a). An isometric view of tetrahedral sheet is shown in Fig.5.5 (b), which illustrates how the oxygen atoms at the base of each tetrahedron are combined to form a sheet structure. The octahedral sheet is a combination of octahedral units. Each octahedral unit consists of six oxygen or hydroxyl, enclosing an aluminium, magnesium, iron or other atom. An octahedral unit is shown in Fig 5.6 (a) while Fig5.6 (b) illustrates how the octahedral units combine to form an octahedral sheet. Substitution of different cations in the octahedral sheet is rather common and is known as isomorphous substitution. Isomorphous substitutions lead to different clay minerals. The common octahedral sheet known as gibbsite has all the anions as hydroxyls and two-third of the cations as aluminum. The variation in the stacking of the two basic sheet structure and nature of bonding has given rise to over dozen clay minerals which have been identified. From an engineering point of view three clay minerals of interest are kaolinite, montmorillonite and illite.

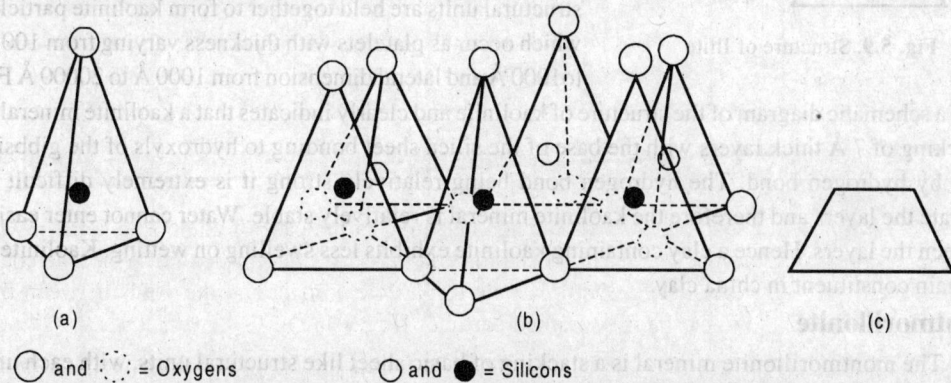

(a)   (b)   (c)

◯ and ⬚ = Oxygens   ◯ and ● = Silicons

Fig. 5.5. Tetrahedral silica unit, silicon sheet and its symbol.

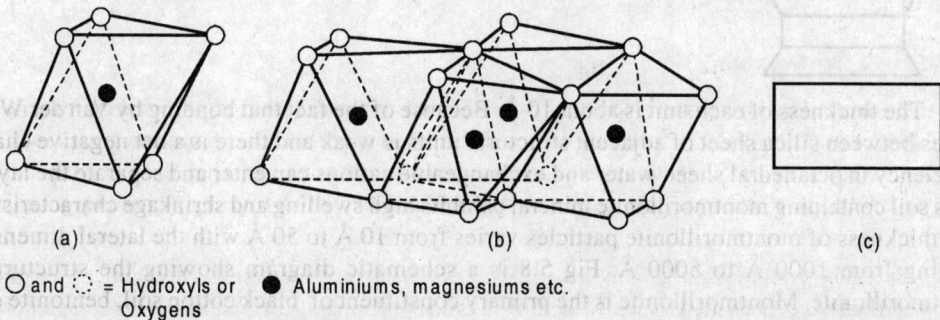

(a)   (b)   (c)

◯ and ⬚ = Hydroxyls or   ● Aluminiums, magnesiums etc.
Oxygens

Fig. 5.6. Octahedral gibbsite unit, gibbsite sheet and its symbol.

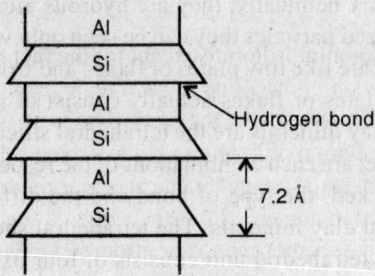

Fig. 5.7. Structure of Kaolinite.

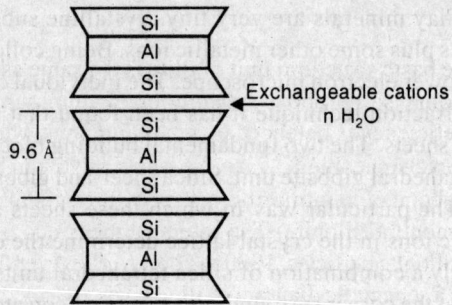

Fig. 5.8. Structure of Montmorillonite

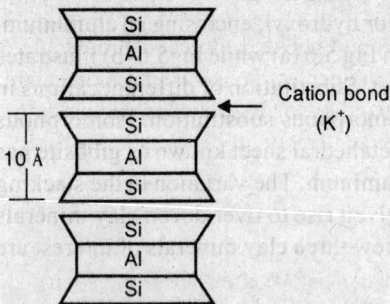

Fig. 5.9. Structure of Illite

Kaolinite

This is the most common mineral of the kaolin group. Each structural unit of kaolinite is a combination of two layers with apeces of a silica layer joined to one of a gibbsite layer. The structural unit is represented by the symbol.

It is about 7 Å thick and successive layers of structural units are held together to form kaolinite particles which occur as platelets with thickness varying from 100 Å to1000 Å and lateral dimension from 1000 Å to 20000 Å Fig 5.7 is a schematic diagram of the structure of kaolinite and clearly indicates that a kaolinite mineral is a stacking of 7 Å thick layers with the base of the silica sheet bonding to hydroxyls of the gibbsite sheet by hydrogen bond. The hydrogen bond being relatively strong it is extremely difficult to separate the layers and therefore the kaolinite mineral is relatively stable. Water cannot enter easily between the layers. Hence a clay containing kaolinite exhibits less swelling on wetting. Kaolinite is the main constituent in china clay.

Montmorillonite

The montmorillonite mineral is a stacking of basic sheet like structural units, with each unit made up of gibbsite sheet sandwiched between two silica sheets and is represented by the symbol

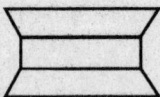

The thickness of each unit is about 10 Å. Because of the fact that bonding by Van der Waals forces between silica sheet of adjacent structural units is weak and there is a net negative charge deficiency in octahedral sheet, water and exchangeable cations can enter and separate the layers. Thus soil containing montmorillonite mineral exhibits high swelling and shrinkage characteristics. The thickness of montmorillonite particles varies from 10 Å to 50 Å with the lateral dimension ranging from 1000 Å to 5000 Å. Fig 5.8 is a schematic diagram showing the structure of montmorillonite. Montmorillonite is the primary constituent of black cotton soil, bentonite clay and other expansive clays.

Illite

The basic structural unit of illite is the same as that of montmorillonite except for the fact that there is some isomorphous substitution of aluminum for silicon in the silica sheet and the resultant charge deficiency is balanced by potassium ions, which bond the layers in the stack. The bond with the non- exchangeable K^+ ions are weaker than the hydrogen bond of kaolinite but is stronger than the water bond of montmorillonite. The illite crystal does not swell so much in the presence of water as does montmorillonite, The lateral dimensions of illite particles are about the same as that of montmorillonite particles, 1000 to 5000 Å. However the thickness is more and varies from 50 to 500 Å, Fig 5.9 is a schematic diagram of illite layer.

[**Note:** *One Angstrom (Å)* $= 10^{-10}m$]

5.7 Primary Valence Bond and Secondary Valence Bond

Primary valence bond is intramolecular bond, that is the bond between atoms in a molecule. Secondary valence bond is the bond between atoms of different molecules. Thus it is intermolecular bond. Primary valence bond being very strong, they are not affected by loads applied in engineering practice. The secondary valence bond is relatively weak and its understanding is essential to fully assess soil structure. A brief summary follows :

The two types of secondary valence bonds recognized as very important in the case of fine-grained soils are Van der Waals forces and the hydrogen bond.

Secondary valence forces acting between molecules are thought to develop from electrical moments in the individual molecules. Although a molecule is electrically neutral, the centre of action of positive charges may not coincide with that of negative charges. Due to this an electrical moment is developed. A system with an electrical moment is said to be polar and is referred to as a dipole. Water molecule is an example of a dipole (Fig 5.10). The solid lines in Fig 5.10 represents primary valence bonds.

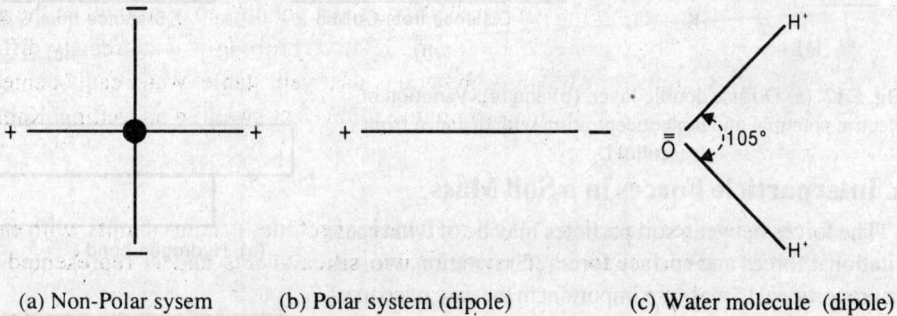

(a) Non-Polar sysem   (b) Polar system (dipole)   (c) Water molecule (dipole)

Fig. 5.10. Non-polar and polar sysems

The three types of dipoles are permanent, induced and fluctuating. The forces of attraction between the oppositely charged ends of permanent dipoles is called the orientation effect. When a non-polar molecule becomes polar in an electrical field it is said to be an induced dipole. The force due to induced polarization is called the induction effect. Since all electrons assume unsymmetrical position periodically as they move in their orbits, temporary fluctuating dipoles result. The intermolecular force due to fluctuating dipoles is called the dispersion effect. Although all the three effects are present, the orientation effect contributes the maximum to the Van der Waals forces.

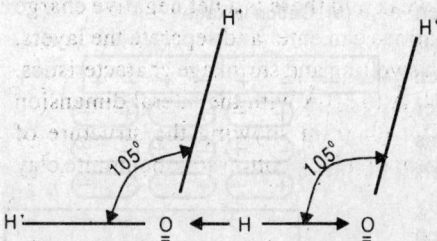

Fig. 5.11. Hydrogen bond

Hydrogen bond occurs when an atom of hydrogen is strongly attracted by two other atoms and the hydrogen atom oscillates between them. In Fig. 5.11 is illustrated the linkage between water molecules and is the best example of a hydrogen bond.

The hydrogen bonds are considerably weaker than primary valence bonds but are strong enough to resist breaking under the normally applied stresses in civil engineering practice. Van der Waals forces are much weaker than the hydrogen bond. They being greatly influenced by applied stresses and by the changes in the nature of soil water system have unique bearing on clay strength and water holding capacity.

5.8 Diffuse Double Layer

Clay particles carry net residual negative charges on their faces. When suspended in water cations from nearest water molecules are attracted towards the surface of a clay particle. Thus each particle will be surrounded by cations plus some anions and these are called counter ions or exchangeable ions since they can be replaced. The swarm of counter ions and the surface charges of the particle together constitute the diffuse double layer. The electric potential decreases with increase in distance from the surface of particle, till at some distance free water exists. [see Fig 5.12].

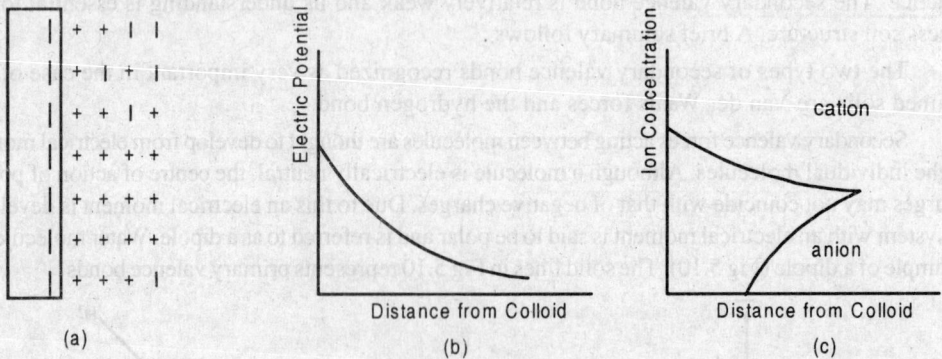

| | | |
| (a) | (b) | (c) |

Fig. 5.12. (a) Diffuse double layer, (b) and (c): Variation in electric potential and ion concentration with distance from colloid.

5.9 Interparticle Forces in a Soil Mass

The forces between soil particles may be of two types: gravitational forces and surface forces. Gravitational forces being proportional to mass are important in the case of coarse–grained soils only.

Surface forces dominate over gravitational forces in the case of clay particles which behave like colloids. A colloid is a particle with specific surface (surface area per unit mass or unit volume) so high that its behaviour is influenced by surface energy than mass energy. When the particles are very close the surface forces can be (*i*) attractive or (*ii*) repulsive. Van der Waals force, hydrogen bond, cation linkage, dipole–cation-dipole linkage, water dipole linkage and ionic bond are the possible mechanisms for attractive forces between particles. The Van der Waals force is the universal attractive force and may be the only attractive force in some soils. The repulsive force between particles is mainly due to similar charges on particle

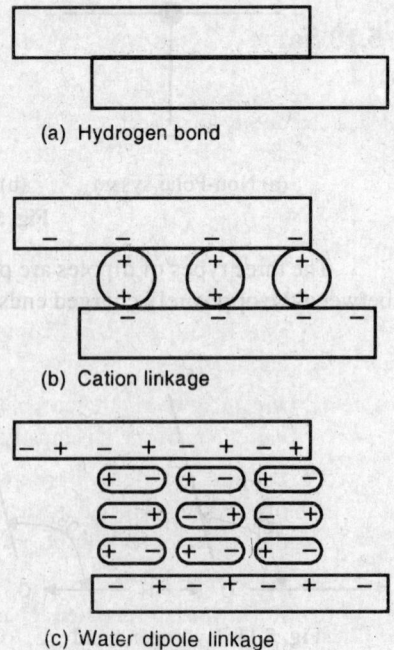

(a) Hydrogen bond

(b) Cation linkage

(c) Water dipole linkage

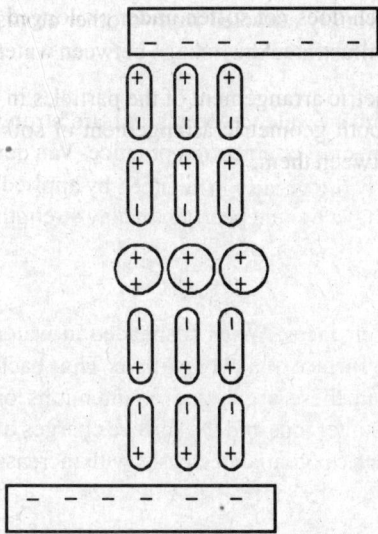

(d) Dipole - Cation - dipole linkage

Fig. 5.13. Interparticle attractive forces

(a) Particle repulsion due to similar charges

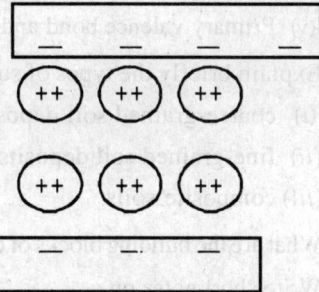

(b) Cation - Cation repulsion

Fig. 5.14. Interparticle repulsive forces (after Lambe).

surfaces. The repulsive forces between two adjacent particles become effective when they approach each other and their double layers just overlap. In Figs. 5.13 and 5.14 are illustrated the mechanisms of inter particle forces. When particles are in aqueous surroundings they may be mutually attracted or repulsed. If the total potential energy between two particles decreases as they approach each other the particles will experience attraction and will flocculate. On the other hand if there is increase in the total potential energy, there will be repulsion and the particles will disperse. The various factors affecting flocculation or dispersion are electrolyte concentration, temperature, ion valence, pH value, dielectric constant and anion adsorption.

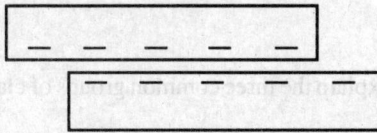

5.10 Some Additional Terms Relating to Soil Formation

1. *Cumulose soils*: Peat and muck are the two types of cumulose soils (organic soils) formed due to decay of vegetable matter under condition of excessive moisture. Peat is partly decayed vegetable matter in which plant forms can still be identified. Muck is thoroughly decomposed vegetable matter in which plant forms cannot be recognized.

2. *Calcareous soil*: A soil containing calcium carbonate which effervesces visibly to the naked eye when treated with hydrochloric acid.

3. *Cemented soil*: The soil in which the particles are bound together by some material acting as cementing agent. Examples of cementing agents are aluminum hydrates and calcium carbonate.

4. *Kankar*: An impure form of limestone with the hard variety occuring in nodular form and soft variety in layers. It is mostly calcium carbonate mixed with siliceous material like soil.

5. *Laterite*: Lateritic soil is considered as a type of residual soil. It contains mainly the hydroxides of iron, alumina and manganese. It is red, brown, yellow, grey and mottled in colors. When dug up laterite is comparatively soft but becomes hard when exposed to air for a few months.

6. *Rock-flour*: Soil material passing 75-micron sieve having little or no plasticity. It has little or no strength when air-dried.

7. *Indurated clay*: It is very strongly cemented clay which does not soften under prolonged wetting.

8. *Soil fabric*: This term is used to refer only to the geometric arrangement of the particles in a soil mass, while the term soil structure considers both geometric arrangement of soil particles and the interparticle forces which may act between them.

EXERCISE–5

5.1. Write a brief note on soil formation.

5.2. Distinguish between the following:

    (*i*) Soil fabric and soil structure

    (*ii*) Residual soil and transported soil

    (*iii*) Primary minerals and secondary minerals

    (*iv*) Coarse grained soil and fine grained soil

    (*v*) Primary valence bond and secondary valence bond

5.3. Explain briefly the types of soil structure recognized in

    (*i*) coarse-grained soil deposits

    (*ii*) fine-grained soil deposits and

    (*iii*) composite soils.

5.4. What are the building blocks of clay minerals? Explain the three common groups of clay minerals.

5.5. Write short notes on :

    (*i*) Van der Waals forces

    (*ii*) Hydrogen bond

    (*iii*) Isomorphous substitution

    (*iv*) Attractive and repulsive forces

5.6. Fill in the blanks :–

    (*i*) Weathering is a …………….. process of ……..................……..of the…...………….rock, it is accomplished through ………..……… and …………… agencies.

    (*ii*) ……………….is the largest group of clay minerals.

    (*iii*) Illite has a structure similar to that of …………………….

    (*iv*) Substitution of different cations in the octahedral sheets will result in……………

    (*v*) Clay particles carry………….....................charges on their faces.

    (*vi*) When clay platelets are in edge to face contact, they are said to have ………....……..structure: but if it is face to face, it is called……………... structure.

    (*vii*) …………and…………..are the types of structure found in composite soils.

5.7. Name the following :–

    (*i*) Very fine grained soil deposited into beds of lakes

    (*ii*) Soil transported by gravity forces

    (*iii*) Wind-drifted, uniformly sized sands

    (*iv*) Water-formed transported soil

(*v*) The material, disintegrated and deposited by glaciers

(*vi*) Under excessive moisture, chemically decompossed vegetable matter

(*vii*) The phenomenon of "replacement of cations".

5.8. Write short notes on :

  (*i*) Specific surface

 (*ii*) Montmorillonite, kaolinite, and illite.

(*iii*) Soil structure

(*iv*) Composite soil

 (*v*) Soil particles in soil mass.

CHAPTER 6

SOIL WATER AND EFFECTIVE STRESS

6.1 Classification

Water present in a soil mass is called soil water. It is broadly divided into two types:

(1) free water or gravitational water and (2) held water.

Water that is free to move through a soil mass under the influence of gravity is known as free water.

Held water is the water that is held within a soil mass by soil particles. It is not free to move under the influence of gravitational forces. Depending on the tenacity with which it is held by soil particles, held water is further classified as (*i*) structural water (*ii*) adsorbed water and (*iii*) capillary water.

Structural water is the water chemically combined in the crystal structure of the soil particle. It cannot be removed without breaking the structure of soil particle. The structure of a soil particle is not broken by loads applied in civil engineering practice or by heating at temperatures (usually upto 110°C) used for drying the soil in laboratory practice. Hence a geotechnical engineer is not much concerned with structural water and treats it as a part of soil particle.

Adsorbed water is the water which is held by fine grained soil particles due to electro chemical forces of adhesion. It can be nearly removed by oven drying (usually at 105 – 110°C) but on exposure to atmosphere the adsorbed layer is again formed due to moisture present in atmosphere. Hence it is also referred to as hygroscopic water. It may be noted that to completely remove the adsorbed water the soil has to be heated to temperature above 200°C.

Capillary water is the water which is held in soil mass due to capillary action. Capillary water can exist on a macroscopic scale compared to other types of held water which can exist on microscopic scale.

Lambe (1953) has suggested that, from the point of view of inter-particle forces, soil water can be classified as adsorbed water and pore water. In this classification capillary water and free water are considered as two types of pore water.

6.2 Adsorbed water

A colloidal soil particle carries a net negative charge on its surface and water molecule is a permanent dipole. Therefore water molecules adjacent to soil particle get attracted by it. (Fig. 6.1). Because of net negative charge on its surface a soil particle can also attract a number of other exchangeable cations like those of sodium, calcium, magnesium, potassium etc and these in turn attract nearby dipolar water molecules. Thus water in the vicinity of a soil particle is subjected to (*i*) attraction by residual negative charge on surface of soil particle and (*ii*) attraction by cations held by soil particle. The thickness of adsorbed layer depends upon mineralogical composition of soil particle, specific surface of soil particle and its environment. The physical properties of water in the adsorbed layer will be different from those of normal water. For example, it has higher boiling point and greater viscosity particularly in layers close to soil particles. The forces of attraction responsible for double layer for-

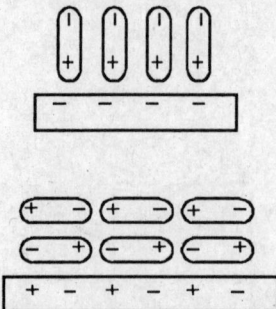

Fig. 6.1. Adsorbed water (after Lambe)

mation decrease exponentially with distance from surface of particle until the adsorbed layer ends meeting normal water free of attractive forces. It is rather difficult to precisely define the thickness of complete adsorbed layer. The innermost part of the double layer with most strongly held water may be about 10 to 15 Å thick. Adsorbed water has significant effect on the cohesion and plasticity characteristics of fine-grained soils.

6.3. Capillary water

Capillary water is the water which is held in a soil mass due to capillary forces.

(a) Derivation of expression for maximum capillary rise

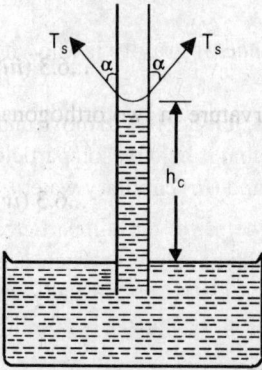

Fig.6.2. Capillary rise

Let a glass tube with narrow bore of diameter d be placed vertically with its bottom end dipped in water taken in a trough as shown in the Fig. 6.2. Formation of concave meniscus causes capillary rise in the tube. With the following notations,

T_s = surface tension (force per unit length)

α = contact angle

h_c = maximum capillary rise

γ_w = unit weight of water

we have for vertical equilibrium,

$$(T_s \cos \alpha)\, \pi d = \left(\frac{\pi d^2}{4}\right) h_c\, \gamma_w$$

$$h_c = \frac{4\,(T_s \cos \alpha)}{d\, \gamma_w} \qquad \ldots 6.3\,(i)$$

Thus capillary rise depends on surface tension T_s, contact angle α, diameter of tube d, and unit weight of water γ_w. The surface tension T_s decreases with temperature. It can be taken as 75×10^{-6} kN/m at 20°C. For pure water and clean glass α can be taken as zero.

(b) Capillary phenomenon in soils

The voids in natural soil formations act as capillary tubes and water rises in the continuous voids to a certain height above ground water table or free water surface. The height to which water rises, the capillary rise, depends on size of voids (particle size and void ratio). All other things being equal the capillary rise is more in a fine grained soil deposit than that in a coarse grained soil deposit as the size of voids is much less in the former case. The zone of soil strata saturated with capillary water is called the capillary fringe (Fig 6.3).

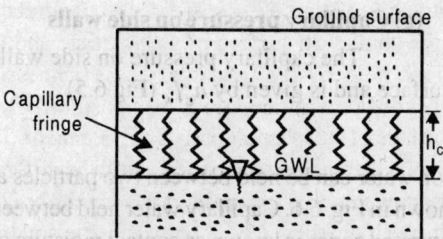

Fig. 6.3. Capillary phenomenon in soil formation

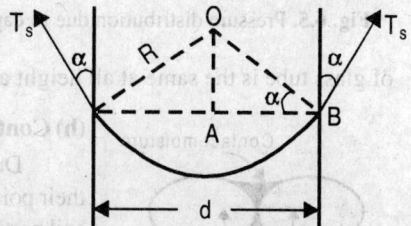

Fig. 6.4. Concave meniscus

(d) Relation between radius of meniscus, R and diameter of bore, d

Referring to Fig 6.4, in triangle OAB,

$$AB = OB \cos \alpha$$

$$\frac{d}{2} = R \cos \alpha$$

$$d = 2R \cos \alpha \qquad\qquad\qquad\qquad\qquad\qquad ...6.3\ (ii)$$

$$\therefore\ h_c = \frac{4T_s \cos \alpha}{d\gamma_w} = \frac{4T_s \cos \alpha}{(2R \cos \alpha)\gamma_w}$$

$$i.e.,\qquad h_c = \frac{2T_s}{R\gamma_w} \qquad\qquad\qquad\qquad ...6.3\ (iii)$$

In case the meniscus in non-uniform with R_1 and R_2 as radii of curvature in two orthogonal planes, we have

$$h_c = \left(\frac{1}{R_1} + \frac{1}{R_2} \right)\frac{T_s}{\gamma_w} \qquad\qquad\qquad ...6.3\ (iv)$$

(e) Capillary rise decreases with increase in temperature

This is illustrated in the following example,

For water at 4°C, $T_s = 75.6 \times 10^{-6}$ kN/m and $\gamma_w = 9.807$ kN/m³.

$$(h_c)_{max} = \frac{4T_s \cos \alpha}{d\gamma_w} = \frac{4(75.6 \times 10^{-6})}{(9.807)(10^{-3})d} = \frac{0.0308}{d}\ \text{m}$$

where α has been taken as zero and d is in mm.

For water at 20°C, $T_s = 72.6 \times 10^{-6}$ kN/m and $\gamma_w = 9.7896$ kN/m³.

$$h_c = \frac{4(72.8 \times 10^{-6})}{d(10^{-3})(9.7896)} = \frac{0.0297}{d}\ \text{m}$$

(f) Capillary tension in water

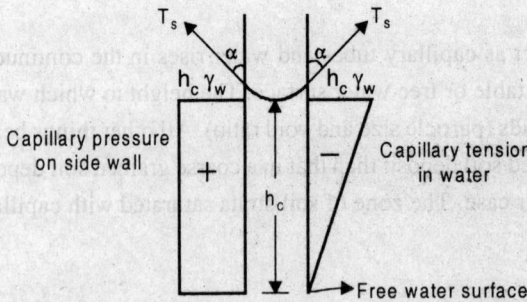

Fig. 6.5. Pressure distribution due to capillary action

When capillary rise takes place, water in the capillary tube above water level in trough (free water surface) will be in a state of tension. The capillary tension in water will increase from zero at free water surface to a maximum of $h_c\gamma_w$ at height (h_c) corresponding to maximum capillary rise (Fig. 6.5). The capillary tension at any height h above free surface is given by $h\gamma_w$.

(g) Capillary pressure on side walls

The capillary pressure on side walls of glass tube is the same at all height above free water surface and is given by $h_c\gamma_w$ (Fig 6.5)

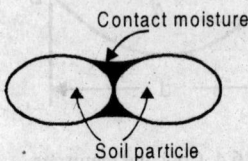

Fig. 6.6. Contact moisture

(h) Contact moisture

Due to surface tension water can be held between two particles at their points of contact as shown in Fig 6.6. Capillary water held between soil particles in partially saturated zones is known as contact moisture or contact capillary water. The tension in the capillary water causes the two particles to tend to press against each other.

(i) Estimation of capillary rise in soil deposit

If D represents the average size of particle, d the size of voids we have

$$\text{Void ratio, } e = \frac{V_v}{V_s} = \frac{d^3}{D^3}$$

$$\therefore d = e^{1/3} . D$$

$$h_c = \frac{4T_s \cos \alpha}{d \gamma_w} \qquad \qquad ...6.3\,(v)$$

Eq. 6.3 (v) can be used to estimate the capillary rise in natural soil formations as well as in other soil deposits.

For a rough estimate of the capillary rise the following equation has been suggested by Allen Hazen

$$h_c = \frac{C}{e \, D_{10}} \qquad \qquad ....6.3\,(vi)$$

where h_c = capillary rise in cm

D_{10} = effective diameter of soil in cm

e = void ratio

C is an empirical constant which is taken between 0.1 to 0.5 (cm^2).

(j) Capillary siphoning

Capillary rise takes place above free water surface in soils. In case of composite earth dams this can lead to a serious problem called capillary siphoning. As shown in Fig. 6.7 if the capillary rise h_c is greater than the height of top of central impervious core above the maximum water level of reservoir then the capillary water will flow over the crest of impervious core towards the down steam shell. This process, known as capillary siphoning, may continue for very long duration. It can cause not only water loss but also damage to down stream slope.

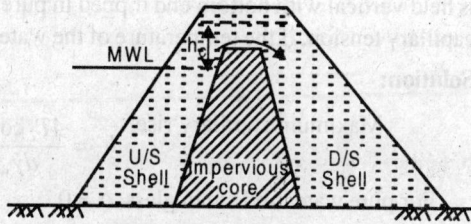

Fig. 6.7. Capillary siphoning

(k) Soil suction

The capillary water held in the capillary fringe in a soil mass will be in a state of tension. The capillary tension at any point is given by the product of capillary rise upto that point and the unit weight of water.

$$p = - h \gamma_w$$

The negative sign indicates that the pressure in capillary column of water is below atmospheric. The maximum pressure deficiency in the capillary water is called soil suction or suction pressure. Soil suction is measured by the height h_c in centimeters to which a water column could be drawn by suction in a soil mass free from external stress.

The common logarithm of h_c (cm) is known as the pF value.

According to Croney and Coleman (1960) soil suction, s and negative pore pressure, p can be related by the equation.

$$p = s + \alpha \sigma \qquad \qquad ...6.3\,(vii)$$

where α = fraction of total stress σ which is effective in changing the suction. For incompressible soils where the applied pressure is resisted by soil grains $\alpha = 0$ and $p = s$. For saturated compressible soil where all the applied pressure is resisted by water, $\alpha = 1$. For example α may be taken as 0.15 for sandy clay and 0.5 for silty clay. It should be noted that in Eq. 6.3 (*vii*) the term soil suction is used to indicate the pressure deficiency measured in a small sample of soil free from external stress and the term negative pore pressure is used to indicate pressure deficiency measured insitu (or in laboratory with the soil subject to stresses associated with the loading condition under consideration).

Table 6.1 Typical ranges of capillary rise in soils

Soil type	Particle size (mm)	Capillary rise (mm)
Coarse gravel	20 to 80	0.4 to 1.7
Fine gravel	4.75 to 20	1.7 to 7
Coarse sand	2 to 4.75	7 to 17
Medium sand	0.425 to 2	17 to 79
Fine sand	0.075 to 0.425	79 to 443
Silt	0.002 to 0.075	443 to 16500
Clay	<0.002	>16500
Colloids	<0.001	>33480

Note: The values of capillary rise in the above table has been obtained for $\alpha = 0$, $T_s = 72.8 \times 10^{-6}$ kN/m (at 20°C), $\gamma_w = 9.79$ kN/m³ (at 20°C) and $e = 0.7$.

Problem 6.1. The internal diameter of a tube is 0.1 mm What will be the maximum capillary rise when it is held vertical with bottom end dipped in pure water taken in a trough? Also compute the maximum capillary tension, if the temperature of the water is 20°C.

Solution:

Maximum capillary rise, $h_c = \dfrac{4T_s \cos \alpha}{d\gamma_w}$

For pure water and clean glass, $\alpha = 0$

For water at 20°C, $T_s = 72.8 \times 10^{-6}$ kN/m

$\gamma_w = 9.79$ kN/m³

$\therefore \quad h_c = \dfrac{4(72.8 \times 10^{-6})}{0.1 \times 10^{-3} \times 9.79} = 0.2974$ m

Maximum capillary tension $= h_c \gamma_w$

$= 0.2974 \times 9.79$

$= 2.91$ kN/m²

Problem 6.2. Compute the range for capillary rise in silt deposits. Assume average value of void ratio as 0.7.

Solution: The particle size range for silt is 0.002 mm – 0.075 mm, according to IS particle size classification chart.

Let $D_1 = 0.002$ mm and $D_2 = 0.075$ mm

If d_1 and d_2 indicate the corresponding size of voids, we have

$d_1 = e^{\frac{1}{3}} D_1 = (0.7)^{\frac{1}{3}} (0.002) = 1.776 \times 10^{-6}$ m

$d_2 = e^{\frac{1}{3}} D_2 = (0.7)^{\frac{1}{3}} (0.075) = 66.593 \times 10^{-6}$ m

Assuming $T_s = 75 \times 10^{-6}$ kN/m and $\gamma_w = 9.81$ kN/m^3

we find h_{c1} and h_{c2} as

$$h_{c1} = \frac{4T_s \cos\alpha}{d_1 \gamma_\omega} = \frac{4(75 \times 10^{-6})}{(1.776 \times 10^{-6})(9.81)} = 17.22 \, m$$

$$h_{c2} = \frac{4T_s \cos\alpha}{d_2 \gamma_\omega} = \frac{4(75 \times 10^{-6})}{(66.593 \times 10^{-6})(9.81)} = 0.46 \, m$$

Answer: Typical range of capillary rise in silt deposit is 0.46 m to 17.22 m

Problem 6.3. The difference in values of capillary rise for fine sand and silt was found to be 4.5 m. If the capillary rise in fine sand is 0.5 m compute the difference in size of voids of the two soils.

Solution:

Capillary rise in fine sand $h_{c1} = 0.5$ m.

Capillary rise in silt $h_{c2} = 0.5 + 4.5 = 5$ m

Let d_1 = size of voids in fine sand

d_2 = size of voids in silt

Assuming,

$$T_s = 75 \times 10^{-6} \text{ kN/m}$$
$$\alpha = 0$$
$$\gamma_w = 9.81 \text{ kN/m}^3$$

we have

$$d_1 = \frac{4T_s \cos\alpha}{h_{c1}\gamma_\omega} = \frac{4(75 \times 10^{-6})}{(0.5)(9.81)} = 61.16 \times 10^{-6} \, m$$

$$d_2 = \frac{4T_s \cos\alpha}{h_{c2}\gamma_\omega} = \frac{4(75 \times 10^{-6})}{(5)(9.81)} = 6.11 \times 10^{-6} \, m$$

Difference in the size of voids = $d_1 - d_2$

$$= 55.04 \times 10^{-6} \, m = 0.055 \, mm$$

Problem 6.4. The capillary rise in soil 1 with average particle size $D_1 = 0.05$ mm is 0.6 m. Estimate the capillary rise in soil 2 with average particle size $D_2 = 0.1$ mm assuming it has the same void ratio as soil 1.

Solution: If d_1 and d_2 denote the size of voids in soils 1 and 2 respectively, we have

$$e = \frac{V_v}{V_s} = \frac{d_1^3}{D_1^3} = \frac{d_2^3}{D_2^3} \qquad (i)$$

For soil 1, $d_1 = \frac{4T_s \cos\alpha}{h_{c1}\gamma_w}$

Assuming

$$T_s = 75 \times 10^{-6} \text{ kN/m}$$
$$\alpha = 0$$
$$\gamma_w = 9.81 \text{ kN/m}^3$$

we get $d_1 = \frac{4(75 \times 10^{-6})}{(0.6)(9.81)} = 50.97 \times 10^{-6} \, m$

From equation (i)
$$d_2 = d_1 \frac{D_2}{D_1} = 50.97 \times 10^{-6} \times \frac{0.1}{0.05}$$

$$= 101.94 \times 10^{-6} \, \text{m}$$

Capillary rise in soil 2,

$$h_{c2} = \frac{4T_s \cos\alpha}{d_2 \gamma_w} = \frac{4(75 \times 10^{-6})}{(101.94 \times 10^{-6})(9.81)} = 0.3 \, \text{m}$$

Problem 6.5. A glass vessel is filled with water. The shape of the vessel is as shown in Fig. 6.8. A fully developed meniscus has formed in the small hole of diameter $d_1 = 0.2$ mm in the upper wall of the vessel. Another hole of diameter d_2 exists in the lower wall. What is the greatest value which d_2 may have?

If $d_2 = d_1$ find the contact angle in the lower hole.

Solution: As it is clear from the given figure, the capillary heights supported by the two holes will be in opposite directions. The algebraic sum of the heights of water supported by them is equal to 0.09 m

$$h_{c1} - h_{c2} = 0.09 \, \text{m}$$

Assuming

$$T_s = 75 \times 10^{-6} \, \text{kN/m}$$

$$\alpha = 0$$

$$\gamma_w = 9.81 \, \text{kN/m}^3$$

Fig. 6.8. For (problem 6.5)

We have

$$h_{c1} = \frac{4T_s \cos\alpha}{d_1 \gamma_w} = \frac{4(75 \times 10^{-6})}{(0.2 \times 10^{-3})(9.81)} = 0.153 \, \text{m}$$

$$h_{c2} = h_{c1} - 0.09 = 0.153 - 0.09 = 0.063 \, \text{m}$$

For full meniscus to be developed in the lower hole,

$$d_2 = \frac{4T_s \cos\alpha_2}{h_{c2} \gamma_w} = \frac{4(75 \times 10^{-6})}{(0.063)(9.81)} = 0.00048 \, \text{m}$$

$$= 0.48 \, \text{mm}$$

If $d_2 = d_1 = 0.2$ mm,

$$\cos\alpha_2 = \frac{d_2 h_{c2} \gamma_w}{4T_s} = \frac{(0.2 \times 10^{-3})(0.063)(9.81)}{4(75 \times 10^{-6})} = 0.412$$

Contact angle for the full meniscus to be developed at second hole, $\alpha_2 = 65.7°$

6.4 Effective Stress, Pore Pressure and Total Stress

Pressure is transmitted through a soil mass by soil particles and pore water. The pressure transmitted through soil mass by soil particles through their points of contact is called effective stress and is denoted by σ'. It is also called intergranular pressure. It is effective in decreasing the void ratio and increasing the shear strength. The pressure transmitted by pore water in a soil mass is called pore water pressure or simply pore pressure. It is denoted by u and is also referred to as neutral pressure.

At any point in soil mass, $u = h_w \gamma_w$ ...6.4 (i)

where h_w = peizometric head at that point.

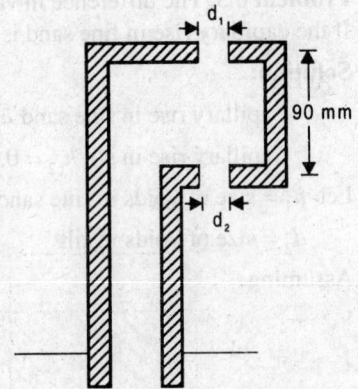

The total stress, denoted by σ, at any point is the sum of effective stress and pore pressure

$$\sigma = \sigma' + u \qquad \qquad \text{...6.4 }(ii)$$

The following discussion illustrates the variation of σ', u and σ with depth in different cases.

Case (i) Submerged soil mass with water table above the surface

At AA

Total pressure,	$\sigma = \gamma_{sat} z_1 + \gamma_\omega z_2$
Pore pressure,	$u = \gamma_\omega (z_1 + z_2)$
Effective stress,	$\sigma' = \sigma - u$
	$= \gamma_{sat} z_1 + \gamma_\omega z_2 - \gamma_\omega z_1 - \gamma_\omega z_2$
	$= (\gamma_{sat} - \gamma_\omega) z_1$
	$= \gamma' z_1$

At BB

$$\sigma = \gamma_\omega z_2$$
$$u = \gamma_\omega z_2$$
$$\sigma' = \sigma - u = 0$$

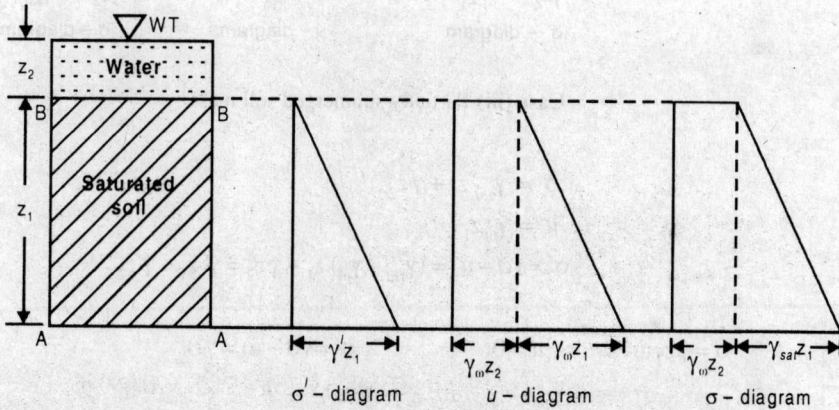

σ' – diagram u – diagram σ – diagram

Case (ii) Submerged soil mass with water table at the surface

At AA

$$\sigma = \gamma_{sat} z$$
$$u = \gamma_\omega z$$
$$\sigma' = \sigma - u = (\gamma_{sat} - \gamma_\omega) z = \gamma' z$$

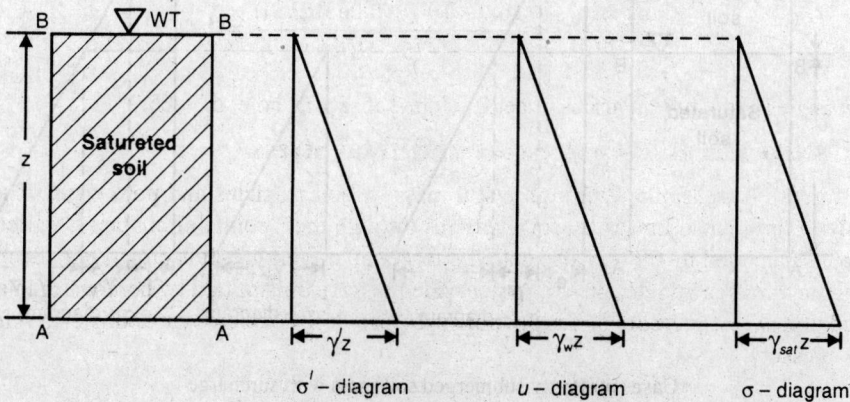

σ' – diagram u – diagram σ – diagram

Case (ii) Fully submerged soil mass

At BB $\qquad$ $\sigma = 0;$ $\qquad$ $u = 0;$ $\qquad$ $\sigma' = 0$

Comparing cases (i) and (ii) we observe that effective stress is independent of depth of water above top surface of soil mass.

Case (iii) Partly submerged soil mass

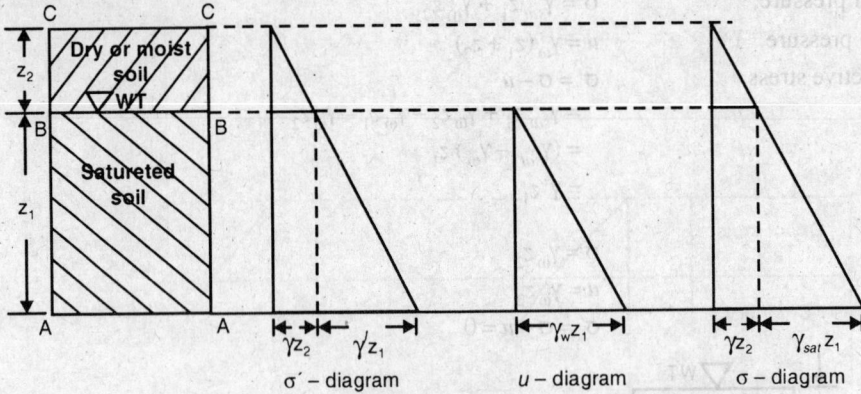

Case (iii) Partially submerged soil mass

At AA

$$\sigma = \gamma_{sat} z_1 + \gamma\, z_2$$
$$u = \gamma_\omega z_1$$
$$\sigma' = (\sigma - u) = (\gamma_{sat} - \gamma_\omega)\, z_1 + \gamma\, z_2 = \gamma'\, z_1 + \gamma\, z_2$$

At BB

$$\sigma = \gamma\, z_2; \qquad u = 0; \qquad \sigma' = (\sigma - u) = \gamma\, z_2$$

At CC

$$\sigma = 0; \qquad u = 0; \qquad \sigma' = 0;$$

Case (iv) Partly submerged soil mass with surcharge

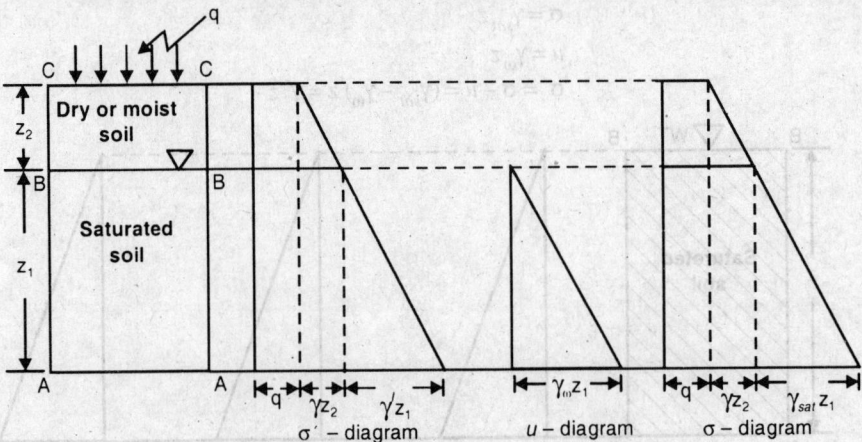

Case (iv) Partly submerged soil mass with surcharge

At AA

$$\sigma = \gamma_{sat} z_1 + \gamma z_2 + q$$
$$u = \gamma_\omega z_1$$
$$\sigma' = (\sigma - u) = (\gamma_{sat} - \gamma_\omega) z_1 + \gamma z_2 + q = \gamma' z_1 + \gamma z_2 + q$$

At BB

$$\sigma = \gamma z_2 + q; \qquad u = 0; \qquad \sigma' = (\sigma - u) = \gamma z_2 + q$$

At CC

$$\sigma = q; \qquad u = 0; \qquad \sigma' = (\sigma - u) = q$$

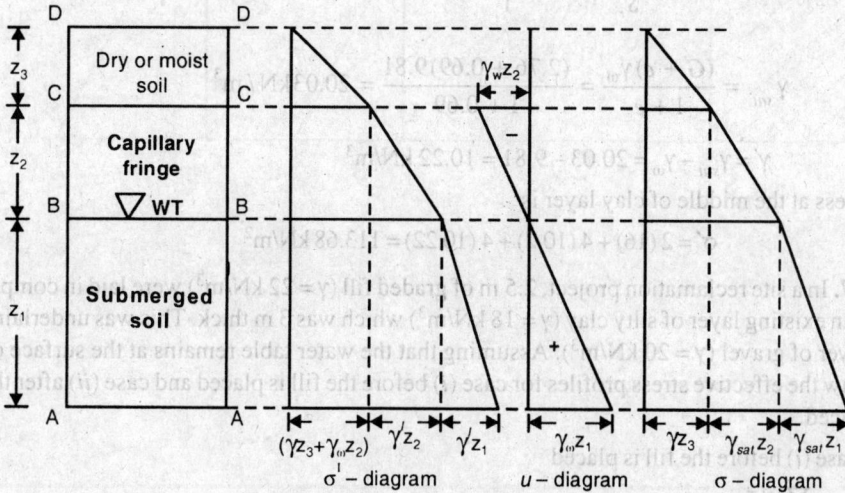

Case (v) Soil mass with capillary fringe

Case (v) Soil mass with capillary fringe

At AA

$$\sigma = \gamma_{sat} z_1 + \gamma_{sat} z_2 + \gamma z_3$$
$$u = \gamma_\omega z_1$$
$$\sigma' = (\sigma - u) = (\gamma_{sat} - \gamma_\omega) z_1 + \gamma_{sat} z_2 + \gamma z_3$$
$$= \gamma' z_1 + \gamma_{sat} z_2 + \gamma z_3$$

At BB

$$\sigma = \gamma_{sat} z_2 + \gamma z_3$$
$$u = 0$$
$$\sigma' = (\sigma - u) = \gamma_{sat} z_2 + \gamma z_3$$

At CC

$$\sigma = \gamma z_3$$
$$u = -\gamma_\omega z_2$$
$$\sigma' = (\sigma - u) = \gamma z_3 + \gamma_w z_2$$

At DD

$$\sigma = 0$$
$$u = 0$$
$$\sigma' = 0$$

Problem 6.6. A clay stratum of thickness 8 m is located at a depth of 6 m below the ground surface. It is overlain by fine sand. The water table is located at a depth of 2 m below ground surface. For fine sand submerged unit weight is 10.2 kN/m³. The moist unit weight of sand located above water table is 16 kN/m³. For clay layer, $G = 2.76$ and water content is 25%. Compute the effective stress at the middle of clay layer.

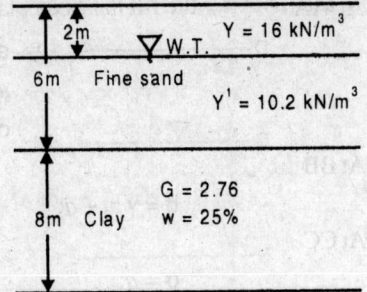

Solution:

For clay layer

$$e = \frac{wG}{S_r} = \frac{(0.25)(2.76)}{1} = 0.69$$

$$\gamma_{sat} = \frac{(G+e)\gamma_\omega}{1+e} = \frac{(2.76+0.69)9.81}{1+0.69} = 20.03\,kN/m^3$$

$$\gamma' = \gamma_{sat} - \gamma_\omega = 20.03 - 9.81 = 10.22\ kN/m^3$$

Effective stress at the middle of clay layer is

$$\sigma' = 2(16) + 4(10.2) + 4(10.22) = 113.68\ kN/m^2$$

Problem 6.7. In a site reclamation project, 2.5 m of graded fill ($\gamma = 22$ kN/m³) were laid in compacted layers over an existing layer of silty clay ($\gamma = 18$ kN/m³) which was 3 m thick. This was underlain by a 2 m thick layer of gravel ($\gamma = 20$ kN/m³). Assuming that the water table remains at the surface of the silty clay draw the effective stress profiles for case (i) before the fill is placed and case (ii) after the fill has been placed.

Solution: Case (i) before the fill is placed

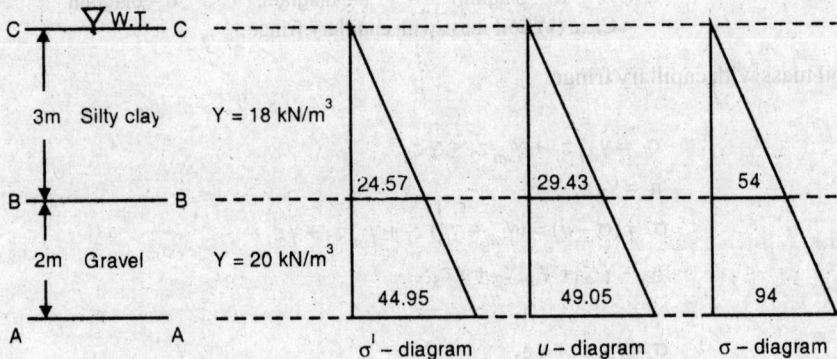

At A–A

$$\sigma = 2(20) + 3(18) = 94\ kN/m^2$$
$$u = 5(9.81) = 49.05\ kN/m^2$$
$$\sigma' = (\sigma - u) = 94 - 49.05 = 44.95\ kN/m^2$$

At B–B

$$\sigma = 3(18) = 54\ kN/m^2$$
$$u = 3(9.81) = 29.43\ kN/m^2$$
$$\sigma' = \sigma - u = 54 - 29.43 = 24.57\ kN/m^2$$

At C–C

$$\sigma = 0 \qquad u = 0 \qquad \sigma' = 0$$

Case (ii) After the fill has been placed

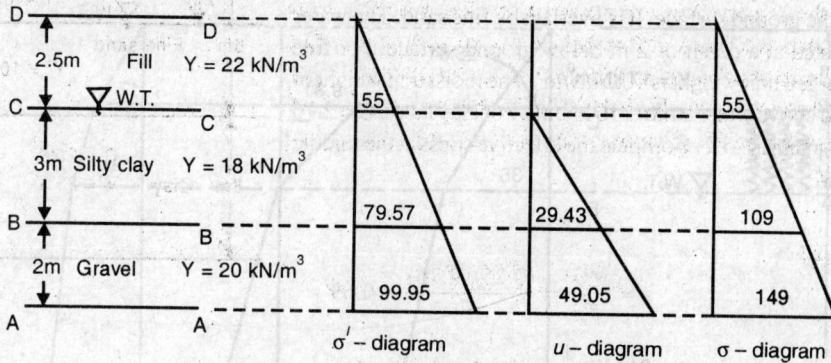

At A–A

$$\sigma = 2\,(20) + 3\,(18) + 2.5\,(22) = 149 \text{ kN/m}^2$$
$$u = 5\,(9.81) = 49.05 \text{ kN/m}^2$$
$$\sigma' = (\sigma - u) = 149 - 49.05 = 99.95 \text{ kN/m}^2$$

At B–B

$$\sigma = 3\,(18) + 2.5\,(22) = 109 \text{ kN/m}^2$$
$$u = 3\,(9.81) = 29.43 \text{ kN/m}^2$$
$$\sigma' = \sigma - u = 109 - 29.43 = 79.57 \text{ kN/m}^2$$

At C–C

$$\sigma = 2.5\,(22) = 55 \text{ kN/m}^2$$
$$u = 0$$
$$\sigma' = \sigma - u = 55 \text{ kN/m}^2$$

At D–D

$$\sigma = 0$$
$$u = 0$$
$$\sigma' = 0$$

Problem 6.8. A sand stratum is 10 m thick. The water table is 2 m below ground level. The unit weights of sand layer above and below water table are 17 kN/m^3 and 21 kN/m^3 respectively. The capillary rise above water table is 1 m. Draw the effective stress, pore pressure and total stress diagrams for the sand stratum.

Solution:

At A–A

$$\sigma = 9\,(21) + 1\,(17) = 206 \text{ kN/m}^2$$
$$u = 8\,(9.81) = 78.48 \text{ kN/m}^2$$
$$\sigma' = (\sigma - u) = 206 - 78.48 = 127.52 \text{ kN/m}^2$$

At B–B

$$\sigma = 1\,(21) + 1\,(17) = 38 \text{ kN/m}^2$$
$$u = 0$$
$$\sigma' = \sigma - u = 38 \text{ kN/m}^2$$

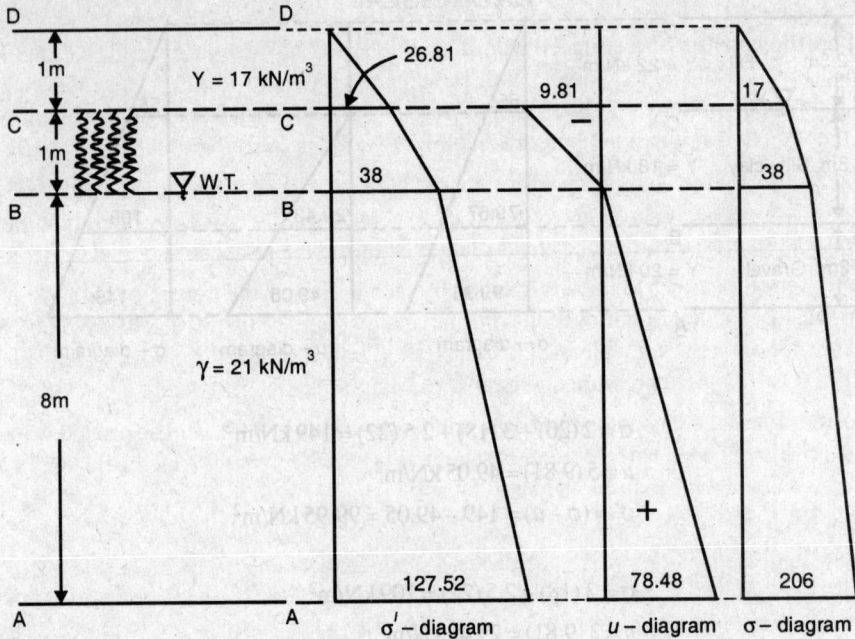

σ' – diagram u – diagram σ – diagram

At C–C

$$\sigma = 1\,(17) = 17\ \text{kN/m}^2$$

$$u = -1\,(9.81) = -9.81\ \text{kN/m}^2$$

$$\sigma' = (\sigma - u) = 17 - (-9.81) = 26.81\ \text{kN/m}^2$$

At D–D

$$\sigma = 0; \qquad u = 0; \qquad \sigma' = 0;$$

Problem 6.9. A deposit of sand has a porosity of 40% and specific gravity of particle 2.7. The ground water table is 2 m below the ground surface. Compute the effective stress at a depth of 6m below ground surface, if capillary rise above water table is 1m.

Solution:

$$n = 40\%$$

$$G = 2.7$$

$$e = \frac{n}{1-n} = \frac{0.4}{1-0.4} = 0.76$$

Assuming the soil to be dry upto 1m depth below ground surface,

$$\gamma_d = \frac{G\gamma_w}{1+e} = \frac{2.7(9.81)}{1+0.67} = 15.86\ \text{kN/m}^3$$

For the soil in the fully saturated zones,

$$\gamma_{sat} = \frac{(G+e)\gamma_\omega}{1+e} = \frac{(2.7+0.67)9.81}{1+0.67} = 19.80\ \text{kN/m}^3$$

Effective stress at A–A (at 6 m depth)

$$\sigma' = 1(15.86) + 1(19.80) + 4(19.80 - 9.81) = 75.65\ \text{kN/m}^2$$

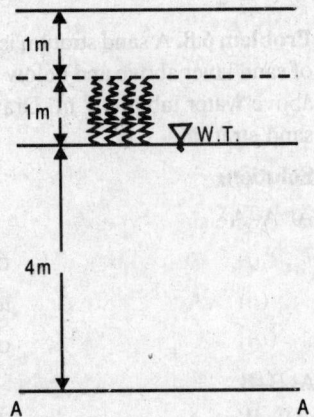

EXERCISE-6

6.1. A soil profile consists of a surface layer of sand 3.5 m thick with unit weight of 16.5 kN/m³, intermediate layer of clayey sand 2.5 m thick with unit weight of 19 kN/m³ and bottom layer of clay 3.5 m thick with unit weight of 19.5 kN/m³. The water table is at top of intermediate layer. Draw the effective stress, pore pressure and total stress diagrams for all the three layers.

6.2. A soil deposit 8 m thick has bulk unit weight of 20 kN/m³. If the water table is at a depth of 3 m below ground surface and soil above water table is saturated by capillary water, plot the effective stress, pore pressure and total stress diagrams for the entire depth of 8 m.

6.3. A layer of saturated clay 5 m thick is overlain by sand 6 m deep, the water table being 2 m below the surface. The saturated unit weights of sand and clay are 20 kN/m³ and 19 kN/m³. The unit weight of sand above water table is 17 kN/m³. If the capillary rise is 1 m above water table, draw the effective stress, pore pressure and total stress diagrams for 11 m depth.

6.4. A deposit of fine sand has porosity of 40% and specific gravity of 2.70. The ground water table is 5 m below the ground surface, sand is saturated by capillary water, upto a height of 1 m due to the water table. The degree of saturation of the first 4 m of moist sand, below the ground surface is 10%. Calculate vertical pressure at a depth of 10 m below ground surface.

6.5. In a saturated soil stratum, water table exists at the surface. The effective stress in the soil, at a depth of 2 m is 20 kN/m³. If the water table rises by 0.5 m, during floods what will be change in the effective stress?

6.6. Name the following

    (i)   Water which is free to move through the soil-mass under the force of gravity.

    (ii)   Water which cannot be removed by heating at ordinary temperatures.

    (iii)   Water attracted and held by the electro chemical forces on the surface of the soil practicles.

    (iv)   Water content of air-dried soil.

    (v)   Pressure-deficiency in held water.

6.7. Fill in the blanks:–

    (i)   With increase in water content, soil suction..

    (ii)   Capillary action is due to .... pressure.

    (iii)   Total soil suction is composed of.............and......................

    (iv)   Water below the water table is called.....................

    (v)   The movement of gravitational water, through soil, is called............

    (vi)   Pressure at the water table is................., and that below it is......................

6.8. Define soil water. What are its main types ?

6.9. Write short notes on:–

    (i)   Structural water.

    (ii)   Adsorbed water

    (iii)   Capillary water

    (iv)   Soil suction.

6.10. Write a note on capillary siphoning.

CHAPTER 7

PERMEABILITY

7.1 Introduction

Permeability is a property of soil by virtue of which the soil mass allows water (or any other fluid) to flow through it. It is an engineering property, which is required to be determined for study of soil engineering problems involving flow of water through soils, such as seepage through body of earth dams and settlement of foundations.

7.2 Darcy's Law and Coefficient of Permeability

According to Darcy's law, for laminar flow conditions the velocity of flow, v is directly proportional to the hydraulic gradient, i.

$$v \, \alpha \, i$$

We can write $v = ki$ \qquad\qquad\qquad ...7.2 (i)

The constant of proportionality, k between v and i is called Darcy's coefficient of permeability. When $i = 1$ we have $k = v$. Therefore the coefficient of permeability can also be defined as the velocity of flow through soil under unit hydraulic gradient and has the same unit as the velocity. It is usually expressed in mm/sec, m/hr or m/day.

Further \quad $q = Av = A \, ki$ \qquad\qquad\qquad ...7.2 (ii)

where \quad q = rate of flow or discharge per unit time

A = total area of cross section of flow perpendicular to direction of flow.

Table 7.1 gives typical values of coefficient of permeability of various soil types.

Table 7.1 - Typical values of k

Soil Type	Coefficient of permeability (cm/sec)
Gravel	1 to 100
Sand	10^{-3} to 1
Silt	10^{-6} to 10^{-3}
Clay	$< 10^{-6}$

We notice that clay has very low value of k and is sometimes described as an impervious soil with silt being described as semi-impervious.

7.3 Discharge Velocity and Seepage Velocity

The velocity of flow of water, v through soil mass is obtained from Darcy's law assuming that the flow takes place through the total cross-sectional area, A of soil mass perpendicular to the direction of flow. This velocity is referred to as discharge velocity or theoretical velocity.

Thus discharge velocity, \quad $v = \dfrac{q}{A}$

The total area A is composed of the area of voids, A_v and area of solids, A_S. But flow can take place only through area of voids, A_v. The actual velocity of flow, referred to as seepage velocity and denoted by v_S is thus greater than the theoretical velocity obtained from Darcy's law.

Seepage velocity $\qquad v_s = \dfrac{q}{A_v}$ ...7.3 (i)

From Eq. 7.2 (ii) and 7.3 (i)

$$q = Av = A_v \cdot v_s$$

$\therefore \qquad v_s = v \cdot \dfrac{A}{A_v} = v \cdot \dfrac{1}{n} \left[\text{since } \dfrac{A_v}{A} = \dfrac{V_v}{V} = n \right]$

$$v_s = \dfrac{v}{n}$$ ...7.3 (ii)

where n = porosity of soil mass.

Further, we have $v = ki$

Similarly we can write $v_s = k_p i$

The constant of proportionality, k_p between v_s and i is referred to as coefficient of percolation

$$\dfrac{v_s}{v} = \dfrac{k_p i}{ki} = \dfrac{k_p}{k}$$

But $\dfrac{v_s}{v} = \dfrac{1}{n}$

$\therefore \dfrac{k_p}{k} = \dfrac{1}{n}$

$$k_p = \dfrac{k}{n}$$ ...7.3 (iii)

7.4 Limitations in the use of Darcy's law

According to Darcy's law the velocity of flow through soil mass is directly proportional to the hydraulic gradient for laminar flow condition only. In practice one can expect the flow to be always laminar in the case of fine-grained soil deposits because of low permeability and hence low velocity of flow. However, in the case of sands and gravels flow will be laminar upto a certain value of velocity for each deposit and investigations have been carried out to find a limit for application of Darcy's law.

According to experimental investigations of Francher, Lewis and Barnes (1933) flow through sands will be laminar and Darcy's law valid so long as the Reynolds number expressed in the form shown below is less than or equal to unity

$$\dfrac{v D_a \gamma_\omega}{\eta g} \le 1$$ ...7.4 (i)

where v = velocity of flow in cm/sec

D_a = size of particles (average) in cm

It is found that the limiting value of Reynolds number taken as 1 is very approximate as its actual value can have wide variation (0.1 to 75) depending partly on the characteristic size of particles used in the equation. Allen Hazen (1982) has suggested that the upper limit of effective size of particles can be taken as 3mm for Darcy's law to be valid.

7.5 Factors affecting permeability

The various factors affecting permeability are listed below:

1. Particle size
2. Properties of pore fluid
3. Void ratio
4. Soil fabric and soil stratification
5. Degree of saturation

6. Presence of foreign matter
7. Adsorbed water

Effect of Particle size

It is found that permeability varies approximately as the square of the grain size. According to Allen Hazen (1911) the permeability of sands can be estimated using the following equation

$$k = CD_{10}^2 \qquad \qquad ...7.5\,(ii)$$

where

$$k = \text{coefficient of permeability (cm/sec)}$$
$$D_{10} = \text{effective size (cm)}$$
$$C = \text{constant, which may be taken as 100}$$

Kozeny (1907) has given the following relation, which reflects one of the attempts made to express permeability in terms of specific surface besides other parameters involved.

$$k = \frac{1}{k_k} \frac{1}{\eta} \frac{n^3}{1-n^2} \frac{1}{s^2} \qquad \qquad ...7.5\,(iii)$$

where

$$k = \text{coefficient of permeability (cm/sec)}$$
$$\eta = \text{coefficient of viscosity (gm-sec/cm}^2)$$
$$n = \text{porosity}$$
$$s = \text{specific surface (cm}^2/\text{cm}^3)$$
$$k_k = \text{constant, taken as 5 for spherical particles}$$

Muskat (1973) suggested a more general coefficient of permeability, called the physical permeability which has nearly same value at all temperatures, for all fluids, for a given soil mass.

$$k_p = k\frac{\eta}{\gamma_l} \qquad \qquad ...7.5\,(iv)$$

where η = coefficient of viscosity

γ_l = unit weight of pore fluid

Effect of properties of pore fluid

It can be shown that permeability is directly proportional to the unit weight of pore fluid and inversely proportional to its viscosity. Therefore, we can write

$$\frac{k_1}{k_2} = \frac{\gamma_{w1}}{\gamma_{w2}} \frac{\eta_2}{\eta_1} \qquad \qquad ...7.5\,(v)$$

where subscripts 1 and 2 are used to refer to quantities at temperatures T_1 and T_2 respectively. As the variation in viscosity with temperature is much greater compared to that of unit weight of water, Eq. 7.5 (v) is approximated as

$$\frac{k_1}{k_2} = \frac{\eta_2}{\eta_1} \qquad \qquad ...7.5\,(va)$$

Effect of void ratio

It has been found that a plot of void ratio on natural scale against permeability on logarithmic scale is approximately a straight line for both coarse-grained soils and fine-grained soils.

For coarse-grained soils the following relation has been established.

$$\frac{k_1}{k_2} = \frac{e_1^3}{1 + e_1} \Bigg/ \frac{e_2^3}{1 + e_2}$$

        ...7.5 (*vi*)

where k_1 = permeability at void ratio e_1

                    k_2 = permeability at void ratio e_2

Effect of soil fabric and stratification

The effect of structural arrangement of soil particles on permeability can be found by determining permeability of undisturbed and disturbed soil samples. The effect of change in structural arrangement of particles on permeability is more in the case of fine grained soils.

Stratified soil masses will have different average permeabilities in directions parallel and perpendicular to their bedding planes. The average permeability parallel to bedding plane will be greater than that perpendicular to bedding plane.

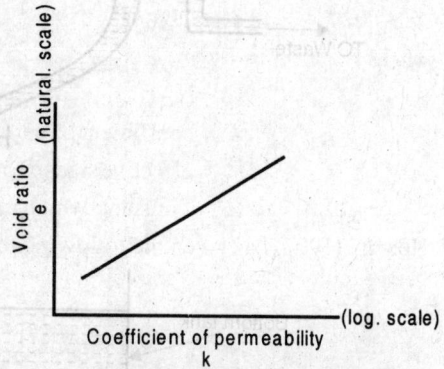

Fig.7.1. Plot of e against log k

Effect of degree of saturation

In partly saturated soils the entrapped air greatly reduces the permeability. Permeability test is always conducted on a fully saturated soil specimen. The water used may contain dissolved air. The use of air-free water is not warranted as the percolating water in the field may contain dissolved gases.

Effect of presence of foreign matter

Organic foreign matter, if present in soil mass, may be carried by flowing water towards critical flow channels and may choke them up, causing reduction in permeability.

Effect of adsorbed water

The adsorbed water, which is held by soil particles, is not free to move and therefore reduces the effective pore space available for the flow of free water. According to Casagrande the net void ratio may be taken approximately as (e-0.1) and permeability assumed to be proportional to the square of the net void ratio.

7.6 Determination of Coefficient of Permeability

The coefficient of permeability can be determined by the following methods:

I. Laboratory Methods

     (*a*) Constant head permeability test

     (*b*) Falling head permeability test

     (*c*) Horizontal capillarity test

II. Field methods

       (*a*) Pumping-out tests

       (*b*) Pumping-in tests

III. Indirect methods involving computation from

       (*i*) grain size

       (*ii*) specific surface

       (*iii*) consolidation test data

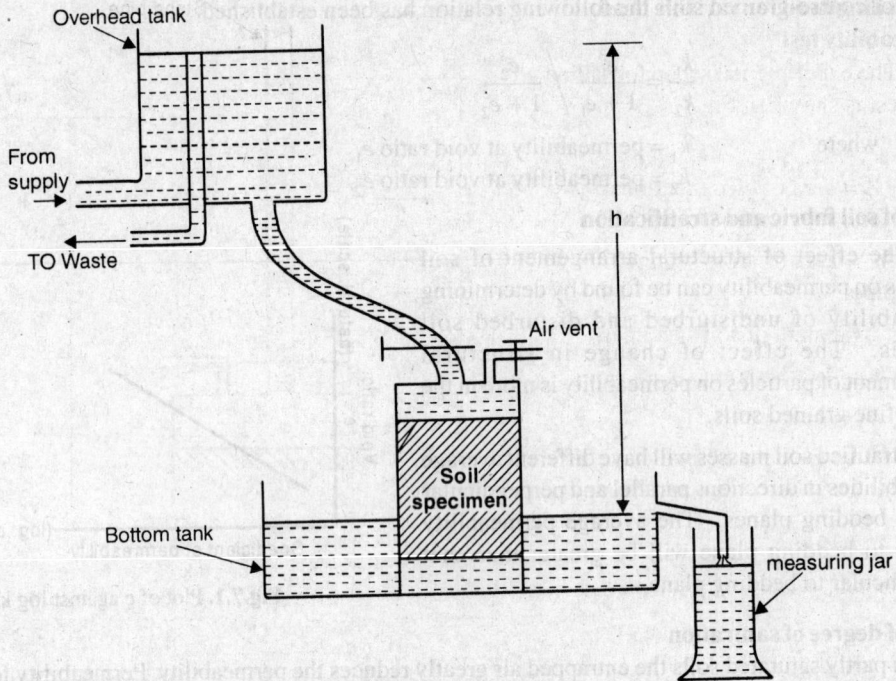

Fig. 7.2. Schematic diagram of constant head test setup

I (a) Constant head permeability test

The experimental set up for the constant head test is shown in Fig.7.2

Let A = area of cross section of soil specimen

L = length of soil specimen

h = constant head

The constant head, h, causing flow is the vertical distance between the water levels in the overhead and bottom tanks.

The air vent is provided to drive out air bubbles, if any, observed in the transparent rubber tubing through which water flows from overhead tank to the soil specimen. After ensuring that the soil specimen is fully saturated and the flow has become steady, the quantity of water, Q flowing through the soil specimen in a known interval of time t is found. The temperature, T of water is noted.

Applying Darcy's law,

$$q = kiA$$

$$\frac{Q}{t} = k \cdot \frac{h}{L} \cdot A$$

$$\therefore k = \frac{QL}{Aht}$$

Usually k is reported at 27° C

$$k_{27°C} = \left[\frac{\gamma_w \text{ at } 27°C}{\gamma_w \text{ at } T°C}\right]\left[\frac{\eta_{at\,T°C}}{\eta_{at\,27°C}}\right]k_{T°C} \qquad ...7.6\,(i)$$

I (b) Falling head (or variable head) permeability test

The experimental set up for falling head test is shown in Fig.7.3

Let

A = area of cross section of soil specimen

a = area of cross section of stand pipe

L = length of soil specimen

h_1 = head at time t_1

h_2 = head at time t_2

If h be the head at time t and 'dh' the fall in head at that instant, then we have

$$q = \frac{-a.dh}{dt} \qquad ...(i)$$

The minus sign indicates fall in head.

Applying Darcy's law

$$q = ki\, A = k.\frac{h}{L}A \qquad ...(ii)$$

From equations (i) and (ii) , we get

$$kdt = \frac{-aL}{A}\frac{dh}{h}$$

Integrating between suitable limits

$$k.\int_{t_1}^{t_2} dt = \frac{-aL}{A}\int_{h_1}^{h_2}\frac{dh}{h}$$

$$k\,(t_2 - t_1) = \frac{aL}{A}\log_e\frac{h_1}{h_2}$$

Replacing $(t_2 - t_1)$ by t, we can write

$$k = \frac{aL}{At}\log_e\frac{h_1}{h_2} \qquad ...7.6\,(ii)$$

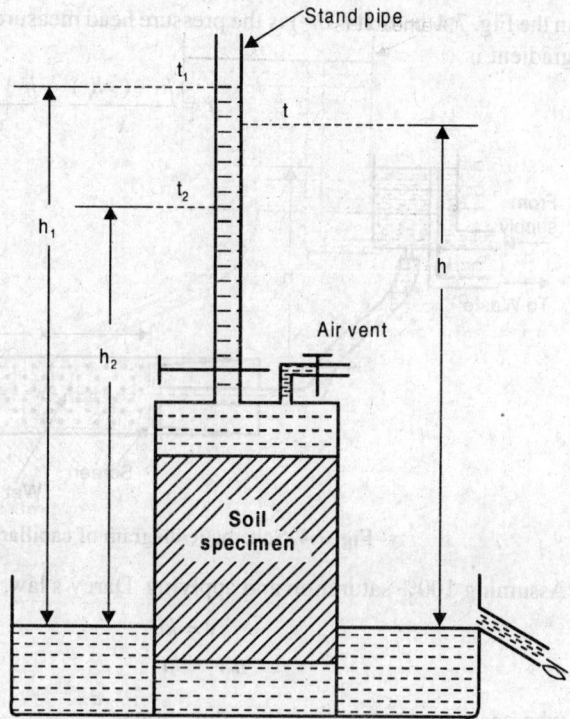

Fig. 7.3. Schematic diagram of falling head test setup

Air bubbles, if any, in water in the stand pipe are driven out by means of air vent.

After ensuring that the sample is fully saturated, the time, t taken for head to fall from h_1 to h_2 is found. The temperature, T of water is noted. Using Eq. 7.6 (ii) the coefficient of permeability is computed. The falling head test is used in the case of soils with low permeability and the constant head test in the case of soils with high permeability.

(c) Capillarity–Permeability test

The coefficient of permeability, k along with capillary height, h_c of soil sample is determined by this test which is also known as horizontal capillarity test.

The set-up for the test essentially consists of a transparent tube (of lucite or glass) about 40 mm in diameter and 0.35 m to 0.5 m long in which dry soil sample is placed at desired density and water is allowed to flow from one end under a constant head h, and the other end is exposed to atmosphere through air vent. Let the time taken by water to saturate the soil for a distance x, shown in the Fig. 7.4 be t. If $(-h_c)$ is the pressure head measured at end of saturated soil, then the hydraulic gradient is

$$i = \frac{h - (-h_c)}{x} = \frac{h + h_c}{x}$$

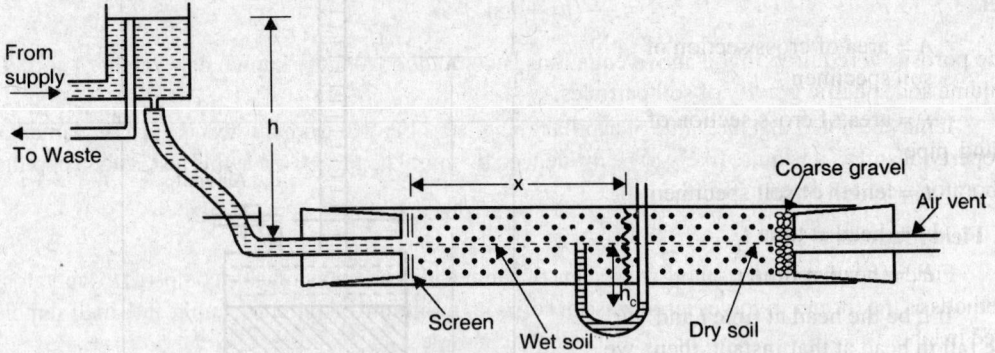

Fig. 7.4. Schematic diagram of capillarity - permeability test setup

Assuming 100% saturation and applying Darcy's law, we have

$$v = ki$$

$$n\,v_s = ki$$

where v_s = seepage velocity in x-direction = $\dfrac{dx}{dt}$

$\therefore \qquad n.\dfrac{dx}{dt} = ki = k\dfrac{(h + h_c)}{x}$

$$x\,dx = \frac{k}{n}(h + h_c)\,dt$$

Integrating between suitable limits,

$$\int_{x_1}^{x_2} x\,dx = \frac{k}{n}(h + h_c)\int_{t_1}^{t_2} dt$$

$$\frac{(x_2^2 - x_1^2)}{2} = \frac{k}{n}(h + h_c)(t_2 - t_1)$$

$$\frac{x_2^2 - x_1^2}{t_2 - t_1} = \frac{2k}{n}(h + h_c) \qquad\qquad ...7.6\,(iii)$$

In order to find the two unknowns k and h_c in the above equation, the first set of observations (for different values of x upto mid-point of tube) are taken under a head h_1 As the capillary saturation progresses the values of x are recorded at different time intervals t. The values of x^2 are plotted against corresponding time intervals t to obtain a straight line whose slope, say m_1 gives the value

of $\dfrac{x_2^2 - x_1^2}{t_2 - t_1}$. The second set of observations (with x beyond mid-point of tube) are taken under an

increased head h_2, and values of x^2 plotted against corresponding values of t to obtain another

straight line, whose slope m_2 will give the value $\dfrac{x_2^2 - x_1^2}{t_2 - t_1}$.

By substitution in Eq. 7.6 (*iii*) we obtain the following two equations, which are solved simultaneously to get k and h_c:

$$m_1 = \frac{2k}{n}(h_1 + h_c)$$

$$m_2 = \frac{2k}{n}(h_2 + h_c)$$

The porosity n required in the above equations is computed from the known dry weight of soil, its volume and specific gravity of soil particles.

It may be noted that the value of capillary height h_c obtained from this test is not a fundamental property of soil as it is quite likely to be influenced by initial head, h. Hence this test is not a popular laboratory test for determining k.

II. Field Methods

Field methods of determining coefficient of permeability are more reliable compared to laboratory methods as the former involves large mass of soil with minimum disturbance unlike the small sample used in laboratory test. The value of coefficient of permeability, k obtained from field test represents an average value of k for the large soil mass over a large area. To obtain such an average value of k in the laboratory, tests will have to be conducted on large number of undisturbed soil samples obtained from different spots at different depths. The degree of disturbance and number of samples can affect the reliability of the average value of k obtained from laboratory tests, for the large soil mass in the field. There are two types of field tests for determining the coefficient of permeability:

(*a*) Pumping-out tests

(*b*) Pumping-in tests

The tests can be conducted in both unconfined aquifer and confined aquifer. An aquifer is a water-bearing stratum in natural ground formations. If it overlies an impervious stratum and the water table is free to fluctuate, it is called unconfined aquifer. On the other hand, if the aquifer is bound by impervious strata both at top and bottom, it is called confined aquifer.

(*i*) Pumping-out test in unconfined aquifer

Figure 7.5 is a schematic diagram illustrating pumping-out test in an unconfined aquifer. Referring to this figure, we let

r = radius of main well

R = radius of zero drawdown known as maximum radius of influence

h = depth of water in the main well, during pumping, measured above impervious layer

H = height of initial water table above impervious layer

q = rate at which water is pumped out of well.

Let P (x, y) be any point on the drawdown curve. The point O at the bottom of central axis of well is chosen as the origin of reference.

Applying Darcy's law, for flow through cylindrical surface of radius x and height y, we have

$$q = k.A_x.i_x$$

$$= k \cdot (2\pi xy)\frac{dy}{dx}$$

Rearranging,

$$q.\frac{dx}{x} = 2\pi kydy$$

Integrating between suitable limits,

$$q\int_r^R \frac{dx}{x} = 2\pi k \int_h^H y \cdot dy$$

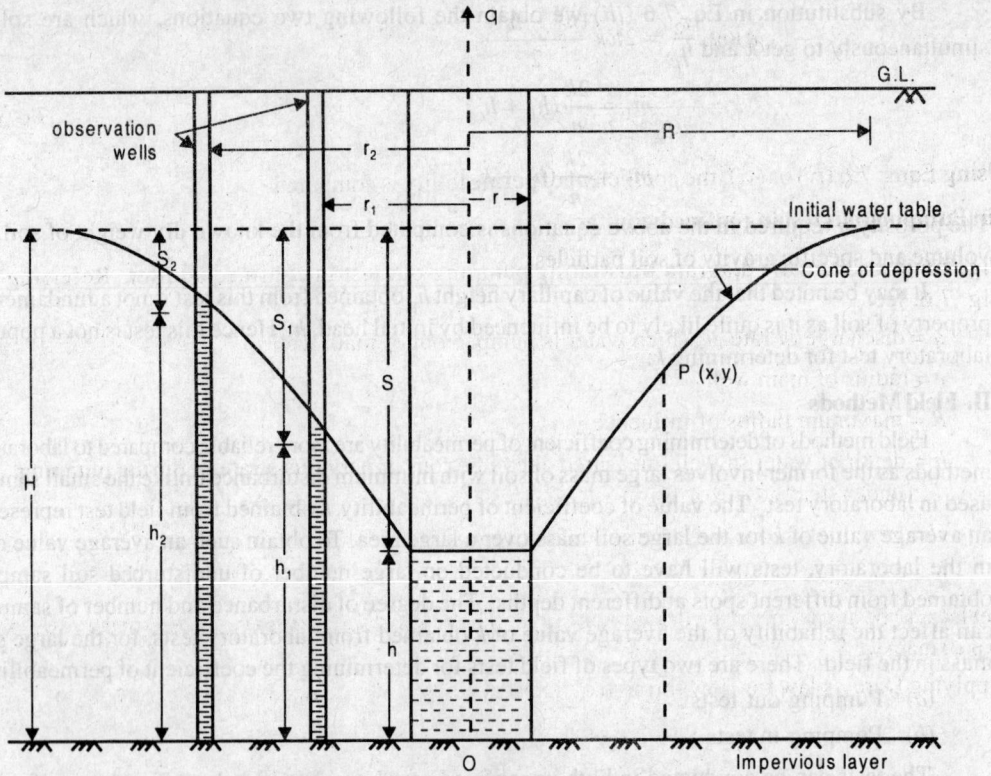

Fig. 7.5. Pumping out test in unconfined aquifer

$$q \log_e \frac{R}{r} = 2\pi k \frac{(H^2 - h^2)}{2}$$

$$\therefore k = \frac{q}{\pi(H^2 - h^2)} \log_e \frac{R}{r} \qquad \qquad ...7.6\,(iv)$$

In Eq. 7.6 (iv) R is found to vary from 150 m to 300 m and can only be estimated crudely, as for example, using the following equation given by Sichardt (1930).

$$R = S\sqrt{k} \qquad \qquad ...7.6\,(v)$$

where k is in m/sec, R and S are in meters. S is the drawdown in the main well ($S = H - h$).

To avoid the use of R, an alternative method is to measure drawdowns S_1 and S_2 in two observation wells located at radial distances r_1 and r_2 from the axis of main well (Fig. 7.5). The depths of water in the two observation wells are

$$h_1 = H - S_1$$
$$h_2 = H - S_2$$

We now have $y = h_1$ at $x = r_1$

$y = h_2$ at $x = r_2$

Therefore, we can write

$$q \int_{r_1}^{r_2} \frac{dx}{x} = 2\pi k \int_{h_1}^{h_2} y\,dy$$

$$q \log_e \frac{r_2}{r_1} = 2\pi k \frac{(h_2^2 - h_1^2)}{2}$$

$$\therefore \quad k = \frac{q}{\pi(h_2^2 - h_1^2)} \log_e \frac{r_2}{r_1} \qquad \qquad ...7.6\,(vi)$$

Using Eqns. 7.6 (iv) or (vi), the coefficient of permeability is computed

(ii) Pumping-out test in confined aquifer

Fig. 7.6 is a schematic diagram illustrating pumping-out test in confined aquifer. Referring to Fig. 7.6, let

q = discharge or rate at which water is pumped out of main well

r = radius of main well

R = maximum radius of influence

h = depth of water in well, measured above bottom impervious stratum, during pumping (under steady state of flow)

H = height of initial piezometric surface above bottom impervious stratum

b = thickness of confined aquifer

Let P (x, y) be any point on the drawdown curve. The origin of reference 0 is chosen at the bottom of axis of main well.

Applying Darcy's law for flow through cylindrical surface of radius x and height b, we have

$$q = k A_x i_x$$

$$= k (2\pi x b) \frac{dy}{dx}$$

Rearranging,

$$q \frac{dx}{x} = 2\pi b k dy$$

Integrating between suitable limits,

$$q \int_r^R \frac{dx}{x} = 2\pi b k \int_h^H dy$$

$$q \log_e \frac{R}{r} = 2\pi b k (H - h)$$

$$\therefore \quad k = \frac{q}{2\pi b (H - h)} \log_e \frac{R}{r} \qquad \qquad ...7.6\,(vii)$$

Alternatively, if h_1 and h_2 are the depths of water measured above bottom impervious stratum in two observation wells located at radial distances r_1 and r_2 from the axis of main well, then we can write

$$q \int_{r_1}^{r_2} \frac{dx}{x} = 2\pi b k \int_{h_1}^{h_2} dy$$

$$q \log_e \frac{r_2}{r_1} = 2\pi b k (h_2 - h_1)$$

$$\therefore \quad k = \frac{q}{2\pi b (h_2 - h_1)} \log_e \frac{r_2}{r_1} \qquad \qquad ...7.6\,(viii)$$

Using Eq. 7.6 (vii) or (viii) the coefficient of permeability is computed.

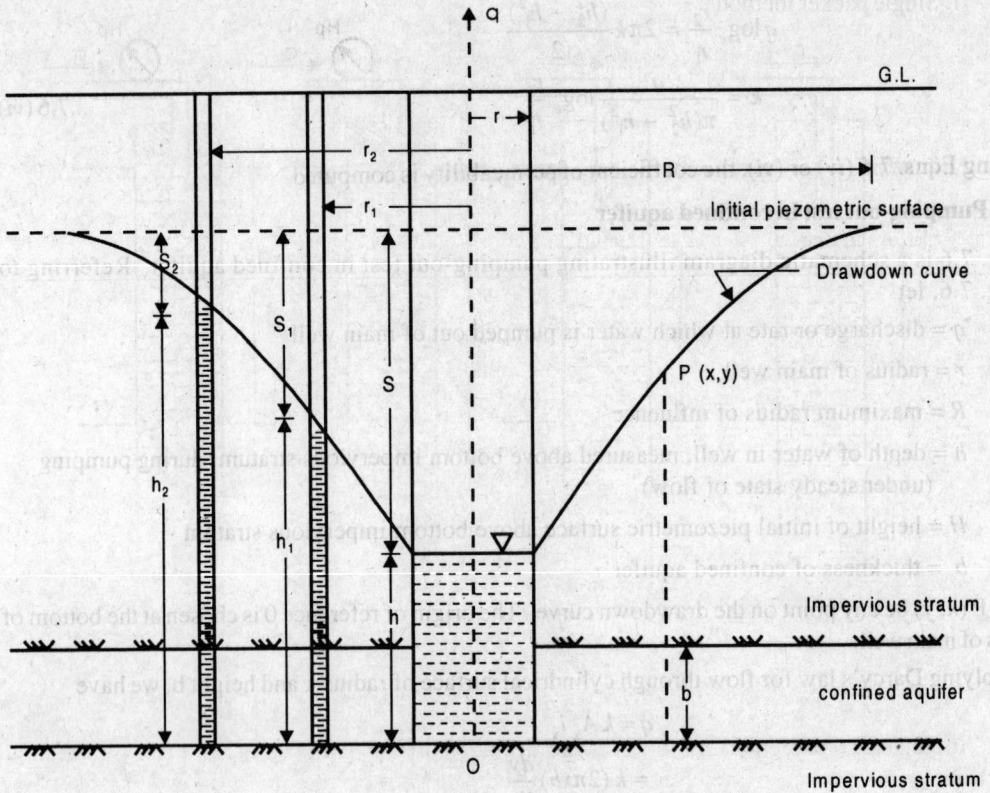

Fig. 7.6. Pumping out test in confined aquifer

(b) Pumping-in tests

The two methods devised by U.S. Bureau of Reclamation are
  (i) Constant water level method (in open-end pipe) and
  (ii) Packer method (in section of borehole)

(i) Constant water level method

An open-end pipe is sunk into the soil to desired depth and the soil is taken out of the pipe till its bottom end. The test is also conducted in a borehole with the pipe casing extending to the desired depth. Fig. 7.7 illustrates the arrangement for the method.

In Fig. 7.7 (a) and (c) the bottom end of pipe is above water table and in Fig. 7.7 (b) and (d) it is below the water table. Water is pumped into the pipe and the rate of flow, q is adjusted to maintain water level constant in the pipe. In the case of soils of low permeability additional pressure head H_p is required to be added to the gravity head H_g in order to maintain constant rate of flow. The coefficient of permeability is computed using the following equation.

$$k = \frac{q}{5.5\,rH}$$

...7.6 (ix)

where r = internal radius of pipe.

(ii) Packer method

A packer is an expandable cylindrical rubber sleeve. Packers are used as a means of sealing of a section of bore hole. Two types of packer methods are used in practice.

1. Single packer method

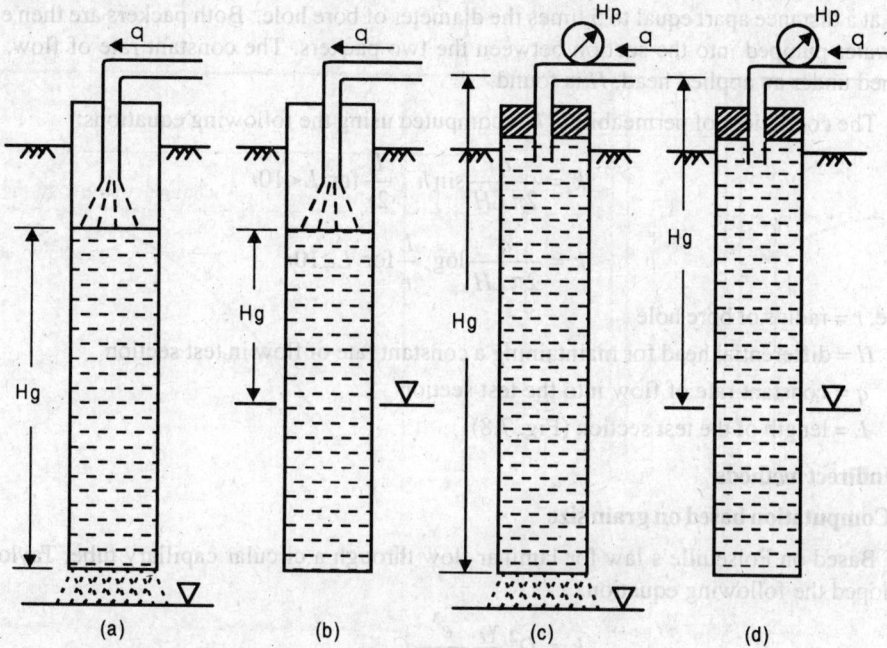

Fig. 7.7. Constant water level method

In single packer method the hole is drilled to the required depth. The packer is fixed at a desired level above the bottom of hole and the water pumped into the section below the packer. The constant rate of flow, q that is attained under an applied head, H is found.

2. Double packer method

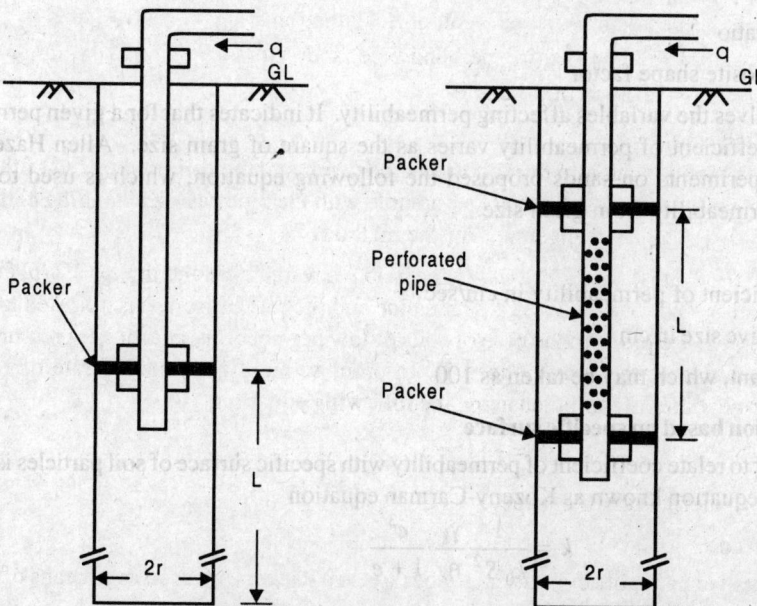

Fig. 7.8. Pumping in test – single and double packer methods

In the double packer method, the hole is drilled to the final depth and cleaned. Two packers are fixed at a distance apart equal to 5 times the diameter of bore hole. Both packers are then expanded and water pumped into the section between the two packers. The constant rate of flow, q that is attained under an applied head, H is found.

The coefficient of permeability, k is computed using the following equations:

$$k = \frac{q}{2\pi LH} \sin h^{-1} \frac{L}{2r} \text{ for } L < 10r \qquad \ldots 7.6\,(x)$$

$$k = \frac{q}{2\pi LH} \log_e \frac{L}{r} \text{ for } L \geq 10r \qquad \ldots 7.6\,(xi)$$

where, r = radius of bore hole

$\quad$ H = differential head for maintaining a constant rate of flow in test section

$\quad$ q = constant rate of flow into the test section

$\quad$ L = length of the test section (Fig. 7.8)

III. Indirect methods

(a) Computation based on grain size

Based on Poiseuille's law for laminar flow through a circular capillary tube, Taylor (1948) developed the following equation.

$$k = D_s^2 \frac{\gamma_\ell}{\eta_\ell} \frac{e^3}{1 + e} C_s \qquad \ldots 7.6\,(xii)$$

in which,

$\quad$ k = Darcy's coefficient of permeability

$\quad$ D_s = effective particle size

$\quad$ γ_l = unit weight of liquid

$\quad$ η_l = viscosity of liquid

$\quad$ e = void ratio

$\quad$ C_s = composite shape factor

Eq. 7.6 (xii) involves the variables affecting permeability. It indicates that for a given permeant and soil mass the coefficient of permeability varies as the square of grain size. Allen Hazen (1911) based on his experiments on sands proposed the following equation, which is used to estimate coefficient of permeability from grain size.

$$k = CD_{10}^2 \qquad \ldots 7.6\,(xiii)$$

where, k = coefficient of permeability in cm/sec

$\quad$ D_{10} = effective size in cm.

$\quad$ C = constant, which may be taken as 100

(b) Computation based on specific surface

An attempt to relate coefficient of permeability with specific surface of soil particles is reflected in the following equation known as Kozeny-Carman equation

$$k = \frac{1}{k_0 S^2} \frac{\gamma_\ell}{\eta_\ell} \cdot \frac{e^3}{1 + e} \qquad \ldots 7.6\,(xiv)$$

in which,

$\quad$ k_0 = factor influenced by particle shape

$\quad$ S = specific surface of particles

$$\gamma_l = \text{unit weight of liquid}$$
$$\eta_l = \text{viscosity of liquid}$$
$$e = \text{void ratio}$$

Loudon (1952) based on his experimental investigations developed the following empirical relation.

$$\text{Log}_{10}\,(kS^2) = a + bn \qquad\qquad ...7.6\,(xv)$$

in which a and b are constants and n is porosity. Suggested values are $a = 1.365$ and $b = 5.15$ for permeability at 10°C.

(c)   Computation of k from consolidation test data

The coefficient of volume change, m_v and coefficient of consolidation, C_v are obtained from laboratory consolidation test. Using the following relation k can be computed.

$$k = C_v\,m_v\,\gamma_w \qquad\qquad \text{(Refer Chapter 11)}$$

Note: A consolidometer with permeability test attachment enables direct determination of k at the end of consolidation of specimen under a particular pressure increment.

7.7 Permeability of Stratified Soil Deposits

In the case of stratified or layered soil deposits each layer may be homogeneous and isotropic, but when we consider flow through the entire deposit the average permeability of deposit will vary with the direction of flow relative to the bedding plane. The average permeability parallel to the bedding plane will be greater than that perpendicular to the bedding plane.

(i)   Average permeability parallel to bedding plane

Let Z_1, Z_2, ... Z_n be the thickness of layers with permeabilities k_1, k_2, ... k_n. For flow parallel to bedding plane the hydraulic gradient i will be same for all layers. The total discharge through the deposit will be the sum of discharges through individual layers

$$q = q_1 + q_2 \;.....\; + q_n \qquad\qquad ...(i)$$

If k_x denotes average permeability parallel to bedding plane, applying Darcy's law, we have

$q = k_x iA \;\; = k_x iZ$ (considering unit thickness perpendicular to plane of figure)
For individual layers, we have

$$q_1 = k_1\,i\,Z_1$$
$$q_2 = k_2\,i\,Z_2$$
. .
. .
. .
$$q_n = k_n\,i\,Z_n$$

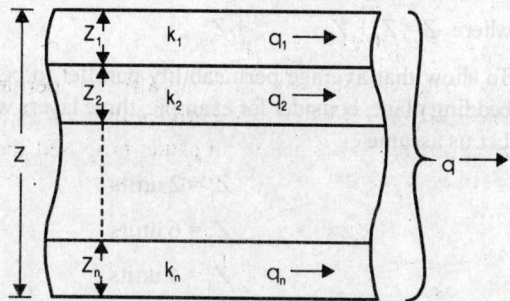

Fig. 7.9. Flow parallel to bedding plane

By substitution in Eq (i), we get

$$k_x iZ = k_1\,i\,Z_1 + k_2\,i\,Z_2 + ... + k_n\,i\,Z_n$$

$$\therefore \qquad k_x = \frac{k_1 Z_1 + k_2 Z_2 + ... + k_n Z_n}{Z} \qquad\qquad ...7.7\,(i)$$

where $Z = Z_1 + Z_2 + ... + Z_n$

(ii) Average permeability perpendicular to bedding plane

For flow perpendicular to bedding plane the discharge, q will be same for all layers. Considering gross area of cross section of flow ($A = A_v + A_s$) the velocity of flow, v will be same for all layers. The total head loss will be sum of head losses in individual layers.

$$h = h_1 + h_2 + \ldots\ldots + h_n \qquad \ldots 7.7\ (i)$$

If k_z denotes average permeability perpendicular to bedding plane, applying Darcy's law we have

$$v = k_z\, i = k_z\, \frac{h}{Z}$$

or
$$h = \frac{vZ}{k_z}$$

For individual layers, we have

$$h_1 = \frac{vZ_1}{k_1}$$

$$h_2 = \frac{vZ_2}{k_2}$$

$$h_n = \frac{vZ_n}{k_n}$$

Fig. 7.10. Flow perpendicular to bedding plane

Substituting in Eq. (i) we get

$$\frac{vZ}{k_z} = \frac{vZ_1}{k_1} + \frac{vZ_2}{k_2} + \ldots\ldots + \frac{vZ_n}{k_n} \qquad \therefore k_z = \frac{Z_1}{\dfrac{Z_1}{k_1} + \dfrac{Z_2}{k_2} + \ldots \dfrac{Z_n}{k_n}} \qquad \ldots 7.7\ (ii)$$

where $Z = Z_1 + Z_2 + \ldots + Z_n$

To show that average permeability parallel to bedding plane is greater than that perpendicular to bedding plane, consider for example, three layers with thickness Z_1, Z_2, Z_3 and permeabilities k_1, k_2, k_3. Let us assume

$$Z_1 = 2 \text{ units} \qquad\qquad k_1 = 5 \text{ units}$$

$$Z_2 = 6 \text{ units} \qquad\qquad k_2 = 3 \text{ units}$$

$$Z_3 = 4 \text{ units} \qquad\qquad k_3 = 7 \text{ units}$$

Then average permeability parallel to bedding plane is given by

$$k_x = \frac{k_1 Z_1 + k_2 Z_2 + k_3 Z_3}{Z_1 + Z_2 + Z_3} = \frac{5(2) + 3(6) + 7(4)}{2 + 6 + 4} = 4.7 \text{ units}$$

The average permeability perpendicular to bedding plane is given by

$$k_z = \frac{Z_1 + Z_2 + Z_3}{\dfrac{Z_1}{k_1} + \dfrac{Z_2}{k_2} + \dfrac{Z_3}{k_3}} = \frac{2 + 6 + 4}{\dfrac{2}{3} + \dfrac{6}{3} + \dfrac{7}{4}} = 2.9 \text{ units}$$

We observe in the above example that $k_x > k_z$

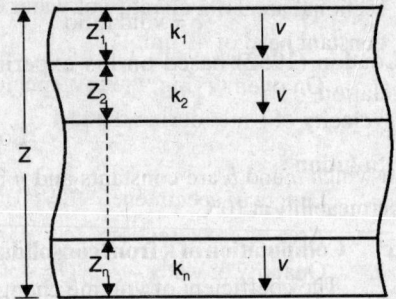

Problem 7.1

Calculate the coefficient of permeability of a soil sample 6 cm in height and 50 cm^2 in cross-sectional area, if a quantity of water equal to 430 cc passed down in 10 minutes under an effective constant head of 40 cm.

On oven drying, the test specimen weighed 4.98 N. Taking G = 2.65, calculate the seepage velocity of water during the test.

Solution :

Length of specimen,　$L = 60$ mm

Area of cross section,　$A = 5000$ mm^2

Quantity of water,　$Q = 430$ cc

Time of flow,　$t = 600$ sec

Constant head,　$H = 400$ mm

$$q = kiA$$

$$\frac{Q}{t} = k\frac{H}{L}A$$

$\therefore$
$$k = \frac{QL}{AHt} = \frac{430 \times 1000 \times 60}{5000 \times 400 \times 600} = 0.0215 \text{ mm/sec}$$

Discharge velocity,
$$v = \frac{q}{A} = \frac{Q}{At} = \frac{430 \times 1000}{5000 \times 600} = 0.143 \text{ mm/sec}$$

Alternatively,
$$v = ki = k\frac{H}{L} = 0.0215 \times \frac{400}{60} = 0.143 \text{ mm/sec}$$

Dry weight of specimen,　$W_d = 4.98$ N

Volume of specimen,　$V = 5000 \times 60 = 3 \times 10^5$ mm^3

$\therefore$
$$\gamma_d = \frac{W_d}{V} = \frac{4.98}{3 \times 10^5} = 1.66 \times 10^{-5} \text{ N/mm}^3$$

$$e = \frac{G\gamma_w}{\gamma_d} - 1 = \frac{(2.65)(9.8 \times 10^{-6})}{1.66 \times 10^{-5}} - 1 = 0.57$$

$$n = \frac{e}{1 + e} = \frac{0.57}{1 + 0.57} \doteqdot 0.36$$

Seepage velocity,
$$v_s = \frac{v}{n} = \frac{0.143}{0.36} = 0.397 \text{ mm/sec}$$

Problem 7.2

A cylindrical mould of diameter 7.5 cm contains 15.0 cm long sample of sand. When water flows through the soil under constant head at a rate of 55 cc/minute, the loss of head between two points 8 cm apart is found to be 12.5 cm. Determine the coefficient of permeability of the soil.

Solution:

Diameter of soil specimen,　$d = 7.5$ cm

Length of soil specimen,　　$L = 15$ cm

Rate of flow,　　$q = 55$ cc/minute

　　　　$= 0.92$ cc/sec

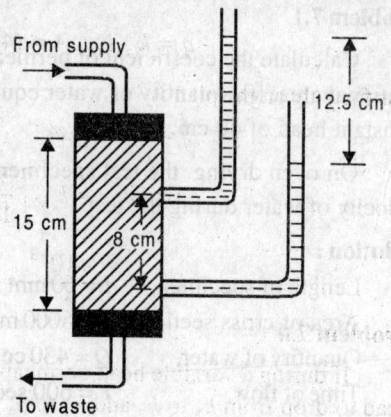

Hydraulic gradient, $\qquad i = \dfrac{12.5}{8} = 1.56$

Area of cross section of specimen,

$$A = \frac{\pi (7.5)^2}{4}$$

$$= 44.18 \text{ cm}^2$$

$$q = kiA$$

$$k = \frac{q}{iA} = \frac{0.92}{(1.56)(44.18)}$$

$$= 0.0133 \text{ cm/sec}$$

Problem 7.3

In the accompanying sketch soil X has a permeability of 4×10^{-3} cm/sec and the head lost in soil Y is 9 times the head lost in soil X. Calculate the coefficient of permeability of soil Y.

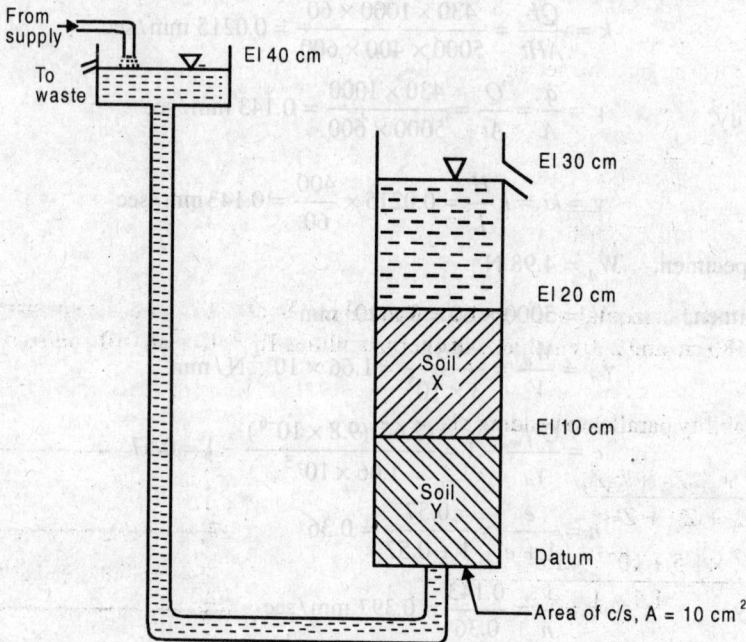

Solution:

Let $h_{(x)}$ and $h_{(y)}$ denote the head losses in soils X and Y respectively.

Then $h_{(y)} = 9\, h_{(x)}$

Also, total head loss $= 40 - 30 = 10$ cm

$\therefore \qquad\qquad h_{(x)} + h_{(y)} = 10$

$$h_{(x)} + 9h_{(x)} = 10$$

$$10h_{(x)} = 10 \text{ or } h_{(x)} = 1 \text{ cm and } h_{(y)} = 9 \text{ cm}$$

For flow through soil X,

$$q = k_{(x)} i_{(x)} A = (4 \times 10^{-3}) \left(\frac{1}{10}\right)(10) = 4 \times 10^{-3} \text{ cm/sec}$$

For flow through soil Y,

$$q = k_{(y)} \cdot i_{(y)} \cdot A$$

$$4 \times 10^{-3} = k_{(y)} \left(\frac{9}{10}\right)(10) = 9k_{(y)}$$

$$\therefore \qquad k_{(y)} = \frac{4 \times 10^{-3}}{9} = 4.44 \times 10^{-4} \text{ cm/sec}$$

Problem 7.4

If during a variable head permeability test on a soil sample, equal time intervals are noted for head to drop from h_1 to h_2 and h_2 to h_3, find the relation between h_1, h_2 and h_3.

Solution : Let t_1 and t_2 be the times taken for head to fall from h_1 to h_2 and h_2 to h_3 respectively.

With usual notations, we can write

$$k = \frac{aL}{At_1} \log_e \frac{h_1}{h_2} = \frac{aL}{At_2} \log_e \frac{h_2}{h_3}$$

If

$$t_1 = t_2, \text{we get} \log_e \frac{h_1}{h_2} = \log_e \frac{h_2}{h_3}$$

i.e.,

$$\frac{h_1}{h_2} = \frac{h_2}{h_3}$$

or

$$h_2 = \sqrt{h_1 h_3}$$

Problem 7.5

Calculate the horizontal and vertical permeabilities of a soil deposit consisting of three layers 150 cm, 180 cm and 200 cm thick with permeabilities 10^{-5}, 10^{-7} and 10^{-9} m/sec respectively.

Solution:

Average permeability parallel to bedding plane,

$$k_x = \frac{k_1 Z_1 + k_2 Z_2 + k_3 Z_3}{Z_1 + Z_2 + Z_3}$$

$$= \frac{10^{-5} \times 1.5 + 10^{-7} \times 1.8 + 10^{-9} \times 2}{1.5 + 1.8 + 2}$$

$$= 0.286 \times 10^{-5} \text{ m/sec}$$
$$= 2.86 \times 10^{-6} \text{ m/sec}$$

$Z_1 = 1.5$ m $\quad k_1 = 10^{-5}$ m/sec

$Z_2 = 1.8$ m $\quad k_2 = 10^{-7}$ m/sec

$Z_3 = 2$ m $\quad k_3 = 10^{-9}$ m/sec

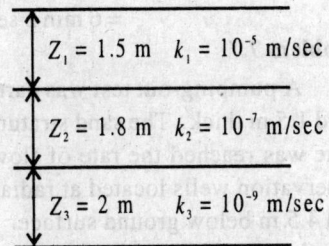

Average permeability perpendicular to bedding plane,

$$k_z = \frac{Z_1 + Z_2 + Z_3}{\dfrac{Z_1}{k_1} + \dfrac{Z_2}{k_2} + \dfrac{Z_3}{k_3}} = \frac{1.5 + 1.8 + 2}{\dfrac{1.5}{10^{-5}} + \dfrac{1.8}{10^{-7}} + \dfrac{2}{10^{-9}}}$$

$$= 0.00026 \times 10^{-5} \text{ m/sec}$$
$$= 2.6 \times 10^{-9} \text{ m/sec}$$

Problem 7.6

In a falling head permeability test the length and area of cross section of soil specimen are 0.17 m and 21.8×10^{-4} m^2 respectively. Calculate the time required for the head to drop from 0.25 m to 0.10 m. The area of cross section of stand pipe is 2×10^{-4} m^2. The sample has three layers with permeabilities 3×10^{-5} m/sec for first 0.06 m, 4×10^{-5} m/sec for second 0.06 m and 6×10^{-5} m/sec for the third 0.05 m thickness. Assume the flow is taking place perpendicular to the bedding plane.

Solution :

Average permeability perpendicular to bedding plane is given by

$$k_z = \frac{Z_1 + Z_2 + Z_3}{\dfrac{Z_1}{k_1} + \dfrac{Z_2}{k_2} + \dfrac{Z_3}{k_3}}$$

$$Z_1 = 0.06 \text{ m} \quad k_1 = 3 \times 10^{-5} \text{ m/sec}$$

$$Z_2 = 0.06 \text{ m} \quad k_2 = 4 \times 10^{-5} \text{ m/sec}$$

$$Z_3 = 0.05 \text{ m} \quad k_3 = 6 \times 10^{-5} \text{ m/sec}$$

$$= \frac{0.06 + 0.06 + 0.05}{\dfrac{0.06}{3 \times 10^{-5}} + \dfrac{0.06}{4 \times 10^{-5}} + \dfrac{0.05}{6 \times 10^{-5}}}$$

$$= 3.92 \times 10^{-5} \text{ m/sec}$$

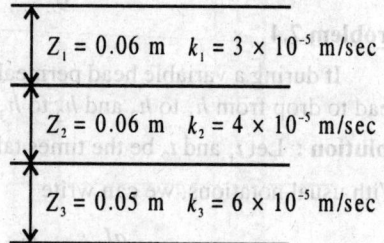

Length of soil sample, $L = 0.17$ m

Area of cross section of soil sample, $A = 21.8 \times 10^{-4}$ m^2

$h_1 = 0.25$ m; $h_2 = 0.10$ m

Time taken for fall from h_1 to h_2, $t = ?$

Area of cross section of stand pipe, $a = 2 \times 10^{-4}$ m^2

$$k = \frac{aL}{At} \log_e \frac{h_1}{h_2}$$

$$t = \frac{aL}{Ak} \log_e \frac{h_1}{h_2} = \frac{(2 \times 10^{-4})(0.17)}{(21.8 \times 10^{-4})(3.92 \times 10^{-5})} \log_e \frac{0.25}{0.10}$$

$$= 365 \text{ sec}$$

$$= 6 \text{ min } 5 \text{ sec.}$$

Problem 7.7

A pumping-out test was carried out at a level site, where 9 m of clay overlies a stratum of sand 1.5 m thick. The sand stratum is underlain by an impermeable rock stratum. When steady-state was reached the rate of flow was found to be 15 litres/second. The water levels in two observation wells located at radial distances of 6 m and 15 m from axis of main well were 5 m and 4.5 m below ground surface. Compute the coefficient of permeability of the sand stratum.

Solution :

Rate at which water is pumped out of well, $q = 15$ litres/sec = 15000 cc/sec

Thickness of confined aquifer, $b = 1.50$ m

Depth of water, measured above bottom impermeable stratum, in the two wells are

$$h_1 = 10.5 - 5.0 = 5.5 \text{ m} \qquad (r_1 = 6 \text{ m})$$

$$h_2 = 10.5 - 4.5 = 6.0 \text{ m} \qquad (r_2 = 15 \text{ m})$$

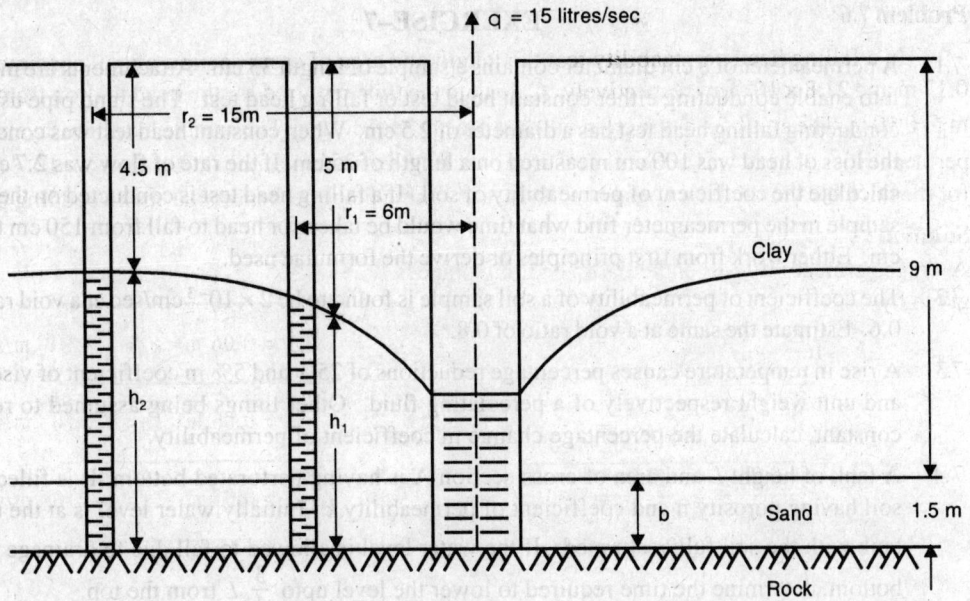

Using
Eq. 7.6 (*vii*),

$$k = \frac{q}{2\pi b (h_2 - h_1)} \log_e \frac{r_2}{r_1}$$

$$= \frac{15000}{2\pi (150)(600 - 550)} \log_e \frac{15}{6}$$

$$= 14.6 \text{ cm/sec}$$

Problem 7.8

Determine the ratio of k_1 to k_2 for flow occurring through two different soil media shown in the following figure when the head loss across L_1 is 25% of that across L_2.

Given, $A_2 = 2 A_1$ and $L_2 = 1.6 L_1$

Let h_1 = head loss across L_1

              h_2 = head loss across L_2

Then, $h_1 = 0.25 h_2$ or $h_2 = 4 h_1$

Applying Darcy's law, we have

$$Q = k_1 i_1 A_1 = k_2 i_2 A_2$$

$$= k_1 (h_1 / L_1) A_1 = k_2 (h_2 / L_2) A_2$$

$$= k_2 (4 h_1 / 1.6 L_1) (2 A_1)$$

$$k_1 = (8 / 1.6) k_2 = 5 k_2$$

$$k_1 / k_2 = 5$$

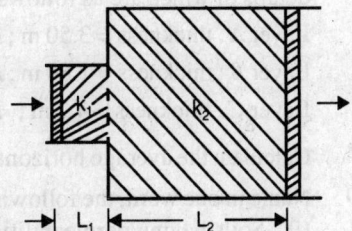

EXERCISE–7

7.1   A permeameter of 8 cm diameter contains a sample of length 35 cm. Attachments are made to
       it to enable conducting either constant head test or falling head test. The stand pipe used for
       conducting falling head test has a diameter of 2.5 cm. When constant head test was conducted
       the loss of head was 100 cm measured on a length of 25 cm. If the rate of flow was 2.7 cc/sec,
       calculate the coefficient of permeability of soil. If a falling head test is conducted on the same
       sample in the permeameter find what time would be taken for head to fall from 150 cm to 100
       cm. Either work from first principles or derive the formulae used.

7.2   The coefficient of permeability of a soil sample is found to be 2×10^{-3} cm/sec at a void ratio of
       0.6. Estimate the same at a void ratio of 0.8.

7.3   A rise in temperature causes percentage reductions of 25% and 5% in coefficient of viscosity
       and unit weight respectively of a percolating fluid. Other things being assumed to remain
       constant, calculate the percentage change in coefficient of permeability.

7.4   A tank of height L and area of cross-section A is having perforated bottom. It is filled with
       soil having porosity n and coefficient of permeability k. Initially water level is at the top of
       tank with the soil fully saturated. If the water level is allowed to fall due to drainage at the
       bottom, determine the time required to lower the level upto $\frac{3}{4} L$ from the top.

7.5   Compute average coefficient of permeability in directions parallel and perpendicular to
       the bedding plane of a layered soil deposit consisting of three layers of total thickness
       3.4 m. The top and bottom layers are each 0.7 m thick. The values of coefficient of
       permeability for the top, middle and bottom layers are k, $2 k$ and $3 k$ respectively, where
       $k = 15 \times 10^{-4}$ cm/sec.

7.6   A test well is installed in a permeable soil to a depth of 15 m below the water table at the site,
       and observation wells are bored at distances of 3m and 6m from axis of test well. When
       pumping at the rate of 2.3 m^3/min reached a steady-state, the drawdown at inner and outer
       wells were 1.5 m and 0.5 m respectively. Compute the coefficient of permeability of soil.

7.7   A falling head permeability test is to be conducted on a soil whose permeability is estimated
       to be 3×10^{-7} cm/sec. What diameter of stand-pipe would you use, if head had to drop from
       27.5 cm to 20.00 cm, in 5 minutes ? (Assume cross section of specimen = 15 cm^2 and length =
       0.5 cm)

7.8   On a certain site, there are three horizontal soil layers, down to an impermeable rock-bed, the
       details of which are as follows –

       Layer A : thickness = 3.50 m ; $k = 2.50 \times 10^{-5}$ m/sec

       Layer B : thickness = 1.80 m ; $k = 1.40 \times 10^{-7}$ m/sec

       Layer C : thickness = 4.2 m ; $k = 5.60 \times 10^{-3}$ m/sec

       Calculate the average horizontal and vertical permeability of the soil.

7.9   Name in one word, the following :
       (i)   Soils having permeability less than $1 x 10^{-6}$ cm/sec (as per USBR)

       (ii) The depth of soil which contains capillary water?

       (iii) Property of any porous media, like soil, to allow water, to pass through it

       (iv) Type of the permeameter to test the permeability of coarse-grained soils

       (v)   Water bearing stratum

7.10   Fill in the blanks :

(*i*) The horizontal permeability is............ than the vertical permeability.

(*ii*) Generally sands have a permeability from....... to......., while, for silts, it varies between ....and.....................

(*iii*) Two soils A and B have 85% and 65% respectively as the degree of saturation. The permeability of A,..…......... than that of B.

(*iv*) The effect of adsorbed water is to…............. the permeability.

(*v*) Either in the field, or in the laboratory, permeability is determined under................. flow conditions.

(*vi*) The effective size of a soil is 0.02 cm. Its permeability is approximately ....

7.11 State Darcy's law and define coefficient of permeability. What are the limitations in the application of Darcy's law to flow through soil media.

7.12 Suppose you are the staff in-charge of a small soils laboratory, which does not have a permeameter, but has basic equipment required for classification tests for all types of soils. Which test, then will you conduct to determine the permeability of a given soil sample?

7.13 Explain the terms discharge velocity and seepage velocity. Derive relation connecting the two.

7.14 Enumerate and briefly explain the factors affecting permeability of a soil.

7.15 What are the different methods to determine the permeability of a soil sample? Discuss briefly their merits and demerits and special applications?

7.16 What are main considerations while determining permeability of stratified soil deposits? Establish the relation between average permeability for flow parallel and that perpendicular to the bedding plane. Establish that the former is greater than the latter.

CHAPTER 8

SEEPAGE ANALYSIS

8.1 Introduction

Seepage or flow of water through a soil mass occurs when there is difference in total head between two points. The total head, h (also known as effective head) at any point in the soil mass in which flow is taking place is the algebraic sum of pressure head, h_w, and datum head or position head, Z.

$$h = h_w + Z \qquad \qquad ...8.1\,(i)$$

where Z will have positive value if the point is above datum and negative value if it is below datum.

Note: (i)   The datum head is the elevation of the point with respect to a chosen datum line.

          (ii)   As the velocity of flow through soil deposits is usually very low, the velocity head $v^2/2g$ being very small is neglected.

        (iii)   The pressure head at a point is also known as peizometric head.

Fig. 8.1 illustrates flow of water through a soil specimen taken in a tube. Referring to this figure we have,

$$\text{total head at } a, h_a = (h_w)_a - Z_a$$
$$\text{total head at } b, h_b = (h_w)_b - Z_b = 0$$

$\therefore$ loss of head for flow through soil mass from a to b $= h_a - h_b = (h_w)_a - Z_a - 0 = H$

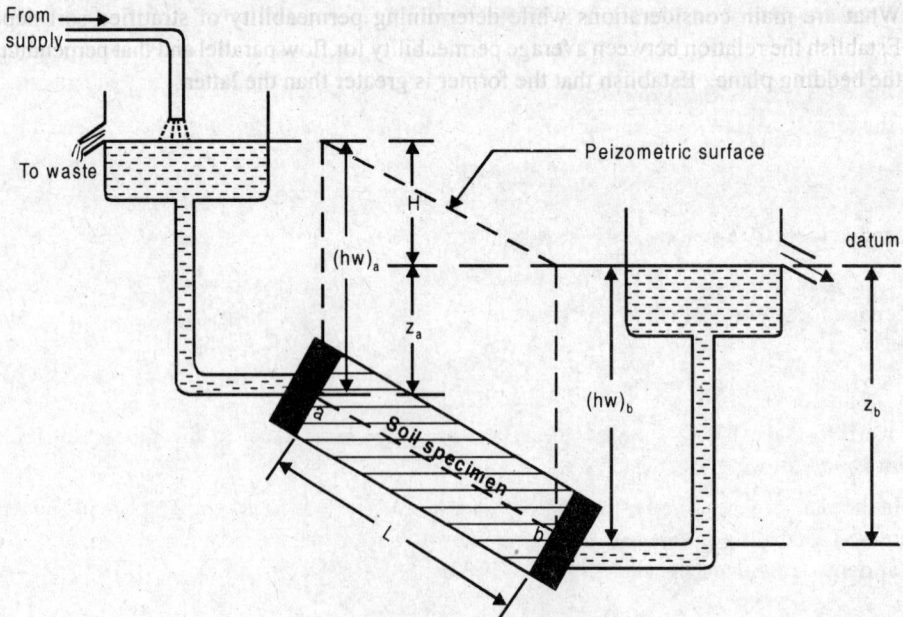

Fig. 8.1. Flow through a soil specimen

where H is the hydraulic head causing flow given by the difference in elevations of water levels in top and bottom tanks.

The loss of head per unit distance of flow through soil mass is called the hydraulic gradient and is denoted by i.

From Fig. 8.1, $$i = \frac{H}{L}$$ ...8.1(ii)

The hydraulic head, h at any point in the soil mass, being the sum of pressure head and position head, may be regarded as potential energy per unit weight of water with respect to datum. It is also referred to as hydraulic potential and is denoted by ϕ. The product of coefficient of permeability, k and hydraulic head h is referred to as velocity potential and is denoted by ϕ. The significance of velocity potential, ϕ, is that when it is differentiated with respect to any coordinate, it gives the velocity component in that direction. For example,

$$\phi = kh$$

$$\frac{\partial \phi}{\partial x} = k \cdot \frac{\partial h}{\partial x} = ki_x = v_x$$

When a piezometer is inserted at any point in the soil mass, water rises in the piezometer upto a height equal to peizometric head at that point. The line joining the water levels in the peizometers is referred to as piezometric surface.

Seepage analysis involves determination of hydraulic head at various points in the soil mass and rate of discharge through the soil mass. It is common practice, to choose the d/s water level (tail water level) as datum.

8.2 Seepage Pressure

The seepage pressure is the pressure exerted by flowing water on the section of soil mass in the direction of flow. It is caused by the force corresponding to energy transfer effected due to frictional drag between water and soil particles.

The seepage pressure at any point in the soil mass is given by

$$p_s = h \gamma_w$$ ...8.2 (i)

where h = total head at that point.
If Z is the length of flow over which head h is lost and i is the hydraulic gradient, we can write

$$p_s = \frac{h}{Z} \cdot Z \gamma_w = iZ\gamma_w$$ ...8.2 (ii)

The seepage force J over total cross-sectional area A of flow in the soil mass is:

$$J = p_s A = iZ \gamma_w A$$ ...8.2 (iii)

The seepage force per unit volume is given by

$$j = \frac{J}{AZ} = \frac{iZ\gamma_w A}{AZ} = i\gamma_w$$ ...8.2 (iv)

Thus, we have shown that seepage force per unit volume is equal to the product of hydraulic gradient i and unit weight of water γ_w.

In the case of vertical flow through soil mass the effective stress at a section will be increased or decreased according as the flow is in the downward direction or upward direction. The effective stress at depth Z is given by

$$\sigma' = \gamma' Z \pm p_s = \gamma' Z \pm iZ\gamma_w$$ ...8.2 (v)

The + sign is used when flow occurs in the downward direction and − sign when it occurs in the upward direction.

Consideration of seepage pressure is of vital importance in the stability analysis of earth structures subjected to seepage. Seepage pressure is responsible for the phenomenon of quick sand.

8.3 Quick Sand or Quick Condition

In the case of upward flow of water through a soil mass, the seepage pressure acts in the upward direction causing reduction in effective stress. In the case of submerged soil mass, the upward seepage pressure may become equal to downward pressure due to submerged weight of soil, at a certain level. When this happens in the case of a cohesionless soil, the soil at that level looses all its shear strength as the effective stress becomes zero.

$$\tau_f = c + \sigma' \tan \phi = 0 + 0 (\tan \phi) = 0$$

Because of this the soil particles have the tendency to be carried away by flowing water. This phenomenon of lifting of soil particles by flowing water is called quick sand or quick condition (or boiling condition). It should be noted that quick sand is not a type of sand but is another name for quick condition. Thus quick condition occurs when

$$\sigma' = \gamma' z - p_s = 0$$

or

$$p_s = \gamma' z$$

Also $p_s = i z \gamma_w$. Therefore, we can write $i_c z \gamma_w = \gamma' z$

i.e., $i_c = \dfrac{\gamma' z}{\gamma_w z} = \dfrac{\gamma'}{\gamma_w} = \dfrac{G - 1}{1 + e}$ ...8.3 (i)

where i_c denotes critical hydraulic gradient.

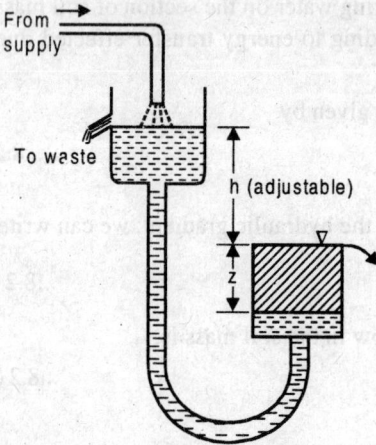

Fig. 8.2. Schematic diagram of experimental setup to demonstrate quick condition

The critical hydraulic gradient is the hydraulic gradient at which quick condition occurs. In Eq.8.3 (i) if we put $G = 2.67$ and $e = 0.67$, i_c will become unity. For most cohesionless soil deposits i_c will be less than unity. An experimental set-up to demonstrate quick condition is illustrated in Fig. 8.2, in which water flows through soil mass of thickness Z, under hydraulic head h. This head is adjustable by moving the supply tank up or down.

The head h is gradually increased until the quick condition is noticed in the soil. At this condition, the upward force at the bottom of the soil mass becomes equal to the downward force due to saturated soil above that level. If A is the area of cross-section of soil sample, we have

$$(h + z) \gamma_w A = \gamma_{sat} Z A$$

$$h \gamma_w = Z (\gamma_{sat} - \gamma_w) = Z \gamma'$$

$$\therefore \quad \dfrac{h}{Z} = i_c = \dfrac{\gamma'}{\gamma_w} = \dfrac{G - 1}{1 + e}$$

The velocity of flow, v required to maintain the critical hydraulic gradient, i_c is directly proportional to coefficient of permeability, k ($v = k i_c$). It is due to this reason that quick sand commonly occurs in the case of fine sands for which coefficient of permeability is low. For quick sand phenomenon to occur in the case of cohesionless soil with greater value of permeability, velocity of flow will have to be much higher.

Problem 8.1

Calculate the critical hydraulic gradient for a coarse-grained soil deposit with void ratio of 0.7. Take $G = 2.67$.

Solution: $e = 0.7$,

$\qquad\qquad G = 2.67$

Critical hydraulic gradient, $i_c = \dfrac{G-1}{1+e} = \dfrac{2.67-1}{1+0.7} = 0.98$

Problem 8.2

An excavation was being carried out in a stiff clay deposit with a saturated unit weight of 17.8 kN/m³. When the depth of excavation reached 6.7 m the bottom of the pit started rising, cracked and a mixture of sand and water was noticed flowing out of the cracks. On enquiry it was found that the clay stratum extended upto a depth of 11.2 m below ground level and was underlain by a sand stratum. Compute the height to which the water would have risen in the bore hole above the surface of sand stratum, prior to excavation.

Solution:

Let h_w be the height to which water will have risen in bore hole, above surface of sand stratum, prior to excavation. Quick condition occurs at the surface of sand stratum when effective stress becomes zero at that level.

$$\sigma' = \gamma_{sat} Z - \gamma_w h_w = 0$$

$$\therefore \qquad h_w = \frac{\gamma_{sat} Z}{\gamma_w} = \frac{(17.8)(4.5)}{9.81} = 8.16 \text{m}$$

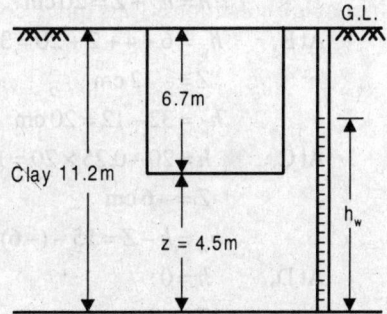

Problem 8.3

A soil sample taken in a container of internal diameter 8 cm has a length of 12.5 cm. The coefficient of permeability of soil is 1.5×10^{-3} cm/sec with void ratio of 0.7. If water is made to flow through the soil sample in upward direction and the rate of discharge is 0.04 ml/sec compute the effective stresses at the bottom and middle sections of the sample. Take $G = 2.65$

Solution :

$$\gamma' = \frac{(G-1)\gamma_w}{1+e} = \frac{(2.65-1)9.81}{1+0.7} = 9.52 \text{ kN/m}^3$$

$$k = 1.5 \times 10^{-3} \text{ cm/sec}, \quad q = 0.04 \text{ cm}^3/\text{sec}$$

Using Darcy's law $q = kiA$ where $A = \dfrac{\pi(8)^2}{4} = 50.26 \text{ cm}^2$

$$i = \frac{q}{kA} = \frac{0.04}{1.5 \times 10^{-3} \times 50.26} = 0.53$$

When flow of water is in upward direction, the effective stress at a point is given by

$$\sigma' = \gamma' Z - p_s = \gamma' Z - iZ\gamma_w$$

The effective stress at the middle section of sample,

$$\sigma' = (9.52)(0.0625) - (0.53)(0.0625)(9.81) = 0.27 \text{ kN/m}^2$$

The effective stress at the bottom section of the sample,

$$\sigma' = (9.52)(0.125) - (0.53)(0.125)(9.81) = 0.54 \text{ kN/m}^2$$

Problem 8.4

Two different types of soil are placed in a permeameter and water is allowed to flow under a constant head as shown in accompanying figure. The internal diameter of permeameter is 8 cm (a) Find the pressure head, datum head and total head at points A, B, C and D, if 25% of the total head causing flow is lost as water flows through lower layer. (b) Find the rate of discharge if the permeability of lower layer is 2.8×10^{-2} cm/sec. (c) Find permeability of top layer.

Solution :

(a)   At A, $h_w = 0$

                     $Z = 20$ cm

                     $h = h_w + Z = 20$ cm

     At B, $h_w = 6 + 4 + 2 + 20 = 32$ cm

                     $Z = -12$ cm

                     $h_w = 32 - 12 = 20$ cm

     At C, $h = 20 - 0.25 \times 20 = 15$ cm

                     $Z = -6$ cm

                     $h_w = h - Z = 15 - (-6) = 21$ cm

     At D, $h = 0$

                     $Z = -2$ cm

                     $h_w = h - Z = 0 - (-2) = 2$ cm

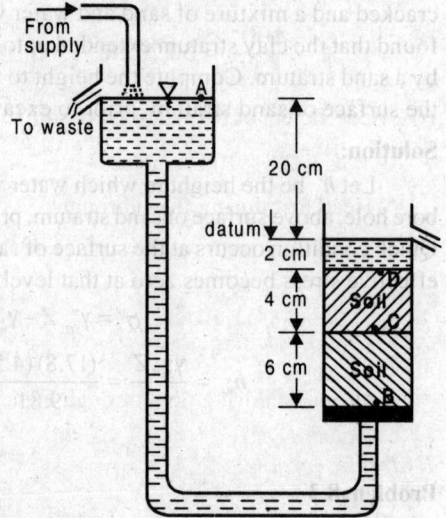

(b)   Area of cross section of soil specimen,

$$A = \frac{\pi (8)^2}{4} = 50.26 \text{ cm}^2$$

Head loss during flow through lower layer = $(0.25)(20) = 5$ cm

Hydraulic gradient, $i = \dfrac{5}{6} = 0.83$

∴   Rate of discharge, $q = kiA = (2.8 \times 10^{-2})(0.83)(50.26)$

           = 1.17 cm³/sec

(c)   Head lost during flow through top layer = $(0.75)(20) = 15$ cm

(Also given by : total head at C – total head at D = $15 - 0 = 15$ cm)

Hydraulic gradient, $i = \dfrac{15}{4} = 3.75$

Coefficient of permeability of soil in top layer,

$$k = \frac{q}{iA} = \frac{1.17}{(3.75)(50.26)} = 6.2 \times 10^{-3} \text{ cm/sec}$$

Problem 8.5

Soils of two types 1 and 2 are taken in a permeameter of diameter 8 cm and water is allowed to flow through them under a head of 30 cm as shown in accompanying figure. (page 113)

(i)   If the permeability of soil 1 is 2.8×10^{-2} cm/sec and 35% of total head causing flow is lost during flow through this layer, calculate the rate of discharge and permeability for soil 2.

(ii)   If the void ratio is 0.55 for soil 1 and 0.70 for soil 2, compute the seepage velocity for flow through each layer.

(iii) If the total head is increased, determine its value at which either of the soil will become quick. Assume G as 2.65 for soil 1 and 2.7 for soil 2.

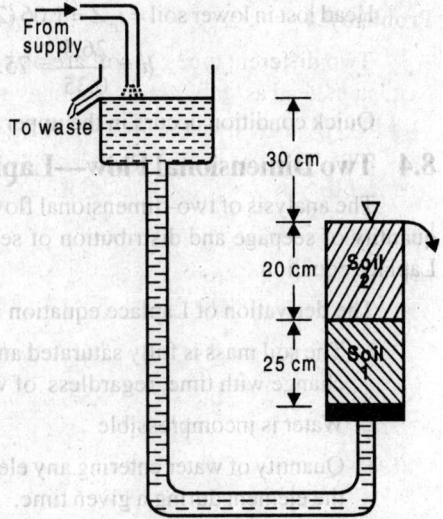

Solution: Area of cross section of soil sample,

$$A = \frac{\pi(8)^2}{4} = 50.26 \text{ cm}^2$$

(i)   Coefficient of permeability of soil 1,

$$k_1 = 2.8 \times 10^{-2} \text{ cm/sec.}$$

Head lost during flow through

lower layer = $0.35 \times 30 = 10.5 \text{ cm}$

$\therefore$ Rate of discharge, $q = k_1 i_1 A = (2.8 \times 10^{-2})\left(\dfrac{10.5}{25}\right)(50.26) = 0.59 \text{ cm}^3/\text{sec}$

Head lost during flow through upper layer = $0.65 \times 30 = 19.5 \text{ cm}$

Hydraulic gradient, $i_2 = \dfrac{19.5}{20} = 0.975$

$\therefore$ Coefficient of permeability of soil 2, $k_2 = \dfrac{q}{i_2 A} = \dfrac{0.59}{0.975 \times 50.26} = 1.2 \times 10^{-2} \text{ cm/sec}$

(ii)   Discharge velocity, $v = \dfrac{q}{A} = \dfrac{0.59}{50.26} = 0.012 \text{ cm/sec}$

Porosity, $n = \dfrac{e}{1+e} = \dfrac{0.7}{1+0.7} = 0.41$ for upper layer and

$$n = \dfrac{0.55}{1+0.55} = 0.36 \text{ for lower layer}$$

Seepage velocity for flow through upper layer,

$$v_s = \dfrac{v}{n} = \dfrac{0.012}{0.41} = 0.029 \text{ cm/sec}$$

Seepage velocity for flow through lower layer,

$$v_s = \dfrac{0.012}{0.36} = 0.033 \text{ cm/sec}$$

(iii) Critical hydraulic gradient, $i_c = \dfrac{G-1}{1+e}$

For upper soil $i_c = \dfrac{2.7-1}{1+0.7} = 1$

For lower soil $i_c = \dfrac{2.65-1}{1+0.55} = 1.06$

At quick condition,

Head lost in upper soil = $i_c Z = 1(20) = 20 = 0.65h$ $\therefore$ $h = \dfrac{20}{0.65} = 30.8 \text{ cm}$

Head lost in lower soil $= i_c Z = 1.06\,(25) = 26.5 = 0.35\text{h}$

$\therefore$ $\qquad\qquad h = \dfrac{26.5}{0.35} = 75.7$ cm

Quick condition occurs in the upper soil when total hydraulic head reaches 30.8 cm.

8.4 Two Dimensional Flow—Laplace equation

The analysis of two-dimensional flow through a saturated soil mass involves determination of quantity of seepage and distribution of seepage pressure and can be done with methods based on Laplace equation.

The derivation of Laplace equation is based on the following assumptions:

1. The soil mass is fully saturated and is incompressible. The size of the pore spaces does not change with time, regardless of water pressure.

2. Water is incompressible

3. Quantity of water entering any element of soil mass is equal to the quantity of water leaving the element during a given time.

4. Darcy's law for flow through soil medium is valid.

5. The hydraulic boundary conditions at entry and exit are known.

Derivation of Laplace equation:

We consider an element of size Δx, Δy and of unit thickness perpendicular to plane of figure. Let v_x and v_y be the velocity components at entry in x and y directions. Then the corresponding velocity components at exit will be

$$\left(v_x + \frac{\partial v_x}{\partial x}\Delta x\right) \text{ and } \left(v_y + \frac{\partial v_y}{\partial_y}\Delta y\right)$$

According to the assumption that quantity of water entering an element is equal to the quantity of water leaving the element in any given time, we have

$$v_x\,(\Delta y \cdot 1) + v_y\,(\Delta x \cdot 1) = \left(v_x + \frac{\partial v_x}{\partial x}\Delta x\right)(\Delta y \cdot 1) + \left(v_y + \frac{\partial v_y}{\partial y}\cdot\Delta y\right)(\Delta x \cdot 1)$$

$i.e.,$ $\qquad \dfrac{\partial v_x}{\partial x} + \dfrac{\partial v_y}{\partial y} = 0$ $\qquad$ ...8.4 (i)

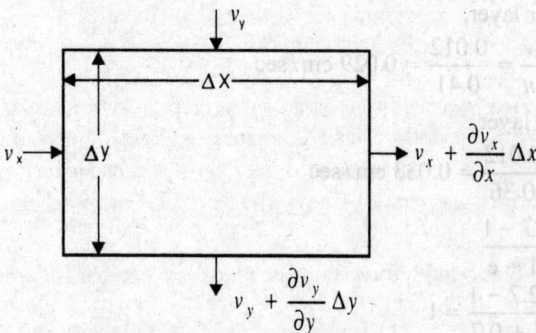

Fig. 8.3. Two-dimensional flow

This is the continuity equation for two-dimensional flow.

According to the assumption that Darcy's law is valid for flow through soil medium, we have

$$v_x = k_x \cdot i_x = k_x \frac{\partial h}{\partial x}$$

$$v_y = k_y \cdot i_y = k_y \frac{\partial h}{\partial y}$$

where k_x and k_y are coefficients of permeability in x and y directions.

By substitution in Eq.8.4 (i), we get

$$k_x \frac{\partial^2 h}{\partial x^2} + k_y \frac{\partial^2 h}{\partial y^2} = 0 \qquad\qquad\qquad ...8.4\,(ii)$$

For an isotropic soil medium, $k_x = k_y = k$ (say)

Then Eq.8.4 (*ii*) reduces to the form

$$\frac{\partial^2 h}{\partial x^2} + \frac{\partial^2 h}{\partial y^2} = 0$$

...8.4 (*iii*)

This is Laplace equation for two-dimensional flow.

8.5   Flow net

The solution of Laplace equation [Eq. 8.4 (*iii*)] gives two sets of curves, namely, flow lines and equipotential lines. Flow lines and equipotential lines together constitute a flow net. A flow line represents the path traced by an individual water particle. An equipotential line is a contour or line joining points of equal potential or head.

The flow lines and equipotential lines cut each other at right angles *i.e.* they are mutually orthogonal as shown in Fig. 8.4.

The space between any two adjacent flow lines is called flow channel. The space enclosed between two adjacent flow lines and two successive equipotential lines is called a field.

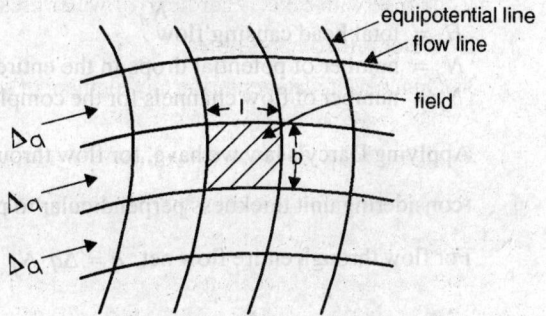

Fig. 8.4. Portion of a flow net

8.6   Properties of Flow net

Note the following properties of flow net before proceeding for construction and application of flow nets.

1. Flow lines and equipotential lines cut each other at right angles i.e. they are mutually orthogonal.
2. Each field is an approximate square and in a well-constructed flow net one should be able to draw a circle in a field touching all the four sides.
3. In a homogeneous soil, every transition in the shape of the two types of curves will be smooth, being either elliptical or parabolic in shape.
4. The rate of flow through each flow channel is same.
5. The same potential drop occurs between two successive equipotential lines.

8.7   Flow net by Graphical Method

The graphical method of flow net construction involves sketching by trial and error. It was first given by Forchheimer (1930). The hydraulic boundary conditions are examined and keeping in mind the properties of flow net initial sketching is done and by trial and error the flow net is improved to make it acceptable for practical applications. For beginners, A. Casagrande (1937) has given the following excellent hints.

1. Well constructed flow nets should be studied and effort should be made to retain the salient features in mind.
2. About four or five flow channels are sufficient for the first trial, as too many flow channels will distract attention from essential features.
3. After initial sketching the flow net should be observed as a whole while adjusting the finer details.
4. All transitions should be made smooth being either elliptical or parabolic in shape.

5. But for a few exceptional fields, all fields should be approximate squares.

Figures 8.5, 8.6, 8.7, 8.8 and 8.9 illustrate flow nets for seepage under sheet pile wall and weirs.

8.8   Applications of Flow net

A flow net can be used to determine (*i*) quantity of seepage, (*ii*) seepage pressure at a point, (*iii*) hydrostatic pressure at a point and (*iv*) exit gradient.

(i)   Determination of seepage

Let us consider the field marked in Fig.8.4, which illustrates a portion of a flow net. (Let *l* and *b* denote the length and breadth of the field. (*l* = *b* = 1)

If Δq = rate of discharge through each flow channel

Δh = head drop per field = $\dfrac{H}{N_d}$

H = total head causing flow

N_d = number of potential drops in the entire flow net

N_f = number of flow channels for the complete flow net,

Applying Darcy's law we have, for flow through the field, $\Delta q = k \cdot \dfrac{\Delta h}{l}(b \times 1) = k \cdot \dfrac{H}{N_d} \cdot \dfrac{b}{l}.$

(considering unit thickness perpendicular to plane of figure).

For flow through entire flow net, $q = \Delta q \cdot N_f = k \cdot \dfrac{H}{N_d} \cdot N_f$ $\left(\dfrac{b}{l} = 1\right)$

i.e., $q = kH \dfrac{N_f}{N_d}$

This equation is used to find the discharge through an isotropic soil for which $k_x = k_y = k$.

(ii)   Determination of seepage pressure at a point

Seepage pressure at a point, p_s is given by

$p_s = h\gamma_w$

where h = total head at that point = $(H - n.\Delta h)$

H = total head causing flow

n = number of potential drops upto the point under consideration.

Δh = potential drop per field = $\dfrac{H}{N_d}$

(iii)   Determination of hydrostatic pressure at a point

The hydrostatic head at a point is given by

$h_w = h - z$

where h = total head at that point = $(H - n.\Delta h)$

n = number of potential drops upto that point

Δh = potential drop per field = $\dfrac{H}{N_d}$

z = datum head at that point.

z will have positive value when the point is above datum and negative value when it is below datum. The hydrostatic pressure at the point is given by

$u = h_w.\gamma_w$

(iv)   Determination of exit gradient.

The exit gradient, i_e is given by

$$i_e = \frac{\Delta h}{l_e}$$

where $\Delta h =$ head drop per field $= \dfrac{H}{N_d}$

$l_e =$ average length of smallest exit field (which will be adjacent to structure).

Problem 8.6

A soil stratum with permeability, $k = 5 \times 10^{-7}$ cm/sec overlies an impermeable stratum. The impermeable stratum lies at a depth of 18 m below the ground surface (surface of soil stratum). A sheet pile wall penetrates 8 m into the permeable soil stratum. Water stands to a height of 9 m on upstream side and 1.5 m on downstream side, above the surface of soil stratum. Sketch the flow net and determine (i) quantity of seepage, (ii) the seepage pressure at a point P located 8 m below surface of soil stratum and 4 m away from the sheet pile wall on its upstream side, (iii) the pore pressure at point P and (iv) the maximum exit gradient.

Solution : The flow net is shown in Fig. 8.5. This type of trial sketching requires about 10 minutes. From the flow net, we have

No of flow channels, $N_f = 4$

Number of potential drops, $N_d = 8$

(i) Head causing flow, $H = 9 - 1.5 = 7.5$ m

Quantity of seepage, $q = kH \dfrac{N_f}{N_d}$

$$= (5 \times 10^{-9})(7.5)\left(\frac{4}{8}\right)$$

$$= 18.75 \times 10^{-9} \text{ m}^3/\text{sec per metre length.}$$

(ii) Potential drop per field, $\Delta h = \dfrac{H}{N_d} = \dfrac{7.5}{8} = 0.9375$ m

Number of potential drops upto point P, $n = 2.5$

∴ Total head at point P, $h = (H - n.\Delta h)$

$$= (7.5 - 2.5 \times 0.9375)$$

$$= 5.16 \text{ m.}$$

Seepage pressure at P, $p_s = h\gamma_w = (5.16)(9.81)$

$$= 50.62 \text{ kN/m}^2$$

(iii) Hydrostatic head at P, $h_w = h - Z$

$$= 5.16 - (-9.5)$$

$$= 14.66 \text{ m}$$

Pore pressure at P, $u = h_w \gamma_w = (14.66)(9.81)$

$$= 143.81 \text{ kN/m}^2$$

(iv) Average length of exit field adjacent to sheet pile wall, $l_e = 2.6$ m

Maximum exit gradient, $i_e = \dfrac{\Delta h}{l_e} = \dfrac{0.9375}{2.8} = 0.33$

Note : In the flow net, the following have been clearly indicated.

    (a) Boundary flow lines : BC, DE and JK

    (b) Boundary equipotential lines : AB and EF

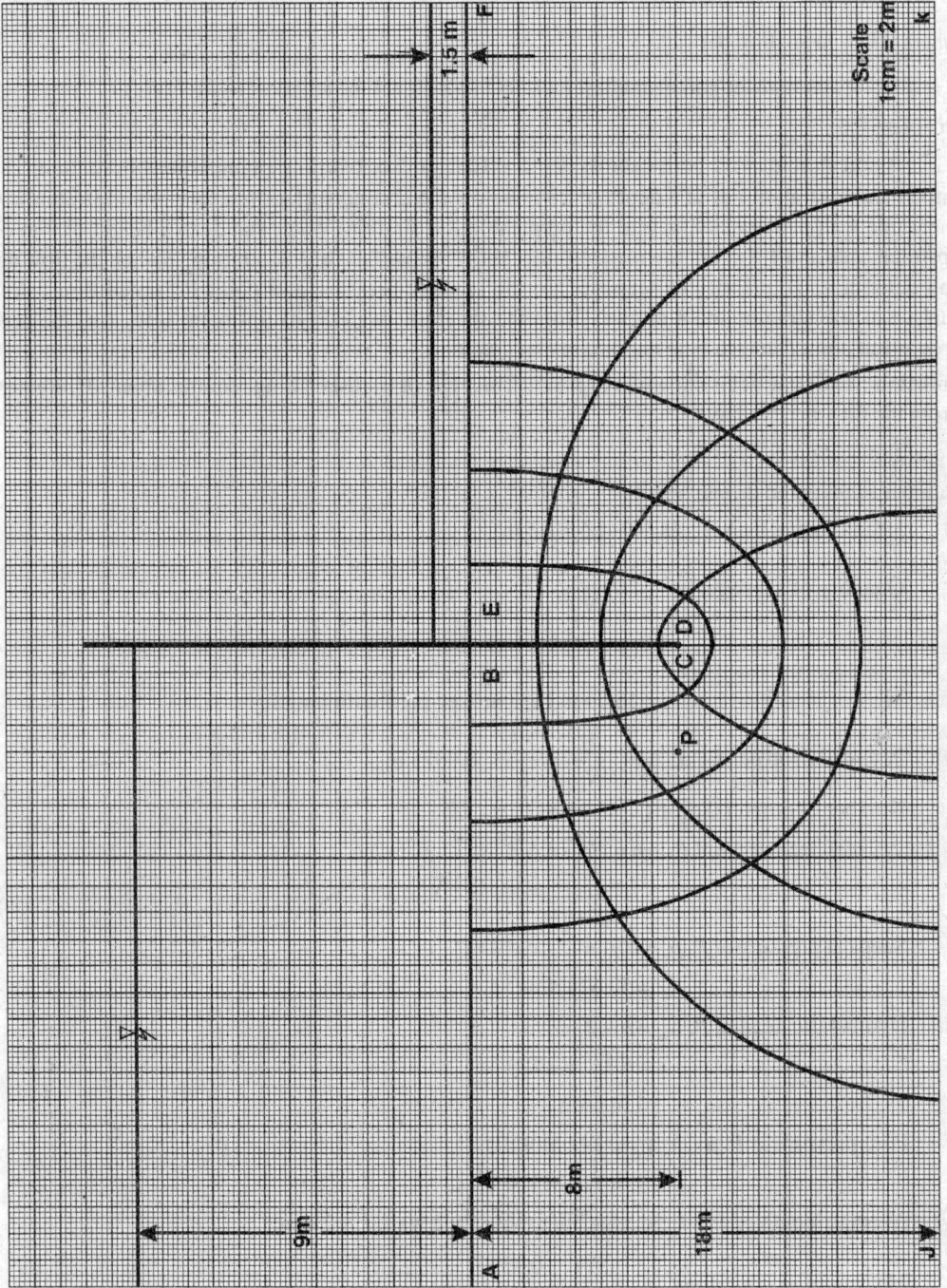

Fig. 8.5 Flow net for flow under sheet pile wall

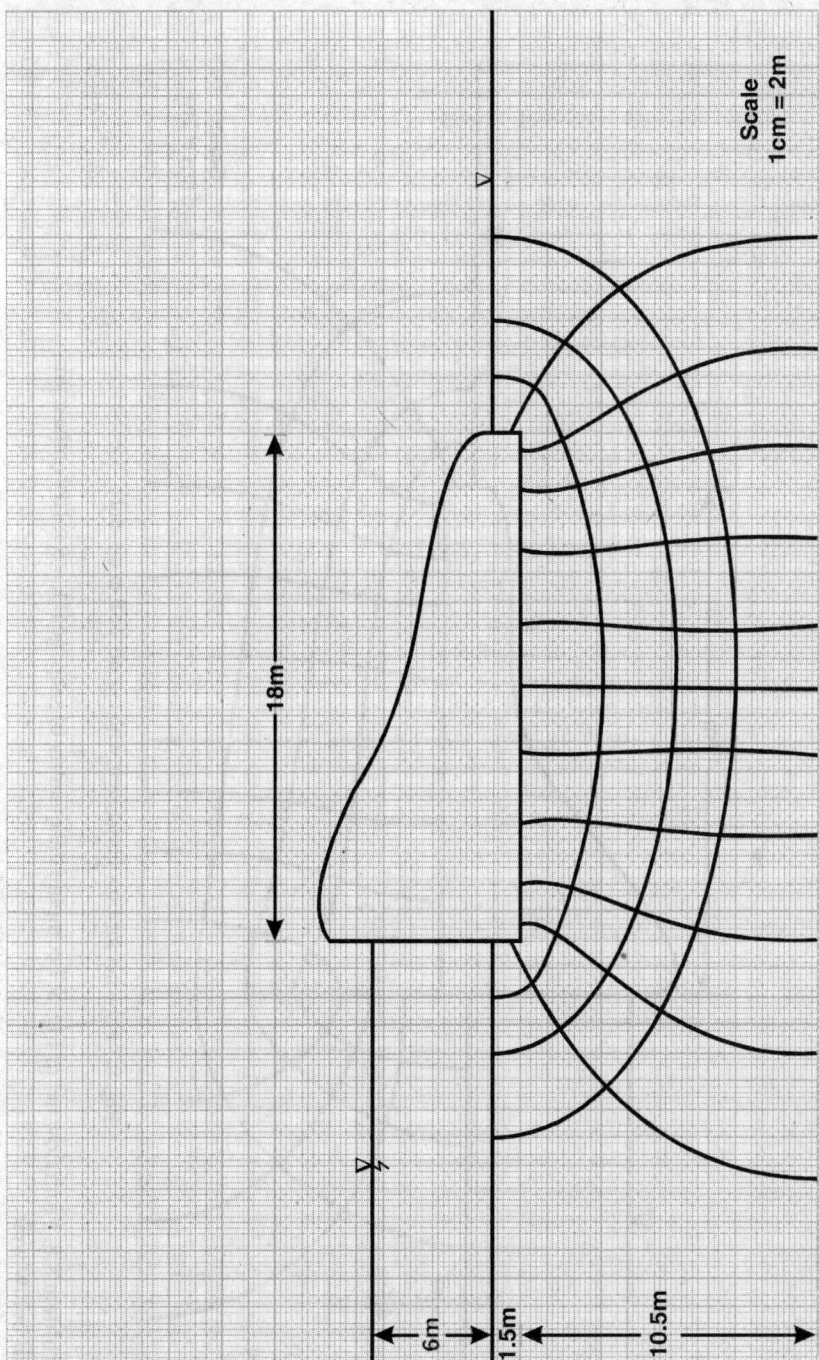

Fig. 8.6 Flow net for flow under weir

Problem 8.7 From flow net of Fig. 8.6 compute the quantity of seepage if $k = 5 \times 10^{-4}$ cm/sec for soil below weir.

Solution: $H = 6$ m $N_f = 4$ $N_d = 12$

$$\therefore q = KH \frac{N_f}{N_d} = (5 \times 10^{-6})(6)\left(\frac{4}{12}\right) = 10 \times 10^{-6} \text{ m}^3/\text{sec per metre run}$$

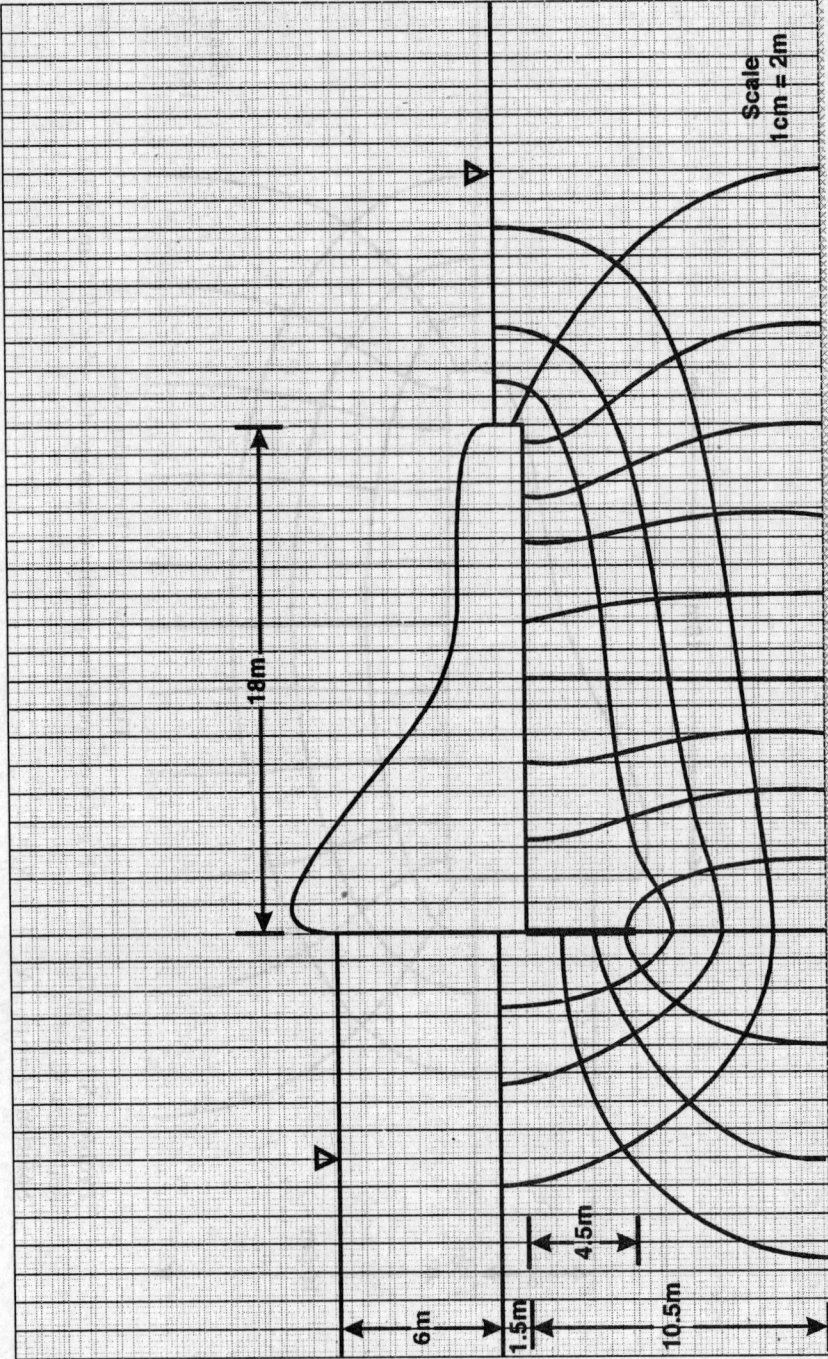

Fig. 8.7 Flow net for flow under weir with cut off wall at u/s end

Problem 8.8. From flow net of Fig. 8.7 compute the quantity of seepage if $k = 5 \times 10^{-4}$ cm/sec for soil below weir.

Solution: $H = 6$ m $N_f = 4$ $N_d = 14$

$$q = kH \frac{N_f}{N_d} = (5 \times 10^{-6})(6)\left(\frac{4}{14}\right) = 8.6 \times 10^{-6} \text{ m}^3 / \text{sec per metre run}$$

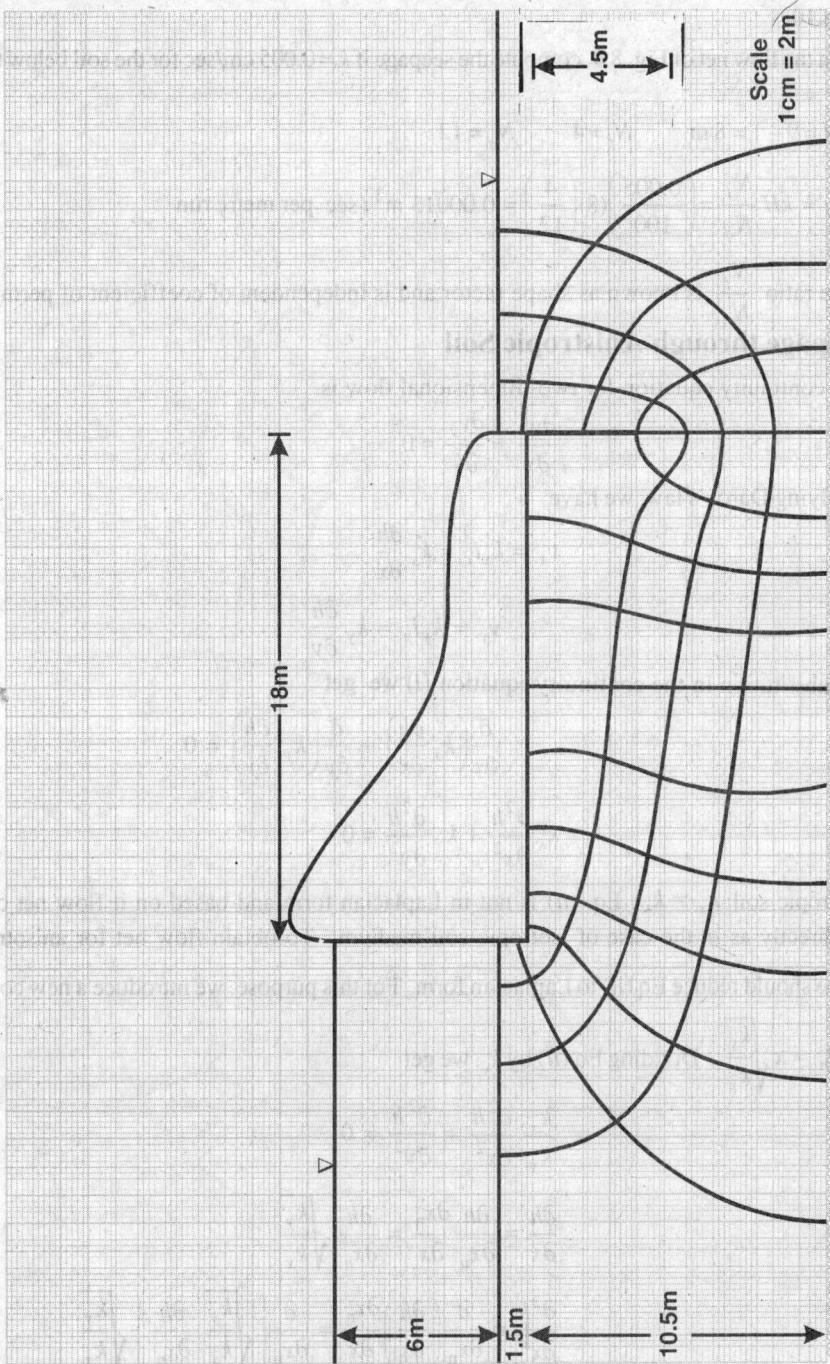

Fig. 8.8 Flow net for flow under weir with cut off wall at d/s side

Problem 8.9. From flow net of Fig 8.8 compute the quantity of seepage if $k = 5 \times 10^{-4}$ cm/sec for soil below weir

Solution : $H = 6$m $\quad N_f = 4 \quad N_d = 14$

$$\therefore \quad q = kH \frac{N_f}{N_d} = (5 \times 10^{-6})(6)\left(\frac{4}{14}\right) = 8.6 \times 10^{-6} \text{ m}^3 / \text{sec per metre run}$$

Problem 8.10

From the flow net of Fig. 8.9, compute the seepage if $k = 0.005$ cm/sec for the soil below the weir.

Solution :

$$H = 9 - 1 = 8 \text{ m} \qquad N_f = 4 \qquad N_d = 12$$

$$q = kH \frac{N_f}{N_d} = \left(\frac{0.005}{100}\right)(8)\left(\frac{4}{12}\right) = 0.00013 \text{ m}^3/\text{sec per metre run}$$

Note : The ratio $\dfrac{N_f}{N_d}$ is known as shape factor and is independent of coefficient of permeability.

8.9 Seepage through Anistropic Soil

The continuity equation for two-dimensional flow is

$$\frac{\partial v_x}{\partial x} + \frac{\partial v_y}{\partial y} = 0 \qquad\qquad ...(i)$$

Applying Darcy's law, we have

$$v_x = k_x i_x = k_x \frac{\partial h}{\partial x}$$

$$v_y = k_y i_y = k_y \frac{\partial h}{\partial y}$$

By substitution in the continuity equation (i) we get

$$\frac{\partial}{\partial x}\left(k_x \frac{\partial h}{\partial x}\right) + \frac{\partial}{\partial y}\left(k_y \frac{\partial h}{\partial y}\right) = 0$$

$$k_x \frac{\partial^2 h}{\partial x^2} + k_y \frac{\partial^2 h}{\partial y^2} = 0 \qquad\qquad ...(ii)$$

For anisotropic soil $k_x \neq k_y$. Eq. (ii) is not in Laplacian form and based on it flow net cannot be obtained directly as in the case of isotropic soil medium. To obtain flow net for anisotropic soil medium we should reduce Eq (ii) to Laplacian form. For this purpose, we introduce a new co-ordinate variable $x_n = x\sqrt{\dfrac{k_y}{k_x}}$. Dividing Eq (ii) by k_y, we get

$$\frac{k_x}{k_y} \frac{\partial^2 h}{\partial x^2} + \frac{\partial^2 h}{\partial y^2} = 0 \qquad\qquad ...(iii)$$

Also

$$\frac{\partial h}{\partial x} = \frac{\partial h}{\partial x_\eta} \frac{\partial x_\eta}{\partial x} = \frac{\partial h}{\partial x_n} \cdot \sqrt{\frac{k_y}{k_x}}$$

$$\frac{\partial^2 h}{\partial x^2} = \frac{\partial}{\partial x_\eta}\left(\frac{\partial h}{\partial x}\right)\frac{\partial x_\eta}{\partial x} = \frac{\partial}{\partial x_\eta}\left(\sqrt{\frac{k_y}{k_x}} \cdot \frac{\partial h}{\partial x_\eta}\right)\cdot\sqrt{\frac{k_y}{k_x}}$$

$$= \frac{k_y}{k_x} \cdot \frac{\partial^2 h}{\partial x_n^2}$$

By substitution in Eq. (iii)

$$\frac{k_x}{k_y} \cdot \frac{k_y}{k_x} \cdot \frac{\partial^2 h}{\partial x_n^2} + \frac{\partial^2 h}{\partial y^2} = 0$$

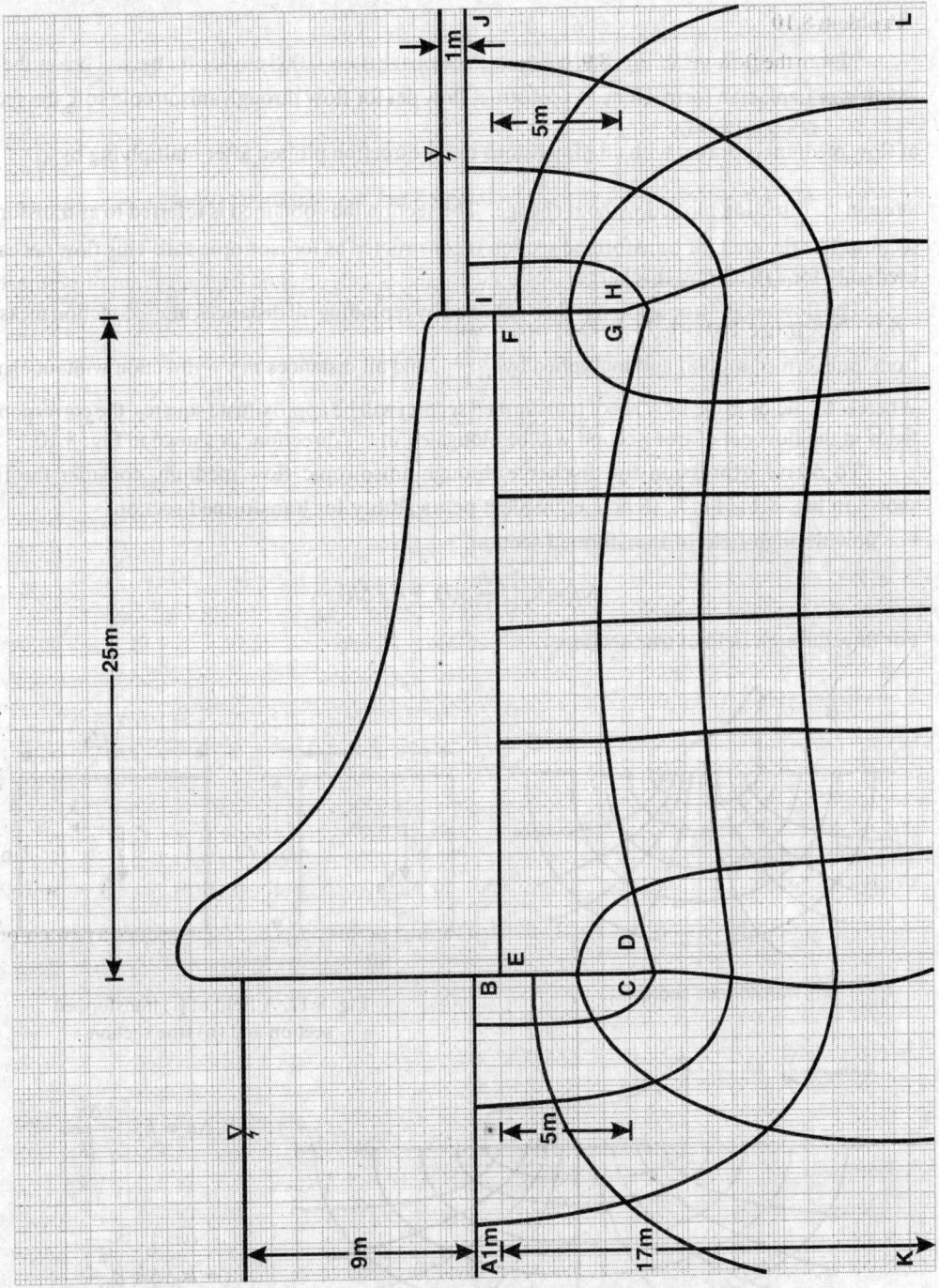

Fig. 8.9. Flow net for flow under weir with sheet pile cut off walls at both u/s and d/s ends.

i.e.
$$\frac{\partial^2 h}{\partial x_n^2} + \frac{\partial^2 h}{\partial y^2} = 0 \qquad \qquad ...(iv)$$

which is in Laplacian form. Hence to construct flow net for flow through anisotropic soil, the section

of flow medium is drawn with all dimensions in x – direction plotted after multiplying by $\sqrt{\dfrac{k_y}{k_x}}$ and

those in y – direction plotted without change. The section thus obtained is referred to as transformed
section. The flow net for transformed section is constructed as for isotropic soil. This flow net can be
used only for calculating discharge.

The flow net for true section is then obtained by replotting all distances in x – direction measured

from transformed section and multiplied by $\sqrt{\dfrac{k_x}{k_y}}$, and all distances in y – direction without change.

Whereas the fields in the flow net of transformed section will be approximate squares, the corresponding
fields in the flow net of true section will be elongated in x – direction, as shown in Fig. 8.10.

To derive expression for discharge through anisotropic flow medium, consider the fields
shown in Fig. 8.11. Let k_e denote equivalent permeability for transformed section.
For flow through field of transformed section,

$$\Delta q = k_e \left(\frac{\Delta h}{l} \right)(l) = k_e \cdot \Delta h \qquad \qquad ...(i)$$

For flow through field of true section,

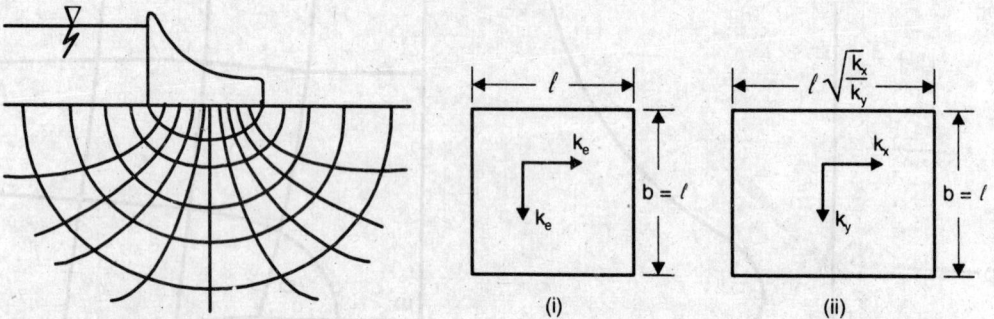

(a) Transformed section

Fig. 8.11. A field of (i) transformed
section and (ii) true section

(i)

(ii)

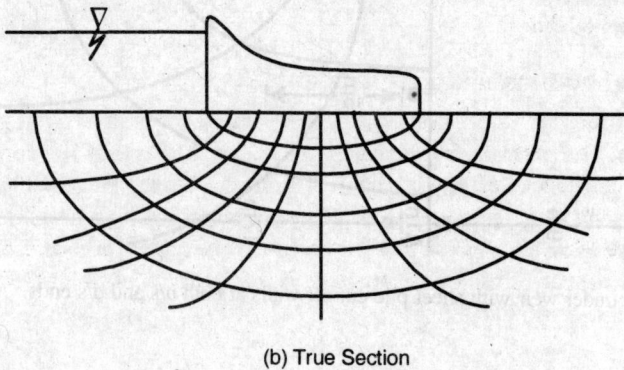

(b) True Section

Fig. 8.10. Flow through anisotropic soil

$$\Delta q = k_x \left[\frac{\Delta h}{l \sqrt{\dfrac{k_x}{k_y}}} \right] \cdot l$$

$$= k_x \cdot \Delta h \cdot \sqrt{\frac{k_y}{k_x}}$$

$$= \Delta h \cdot \sqrt{k_x k_y} \qquad \qquad ...(ii)$$

In both cases we have assumed unit thickness perpendicular to plane of figure.

From Eq. (*i*) and (*ii*), we get

$$k_e \cdot \Delta h = \Delta h \cdot \sqrt{k_x k_y}$$

i.e.,

$$k_e = \sqrt{k_x k_y} \qquad \qquad ...8.9\,(i)$$

$$\therefore \qquad q = k_e \cdot H \cdot \frac{N_f}{N_d} = \sqrt{k_x k_y} \cdot H \cdot \frac{N_f}{N_d} \qquad \qquad ...8.9\,(ii)$$

8.10 Deflection of flow lines at interface between two soils with different permeabilities

At the interface between two soils having different permeabilities the flow lines get deflected while passing from one soil to another.

Fig 8.12 shows two adjacent flow lines AB and DE inclined to the normal to the interface at angle θ_1. After crossing the interface, the flow lines get deflected to new positions BC and EF. Let the deflected flow lines be inclined at angle θ_2 to the normal to the interface. The equipotential lines EE_1 and BB_1 also get deflected at interface to new positions EE_2 and BB_2 as shown in Fig. 8.12. The rate of flow in the fields before and after interface, in the same flow channel, being same we have

$$\Delta q = k_1\, i_1\, (EE_1) = k_2\, i_2\, (BB_2) \text{ [considering unit thickness perpendicular to plane of figure]. } k_1$$
and k_2 denote permeabilities of soils before and after interface, respectively.

Also $$i_1 = \frac{\Delta h}{E_1 B} \text{ and } i_2 = \frac{\Delta h}{EB_2}$$

By substitution,

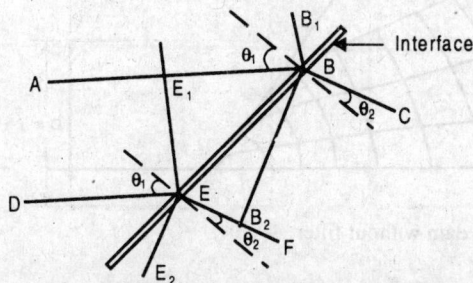

$$k_1 \cdot \frac{\Delta h}{E_1 B} \cdot EE_1 = k_2 \cdot \frac{\Delta h}{EB_2} \cdot BB_2$$

From triangles $EE_1 B$ and $EB_2 B$ we have

$$\frac{E_1 B}{EE_1} = \tan \theta_1 \text{ and } \frac{EB}{BB_2} = \tan \theta_2$$

$$\therefore \quad \frac{k_1}{\tan \theta_1} = \frac{k_2}{\tan \theta_2} \text{ or } \frac{\tan \theta_1}{\tan \theta_2} = \frac{k_1}{k_2} \qquad ...8.10$$

Fig. 8.12. Deflection of flow lines at interface between two soils with different permeabilities.

Eq. 8.10 is helpful in determining the transition when flow lines intersect at interface between two dissimilar soils.

8.11 Seepage through body of homogeneous earth dam

The study of the problem of seepage through body of earth dam involves determining the location and shape of phreatic line. The phreatic line is the top boundary flow line of the zone of seepage in section of earth dam. It can be located by (*i*) graphical method, (*ii*) analytical method or (*iii*) experimental methods. It is useful to note that there will be positive hydrostatic pressures at points below the phreatic line, where as on the phreatic line it is equal to atmospheric pressure. Once the phreatic line is located the flow net can be easily constructed. Figs 8.13 and 8.14 illustrate flow nets for two cases of an homogeneous earth dam.

Casagrande's graphical method

A graphical method for obtaining phreatic line has been suggested by A.Casagrande (1940) based on the findings of Kozeny (1931). According to Kozeny's analytical solution for the case of water flowing above an impervious infinite horizontal plane which at a certain point F becomes permeable, the flow net consists of flow lines and equipotential lines which will be two sets of confocal parabolae with the point F as common focus (Fig.8.15). In Casagrande method, to obtain the phreatic line the base parabola is constructed and is then corrected at entry and exit ends to suit the boundary conditions. This is illustrated in the following two cases.

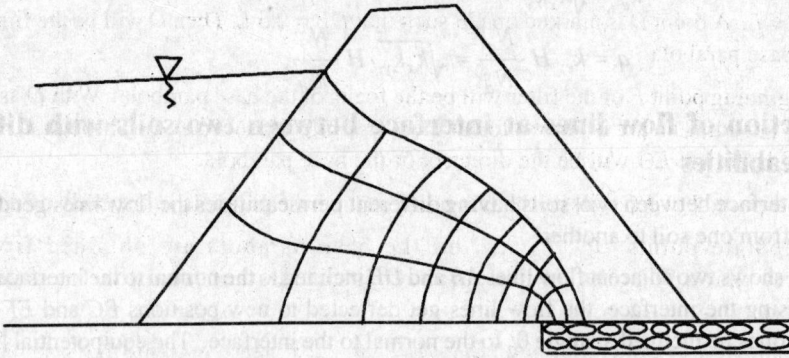

Fig. 8.13. Flow through earth dam with filter

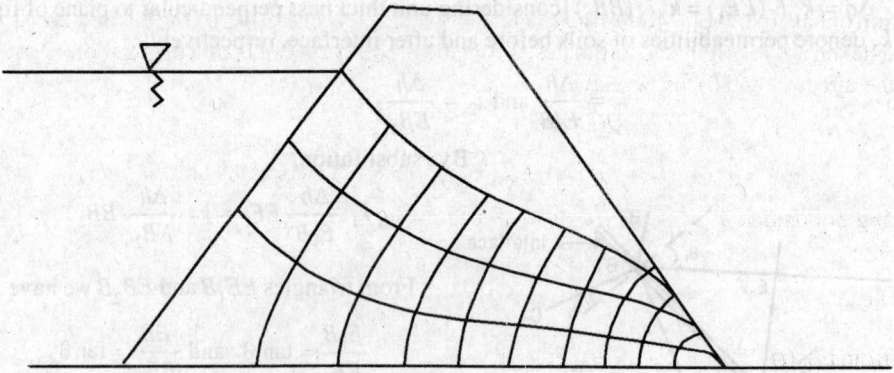

Fig. 8.14. Flow through earth dam without filter

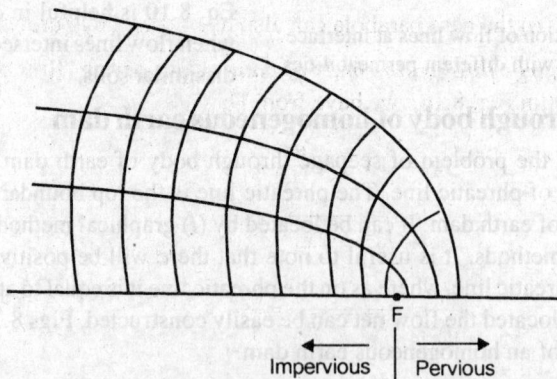

Impervious | Pervious

Fig. 8.15. Flow lines and equipotential lines as
confocal parabolae (after Kozeny, 1931)

Case (i) Homogeneous earth dam with horizontal filter.

The graphical procedure for obtaining phreatic line consists of the following steps and is illustrated in Fig. 8.16

1. Let AB be the wetted portion of upstream face and CB the projection of AB on the water line.

2. Let $CB = L$. A point D is marked on CB such that $DB = 0.3\,L$. Then D will be the first point of the base parabola.

3. The beginning point F of the filter will be the focus of the base parabola. With D as centre and DF as radius an arc is drawn to cut the water line produced at E. Through E, vertical line EG is drawn. EG will be the directrix of the base parabola.

4. The mid-point J of FG will be the last point of the base parabola in the body of dam.

5. Intermediate points P_1, P_2 ..... on the base parabola are obtained from the property that any point on the parabola is equidistant from the focus and the directrix.

6. A smooth curve is drawn through points $D\ P_1\ P_2 \dots J$, and represents the base parabola. The phreatic line is obtained by correcting this base parabola to meet the u/s face and the filter, both at right angles, at B and J.

Equation of base parabola. Referring to Fig. 8.16, let $P\,(x, y)$ be a point on the base parabola with the focus F taken as the origin of coordinates. From the property that any point on parabola is equidistant from the directrix and focus, we have

$$QG = PF$$

$$(x + s) = \sqrt{x^2 + y^2} \quad \text{where } s = \text{focal distance} \qquad \dots(i)$$

Squaring both sides,

$$(x + s)^2 = x^2 + y^2$$

$$x^2 + s^2 + 2xs = x^2 + y^2$$

i.e.,

$$y^2 = s^2 + 2xs \qquad \dots(ii)$$

Also, from Eq. (i)

$$s = \sqrt{x^2 + y^2} - x$$

At $D, x = t$ and $y = H$

$\therefore$

$$s = \sqrt{t^2 + H^2} - t \qquad \dots(iii)$$

Using Eqs. (iii) and (ii) the base parabola can also be obtained by purely analytical method.

Expression for discharge through the body of dam: Considering flow through unit thickness of vertical section PQ, in Fig. 8.16, we have from Darcy's law,

$$q = kiA = k \cdot \frac{dy}{dx} \cdot y \qquad \dots(iv)$$

From Eq. (ii)

$$y = \sqrt{s^2 + 2xs}$$

$$\frac{dy}{dx} = \frac{s}{\sqrt{s^2 + 2xs}} = \frac{s}{y}$$

By substitution is Eq. (iv) $q = k \cdot \dfrac{s}{y} \cdot y$

$\therefore$

$$q = ks \qquad \dots 8.11\,(i)$$

The focal distance, s, can be obtained by measurement or calculated using Eq. (iii)

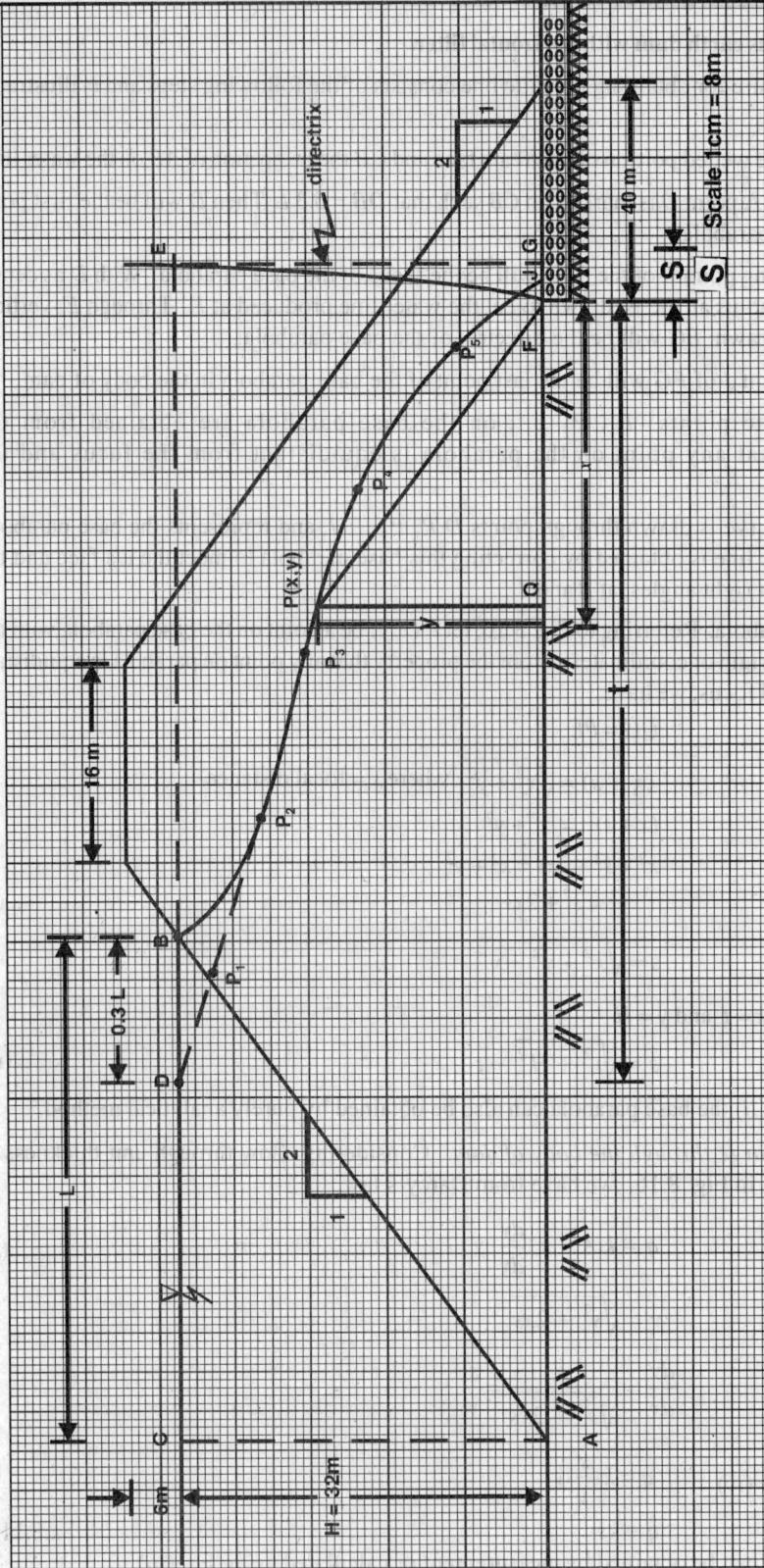

Fig. 8.16. Phreatic line in earth dam (with filter)

Problem 8.11 For the section of dam shown in Fig. 8.16 compute the quantity of seepage if $k = 3 \times 10^{-4}$ cm/sec for the soil in body of dam

Solution : By measurement, focal distance, $s = 0.8 \times 8 = 6.4$ m

$\therefore q = ks = (3 \times 10^{-6})(6.4) = 19.2 \times 10^{-6}$ m³/sec per metre length of dam

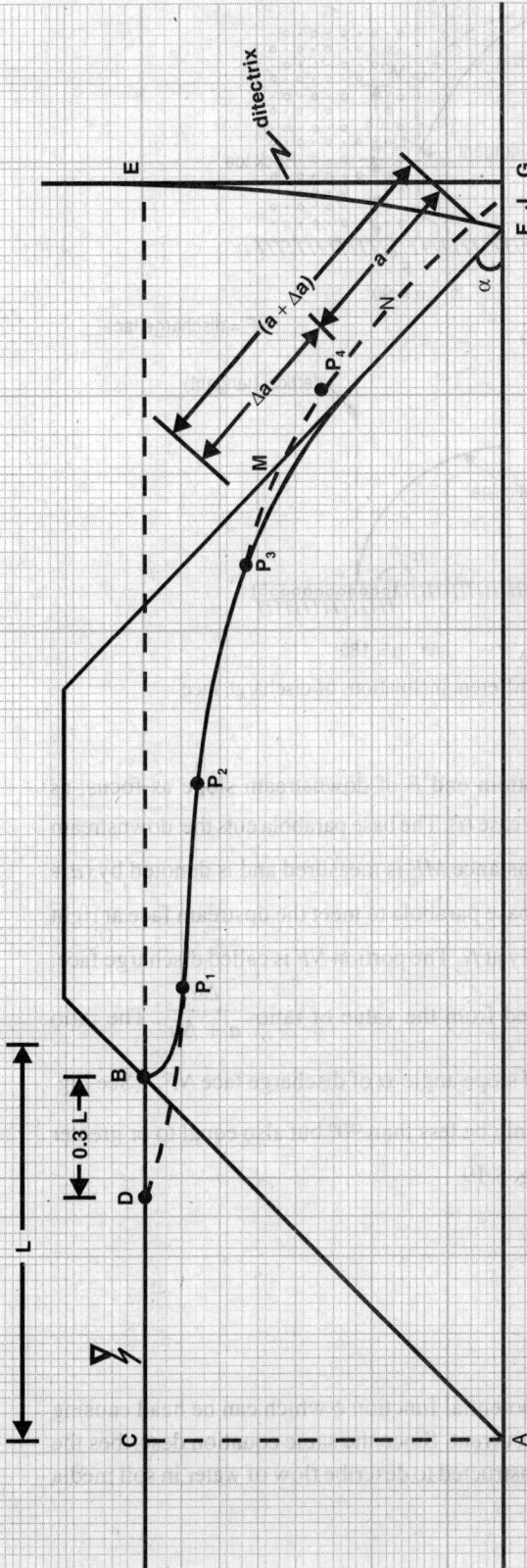

Fig. 8.17. Phreatic line in earth dam (without d/s filter)

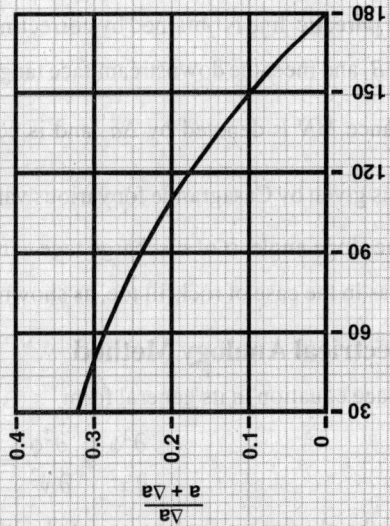

Fig. 8.18 Relation between α and $\dfrac{\Delta a}{a + \Delta a}$

Angle of inclination, α in degrees, of discharge face

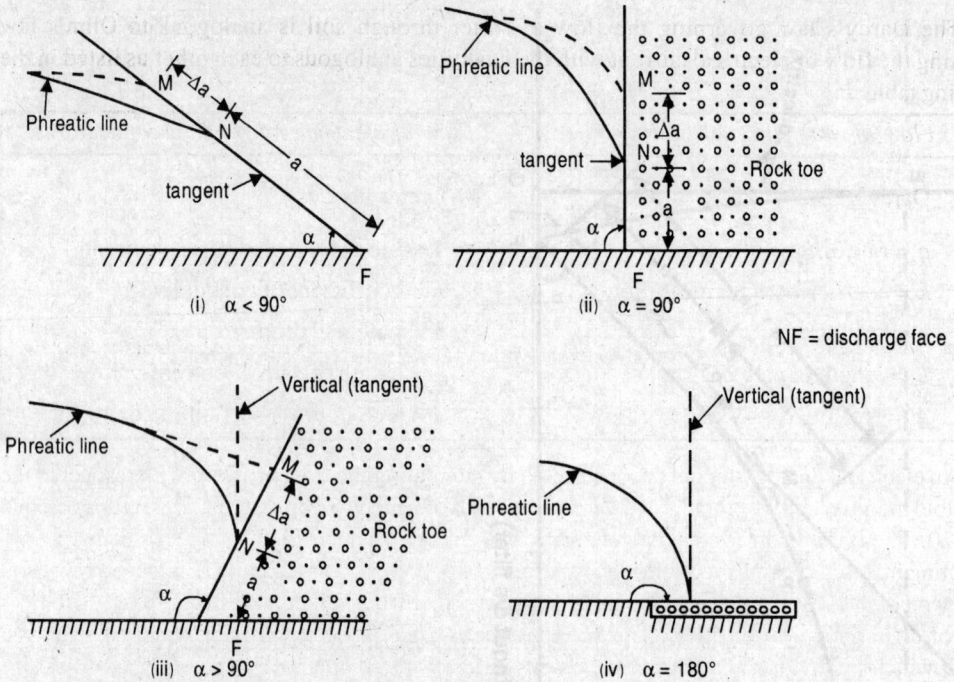

Fig. 8.19. Lower end of phreatic line for different inclinations of discharge face

Case (ii) Homogeneous earth dam without filter.

The base parabola is constructed with the bottom end F of downstream slope as focus, as shown in Fig. 8.17, following the same procedure as in case (i). The base parabola cuts the downstream slope at point M and extends beyond d/s face. The distance MF is measured and is denoted by ($a +$ Δa). The phreatic line is obtained by correcting the base parabola to meet the upstream face at right angles at B, and meet the downstream face tangentially at N. The portion NF is called discharge face.

The distance MN is denoted by Δa and is computed from the value of ratio $\dfrac{\Delta a}{a + \Delta a}$. The ratio $\dfrac{\Delta a}{a + \Delta a}$ is given by Casagrande for various values of slope angle α of discharge face NF. (See Fig. 8.18.) The slope angle α of discharge face can not only be less than 90° but also equal to or greater than 90° as in the case of rockfill toe, as shown in Fig. 8.19.

8.12 Electrical Analogy Method

The Laplace equation in its general form

$$\frac{\partial^2 \phi}{\partial x^2} + \frac{\partial^2 \phi}{\partial y^2} = 0$$

describes the physical quantity associated with the potential function ϕ which can be head causing flow of water or the voltage causing flow of electric current. Since the same equation describes the two phenomena, electrical analogy models can be constructed to describe flow of water in soil media in different types of flow problems.

The Darcy's law governing the flow of water through soil is analogous to Ohm's law governing the flow of electrical current with the quantities analogous to each other as listed in the following table.

Flow of water through soil	Flow of electricity through a conductor
Darcy's law : $q = k \dfrac{h}{L} A$	Ohm's law : $I = K \dfrac{E}{L} a$
q = rate of flow of water	I = rate of flow of electrical current
k = coefficient of permeability	K = coefficient of conductivity
h = head drop	E = potential drop
L = length of path of flow	L = length of path of current
A = area of cross-section of flow	a = area of cross-section of conductor

An electrical analogy model is constructed by simulating the flow problem with respect to the permeable medium and its boundaries. An example for the electrical analogy model is illustrated in Fig. 8.20. The base of the weir and sheet pile wall are represented by EG and EF made of non-conducting material (perspex, ebonite etc.). A special paper coated with graphite is used to represent the flow medium. By changing the coating composition or thickness it is possible to simulate flow media of different permeabilities. The boundary equipotential lines are represented by strips AB and CD which are made conducting by coating with conducting paint (alternatively copper strips can be used). A direct voltage is applied across AB and CD (about 6 to 10 V). A high resistance wire is calibrated to give potential drop as a percentage of the total drop between the terminals. This is connected in series with the voltage source. A sliding contact point S on the high resistance wire is connected through a galvanometer (G) to a probing pencil. This acts as the null-indicator system and is used to locate points on equipotential lines of different percentage drops, say 20%, 30% etc. Having thus obtained the set of equipotential lines, the flow net is completed by sketching the flow lines which will be orthogonal to the equipotential lines.

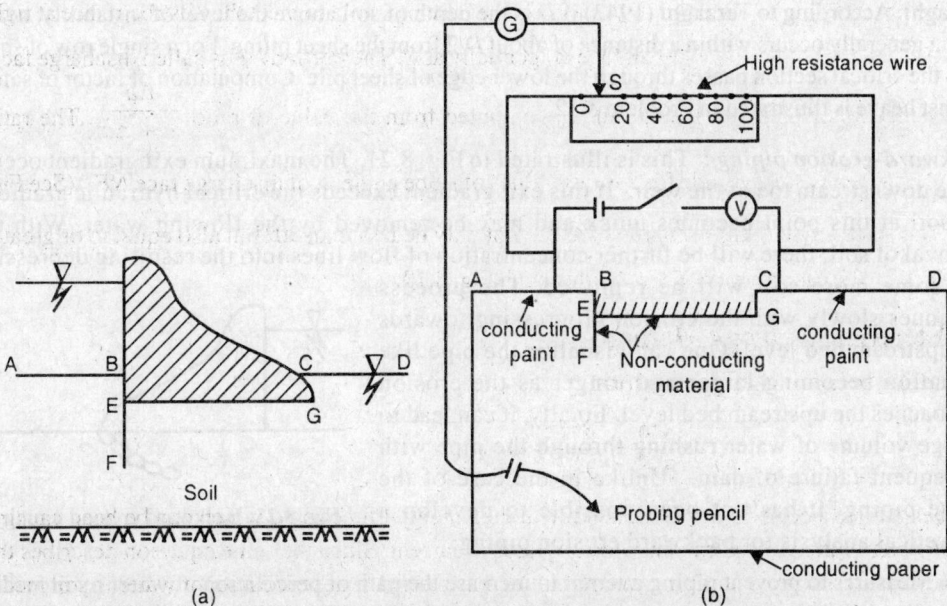

Fig. 8.20. An example for electrical analogy method (a) weir on permeable bed, (b) electrical analogy model

To minimise errors due to scale effect, it has been found experimentally that the length to depth ratio of the conducting medium be kept not less than 4.

An electrical analogy tray made of an insulating material, shallow in depth and filled with salt or copper sulphate solution can be used as a conducting medium instead of conducting paper. However, in this case an alternating current source is required to prevent polarization of electrodes. In the case of composite sections, the different permeabilities can be represented by varying the depths of liquid.

8.13 Scaled Model Method

This method also referred to as sand model method is useful as a direct demonstration of seepage or hydraulic flow. Scaled models of the prototypes are placed in narrow flumes provided with devices for regulating and maintaining appropriate water levels on upstream and downstream sides of the model. The sides of the tank are transparent (glass or perspex). Coarse sand, preferably white in colour, is used to represent the soil medium in the field problem. In the case of hydraulic structure of composite section composed of materials of different permeabilities, the materials used in the model should have same ratio of permeabilities as in the field problem. Under steady state of flow a dye is injected at different points on the upstream face of the flow medium and seepage lines or flow lines observed are traced on the side of the flume. By drawing curves orthogonal to the flow lines to obtain equipotential lines the flow net is completed. It is also possible to obtain pressure heads at various points by inserting piezometers.

8.14 Piping

When movement of soil particles by percolating water takes place leading to the formation of a hole or pipe the phenomenon of piping is said to have occurred. Heave-piping and backward-erosion piping are the two types of piping observed in practice.

Heave piping : When the upward seepage pressure becomes equal to downward pressure due to submerged weight of soil at a certain level, the soil above this level becomes quick. This may cause the entire soil above the level of instability to heave up and be blown out by the flowing water. This phenomenon is known as heave piping. The mechanics of heave piping was first analysed by Terzaghi. According to Terzaghi (1943) if D is the depth of soil above the level of instability, heave piping generally occurs within a distance of about $D/2$ from the sheet piling. For a single row of sheet piles the critical section passes through the lower edge of sheet pile. Computation of factor of safety against heave is illustrated in Problem 8.12.

Backward-erosion piping : This is illustrated in Fig. 8.21. The maximum exit gradient occurs at the downstream toe of the weir. If this exit gradient exceeds the critical hydraulic gradient, the soil at this point becomes quick and may be removed by the flowing water. With the removal of soil, there will be further concentration of flow lines into the resulting depression and some more soil will be removed. The process continues slowly with the erosion progressing towards the upstream bed level. One can visualize the pipe like formation becoming larger and longer as the erosion approaches the upstream bed level. Finally, it can lead to a large volume of water rushing through the pipe with subsequent failure of dam. Unlike in the case of the heave piping, it has not been possible to develop a theoretical analysis for backward erosion piping.

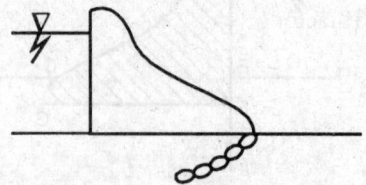

Fig. 8.21. Backward erosion piping

Measures to prevent piping attempt to increase the path of percolation of water by providing sheet pile walls below a structure and reduce seepage in the body of dam by providing impervious core walls. The provision of upstream blanket reduces the exit gradient. Provision of filters will protect the downstream face of dam by preventing piping in the body of dam.

Problem 8.12

From the flow net of Fig. 8.5, compute (i) the factor of safety against heave piping if the saturated weight of soil is 20 kN/m^3 and (ii) the effective filter weight required to be provided on downstream side to increase the factor of safety to 3.

Solution : A part of flow net of Fig. 8.5 is shown in accompanying sketch.

$$H = 9 - 1.5 = 7.5 \text{ m}$$

$$N_d = 8$$

$$\Delta h = \frac{H}{N_d} = \frac{7.5}{8} = 0.9375 \text{ m}$$

According to Terzaghi, heave piping occurs if the upward seepage pressure on the section ae equals (or exceeds) the downward effective stress at that level.

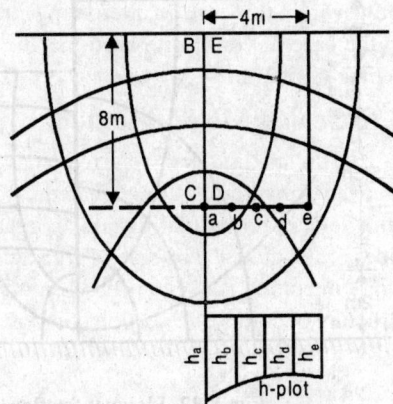

Figure for problem 8.12

We have

$$h_a = (H - n.\Delta h) = (7.5 - 4 \times 0.9375) = 3.75 \text{ m}$$

$$h_b = 7.5 - 4.75 \times 0.9375 = 3.05 \text{ m}$$

$$h_c = 7.5 - 5 \times 0.9375 = 2.81 \text{ m}$$

$$h_d = 7.5 - 5.25 \times 0.9375 = 2.58 \text{ m}$$

$$h_e = 7.5 - 5.5 \times 0.9375 = 2.34 \text{ m}$$

$$h_{av} = \frac{\text{Area of h - plot}}{4} = \frac{12.18}{4} = 3.05 \text{ m}$$

Upward seepage pressure at level $a - e$ = (3.05) (9.81) = 29.92 kN/m^2

Downward effective stress due to soil = 8(20 − 9.81) = 81.52 kN/m^2

Factor of safety against heave piping $= \dfrac{81.52}{29.92} = 2.7$

If factor of safety is to be increased to 3, downward effective vertical stress due to soil should be

$$3 \times p_s = 3 \times 29.92$$

$$= 89.76 \text{ kN/m}^2$$

∴ Effective filter weight required = 89.76 − 81.52 = 8.24 kN/m^2.

Problem 8.13

If in Fig. 8.22, the water level between the sheet piles is kept at trench level at a rate of pumping giving $q = 0.2$ m^3/hour per metre length of sheet pile, compute the coefficient of permeability of soil and exit gradient.

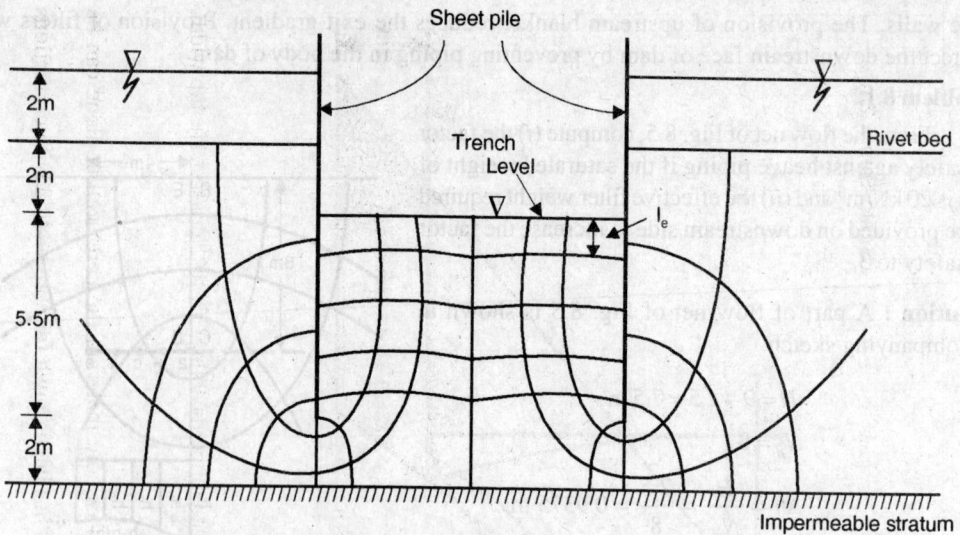

Fig. 8.22. Flownet for flow into trench between two rows of sheet piles

Solution : From flow net, $N_f = 6$, $N_d = 10$

Total head causing flow, $H = 4$ m

$$q = kH \frac{N_f}{N_d}$$

∴

$$k = \frac{q}{H} \times \frac{N_d}{N_f} = \frac{0.2 x 10}{60 \times 60 \times 4 \times 6} = 2.3 \times 10^{-5} \, \text{m/sec}$$

Maximum exit gradient,
$$i_e = \frac{\Delta h}{l_e} = \frac{4}{10 \times 0.9} = 0.44$$

8.15 Protective Filter

A protective filter is designed to provide for drainage of water from a body of soil without allowing movement of soil particles by flowing water. It can be a single-layer filter or a multiple-layer filter. In the latter case each subsequent layer will be increasingly coarser than the previous one and is sometimes referred to as reverse filter.

The materials used in filters provided for preventing piping should satisfy the following two requirements apart from adding weight where required.

(i) The gradation of the filter material should be such that the voids of the filter are small enough to prevent the particles of protected soil from penetrating and clogging the filter.

(ii) The gradation of filter material should be such that rapid drainage of incoming water takes place without allowing development of large seepage forces within the filter.

According to Terzaghi, the two requirements are satisfied if the gradation of filter material is such that

$$\frac{D_{15(\text{filter})}}{D_{85(\text{protected soil})}} < 4 \text{ to } 5 \qquad \qquad ...8.15 \, (i)$$

and
$$\frac{D_{15(\text{filter})}}{D_{15(\text{protected soil})}} > 4 \text{ to } 5 \qquad \qquad ...8.15 \, (ii)$$

The soil protected by filter is referred to as base material. Filters are usually multi-layered and each layer should satisfy the requirements with respect to preceding layer. As a rough guideline the grain size distribution curves of filter material and base material should be roughly parallel.

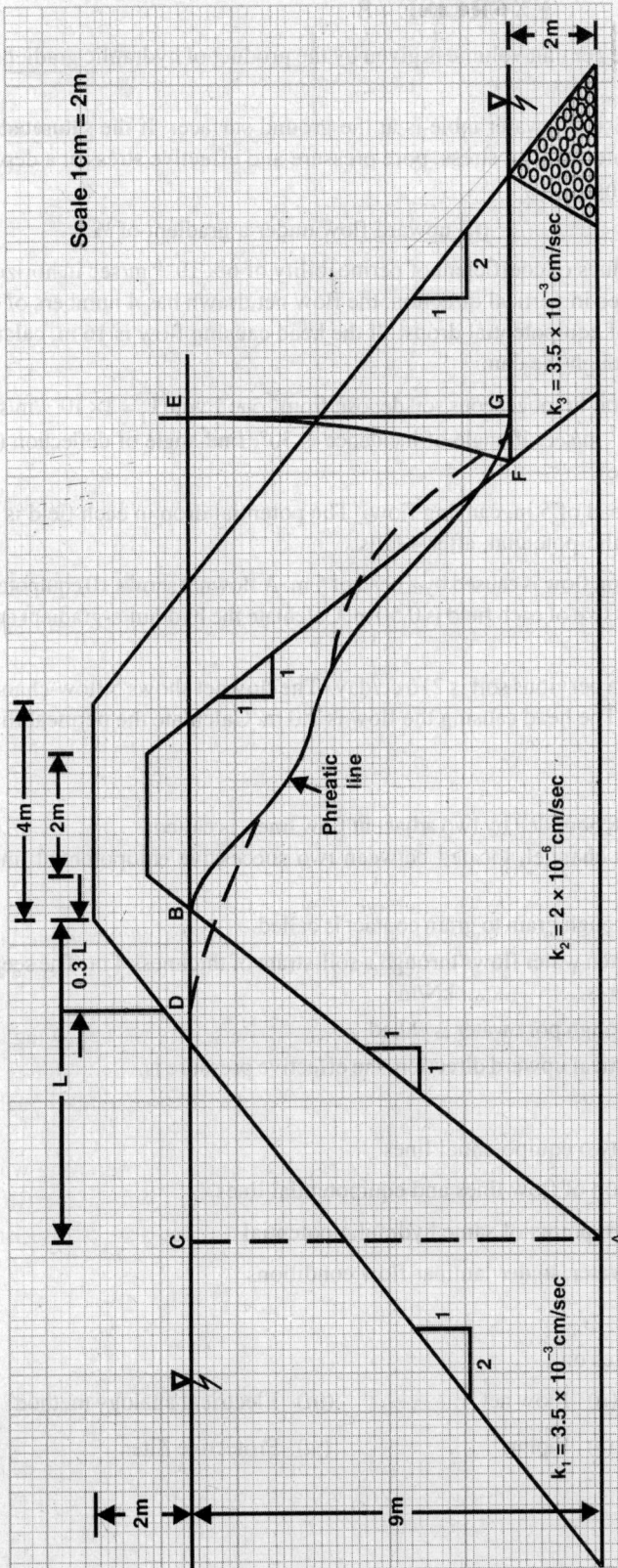

Scale 1cm = 2m

$k_3 = 3.5 \times 10^{-3}$ cm/sec

$k_2 = 2 \times 10^{-6}$ cm/sec

Phreatic line

$k_1 = 3.5 \times 10^{-3}$ cm/sec

Fig. 8.23. Phreatic line in earth dam of composite section

Problem 8.14

Sketch the phreatic line of earth dam with two outer shells of permeability 3.5×10^{-3} cm/sec and core of permeability 2×10^{-6} cm/sec. The top width, side slopes and positions of water levels on u/s and d/s sides are indicated in Fig. 8.23.

Solution :

According to L. Casagrande if permeability of shell material is not less than 10 times that of core material, the core section alone can be considered for drawing the flow net. Since $\dfrac{k_1}{k_2}$ and $\dfrac{k_3}{k_2}$ are both greater than 10, the phreatic line has been drawn for core section with corrections made free-hand at the ends.

EXERCISE – 8

8.1   Show that the seepage force per unit volume is given by the product of hydraulic gradient and unit weight of water.

8.2   In a cohesionless soil deposit the water table is at the ground surface. If the saturated unit weight is 18 kN/m³, compute the total stress, pore pressure and effective stress at a depth of 5 m for each of the following cases:

    (*i*) static ground water   (*ii*) upward flow under a gradient of ½.

8.3   For a zoned dam having values of coefficient of permeability of 6×10^{-6} m/sec in horizontal direction and 4×10^{-6} m/sec in vertical direction, the flow net drawn has 4 numbers of flow channels and 12 numbers of equipotential drops. If the head causing flow is 16 m, calculate the quantity of seepage through the dam.

8.4   Flow takes place through a non-homogeneous soil deposit from zone 1 with $k_1 = 1 \times 10^{-7}$ m/sec to zone 2 having $k_2 = 4 \times 10^{-7}$ m/sec. If angle of incidence is 30°, find angle of deflection (with respect to normal to interface).

8.5   (*a*) A flownet has a total head of 5 m, causing flow. The potential drop in each field is 0.50 m. Calculate the hydraulic-potential, after 4 falls.

    (*b*) In a saturated soil medium, flow is caused by a head of 5 m. A flownet shows 10 equipotential drops. The approximate size of each field is 0.50 m. Calculate the hydraulic-gradient across each field.

    (*c*) The discharge through a pervious soil is 216 cc/day. The flownet shows 5 flow-channels, 10 equipotential drops. The head causing the flow is 2.0 m. Calculate the permeability of the soil.

8.6   Fill in the blanks :

    (*i*) The portion of a flownet, bounded by two adjacent flow lines, is called...................., and every section of a flow-channel, located between two successive equipotential lines is called a..........

    (*ii*) Pressure transmitted through grain to grain contact is called....................

    (*iii*) A head of 2 m is lost, when water flows through a soil-strata of thickness 4 m. The seepage pressure exerted by water is.................kN/m².

    (*iv*) Pressure transmitted through pore water is called....................

    (*v*) If flow occurs in the vertical upward direction, the effective pressure is.................

8.7   Name the following :

    (*i*) The head lost between two equipotential lines

    (*ii*) A pictorial representation of flow lines and equipotential lines

    (*iii*) The line passing through points of same hydraulic potential

    (*iv*) The path followed by water, under laminar flow conditions

    (*v*) The ratio of N_f to N_d.

8.8   Write short explanatory notes on :

    (*i*) Quick sand   (*ii*) Flow net   (*iii*) Electrical analogy method

    (*iv*) Scaled model method   (*v*) Piping   (*vi*) Protective filter.

CHAPTER 9

STRESS DISTRIBUTION IN SOIL MASS

9.1 Introduction

A stress component at a point inside a soil mass can be the sum of corresponding stress components due to self-weight of soil and external loading. Considering the soil mass to be semi-infinite, bound by the ground surface, the stress components on an element inside soil mass due to its self weight can be shown to be as indicated in Fig. 9.1.

The point is considered to be at depth z below the ground surface.

$$\sigma_z = \gamma z \qquad \qquad \text{...9.1 } (i)$$

$$\sigma_x = \sigma_y = \left(\frac{\mu}{1 - \mu} \right) \sigma_z \qquad \qquad \text{...9.1 } (ii)$$

where μ is Poisson's ratio of soil and γ is unit weight of soil

Shear stress on all planes are zero.

i.e., $$\tau_{xy} = \tau_{yx} = \tau_{yz} = \tau_{zy} = \tau_{zx} = \tau_{xz} = 0$$

The vertical stress due to self-weight of soil is referred to as geostatic stress. The rest of the discussion in this chapter is focussed on stress distribution inside soil mass due to surface loading. Boussinesq analysis for homogeneous soil deposits and Westergaard analysis for layered soil deposits have been introduced.

9.2 Boussinesq Analysis

Boussinesq (1885) solved the problem of evaluating stress at a point inside soil mass due to a point load acting on the surface of soil mass. (Fig. 9.2)

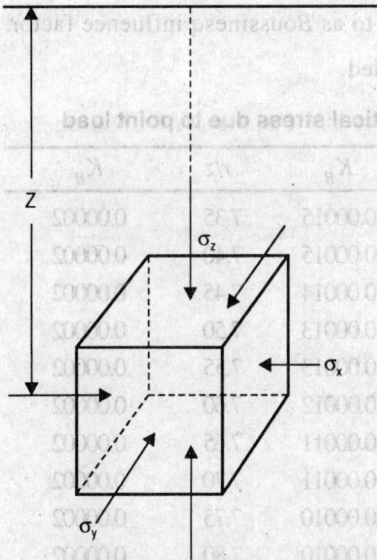

Fig. 9.1. Stress components at a point due to self weight of soil

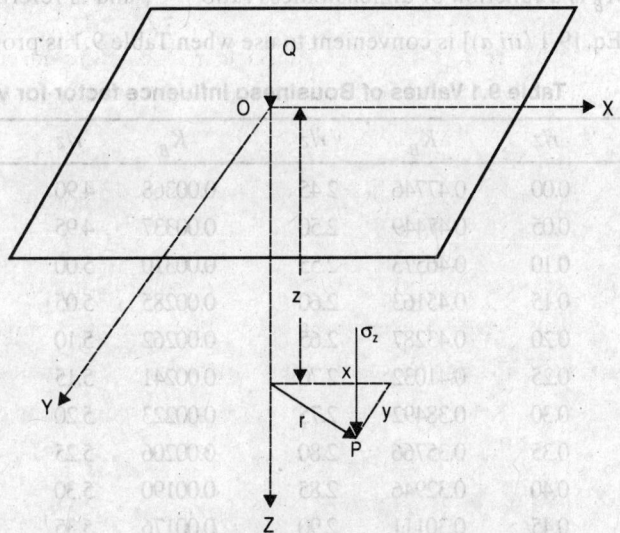

Fig. 9.2. Vertical stress due to point load

Following are the assumptions made in the analysis :

1. The soil mass is an elastic medium.
2. The modulus of elasticity E is constant.
3. The soil mass is homogeneous, that is elastic properties of material are the same at all points in identical directions.
4. The soil mass is isotropic, that is, at any point the elastic properties are same in all directions.
5. The soil mass is semi-infinite, that is, it is bound only by ground surface and extends to infinity in all other directions.
6. The soil mass is weightless.

Referring to Fig 9.2 the vertical stress σ_z at point P inside soil mass, due to point load Q acting on the surface, is obtained from Boussinesq analysis as

$$\sigma_z = \frac{3Q}{2\pi} \frac{z^3}{(r^2 + z^2)^{5/2}}$$

...9.1 (*iii*)

where z = depth of point below surface of soil mass

r = radial distance of point P from axis of load $= \sqrt{x^2 + y^2}$

Eq 9.1 (*iii*) is also written as

$$\sigma_z = K_B \frac{Q}{z^2}$$

...9.1 (*iii a*)

$$K_B = \frac{3}{2\pi} \left\{ \frac{1}{1 + \left(\dfrac{r}{z}\right)^2} \right\}^{5/2}$$

...9.1 (*iii b*)

K_B is a function of dimensionless ratio $\left(\dfrac{r}{z}\right)$ and is referred to as *Boussinesq influence factor*. Eq. [9.1 (*iii a*)] is convenient to use when Table 9.1 is provided.

Table 9.1 Values of Bousinesq Influence factor for vertical stress due to point load

r/z	K_B	r/z	K_B	r/z	K_B	r/z	K_B
0.00	0.47746	2.45	0.00368	4.90	0.00015	7.35	0.00002
0.05	0.47449	2.50	0.00337	4.95	0.00015	7.40	0.00002
0.10	0.46573	2.55	0.00310	5.00	0.00014	7.45	0.00002
0.15	0.45163	2.60	0.00285	5.05	0.00013	7.50	0.00002
0.20	0.43287	2.65	0.00262	5.10	0.00013	7.55	0.00002
0.25	0.41032	2.70	0.00241	5.15	0.00012	7.60	0.00002
0.30	0.38492	2.75	0.00223	5.20	0.00011	7.65	0.00002
0.35	0.35766	2.80	0.00206	5.25	0.00011	7.70	0.00002
0.40	0.32946	2.85	0.00190	5.30	0.00010	7.75	0.00002
0.45	0.30111	2.90	0.00176	5.35	0.00010	7.80	0.00002
0.50	0.27332	2.95	0.00163	5.40	0.00010	7.85	0.00002
0.55	0.24660	3.00	0.00151	5.45	0.00009	7.90	0.00001
0.60	0.22136	3.05	0.00140	5.50	0.00009	7.95	0.00001
0.65	0.19784	3.10	0.00130	5.55	0.00008	8.00	0.00001

r/z	K_B	r/z	K_B	r/z	K_B	r/z	K_B
0.70	0.17619	3.15	0.00121	5.60	0.00008	8.05	0.00001
0.75	0.15646	3.20	0.00113	5.65	0.00008	8.10	0.00001
0.80	0.13862	3.25	0.00105	5.70	0.00007	8.15	0.00001
0.85	0.12262	3.30	0.00098	5.75	0.00007	8.20	0.00001
0.90	0.10833	3.35	0.00091	5.80	0.00007	8.25	0.00001
0.95	0.09564	3.40	0.00085	5.85	0.00006	8.30	0.00001
1.00	0.08440	3.45	0.00080	5.90	0.00006	8.35	0.00001
1.05	0.07449	3.50	0.00075	5.95	0.00006	8.40	0.00001
1.10	0.06576	3.55	0.00070	6.00	0.00006	8.45	0.00001
1.15	0.05809	3.60	0.00066	6.05	0.00006	8.50	0.00001
1.20	0.05134	3.65	0.00062	6.10	0.00005	8.55	0.00001
1.25	0.04543	3.70	0.00058	6.15	0.00005	8.60	0.00001
1.30	0.04023	3.75	0.00054	6.20	0.00005	8.65	0.00001
1.35	0.03568	3.80	0.00051	6.25	0.00005	8.70	0.00001
1.40	0.03168	3.85	0.00048	6.30	0.00005	8.75	0.00001
1.45	0.02816	3.90	0.00045	6.35	0.00004	8.80	0.00001
1.50	0.02508	3.95	0.00043	6.40	0.00004	8.85	0.00001
1.55	0.02236	4.00	0.00040	6.45	0.00004	8.90	0.00001
1.60	0.01997	4.05	0.00038	6.50	0.00004	8.95	0.00001
1.65	0.01786	4.10	0.00036	6.55	0.00004	9.00	0.00001
1.70	0.01600	4.15	0.00034	6.60	0.00004	9.05	0.00001
1.75	0.01436	4.20	0.00032	6.65	0.00003	9.10	0.00001
1.80	0.01290	4.25	0.00030	6.70	0.00003	9.15	0.00001
1.85	0.01161	4.30	0.00028	6.75	0.00003	9.20	0.00001
1.90	0.01047	4.35	0.00027	6.80	0.00003	9.25	0.00001
1.95	0.00945	4.40	0.00026	6.85	0.00003	9.30	0.00001
2.00	0.00854	4.45	0.00024	6.90	0.00003	9.35	0.00001
2.05	0.00774	4.50	0.00023	6.95	0.00003	9.40	0.00001
2.10	0.00701	4.55	0.00022	7.00	0.00003	9.45	0.00001
2.15	0.00637	4.60	0.00021	7.05	0.00003	9.50	0.00001
2.20	0.00579	4.65	0.00020	7.10	0.00003	9.60	0.00001
2.25	0.00528	4.70	0.00019	7.15	0.00002	9.70	0.00001
2.30	0.00481	4.75	0.00018	7.20	0.00002	9.80	0.00001
2.35	0.00440	4.80	0.00017	7.25	0.00002	9.90	0.00000
2.40	0.00402	4.85	0.00016	7.30	0.00002	10.00	0.00000

Again referring to Fig. 9.2 the Boussinesq equation for shear stress, at point P due to point load Q acting on surface is

$$\tau_{rz} = \frac{3Q}{2\pi} \frac{rz^2}{(r^2 + z^2)^{5/2}} \qquad \ldots 9.1\,(iv)$$

or

$$\tau_{rz} = \frac{3Qr}{2\pi z^3} \left\{ \frac{1}{1 + \left(\dfrac{r}{z}\right)^2} \right\}^{5/2} \qquad \ldots 9.1\,(iva)$$

or
$$\tau_{rz} = K_B \frac{Qr}{z^3}$$
...9.1 (ivb)

9.3 Isobar and Pressure Bulb

An isobar is a line or contour joining points inside soil mass at which the vertical stress have same value. An isobar of given vertical stress intensity can be constructed using Boussinesq equation for vertical stress due to point load. All other quantities in the equation being known the radial distance r is computed for different values of z and then plotted to obtain isobar. The following example, in which an isobar of intensity $0.1Q$ is plotted, illustrates the procedure.

The Boussinesq equation for vertical stress due to point load is

$$\sigma_z = \frac{3Q}{2\pi} \frac{z^3}{(r^2 + z^2)^{5/2}}$$

Substituting $\sigma_z = 0.1\,Q$ we have

$$0.1Q = \frac{3Q}{2\pi} \frac{z^3}{(r^2 + z^2)^{5/2}}$$
...(i)

Rearranging,

$$(r^2 + z^2)^{5/2} = \frac{3z^3}{0.2\pi}$$

$$(r^2 + z^2) = \left(\frac{3z^3}{0.2\pi}\right)^{2/5} = 1.869\, z^{1.2}$$

$$r = \sqrt{(1.869 z^{1.2} - z^2)}$$
...(ii)

Putting $r = 0$ in Eq. (i), we get

$$0.1 = \frac{3}{2\pi z^2}$$

$$z = \sqrt{\frac{3}{0.2\pi}} = 2.18\,m$$

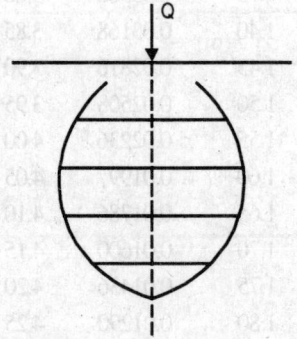

Fig. 9.3. Isobar of intensity 0.1 Q

The isobar cuts the axis of loading at $z = 2.18$ m. For values of z at suitable interval upto 2.18 m, r is evaluated using Eq. (ii) and then plotted to obtain the isobar.

z (m)	r (m)
0.50	0.89
1.00	0.93
1.50	0.89
2.00	0.54

When an isobar is rotated about the axis of loading a volume is generated which is referred to as pressure bulb. The pressure bulb is of great significance in practice. The depth of a pressure bulb of intensity $0.1\,q$ to $0.2\,q$, where q is the loading intensity at the surface, varies generally from one and half times to two times the width of loaded area, and is taken as a guide line for determining the significant depth for site exploration.

9.4 Vertical Stress Distribution on Horizontal Plane

The vertical stress distribution on a horizontal plane at a given depth z can be determined using Boussinesq equation. z being constant, σ_z is calculated for different values of r. The procedure is illustrated in the following example, The Boussinesq equation for vertical stress due to point load acting at surface is

$$\sigma_z = \frac{3Q}{2\pi} \frac{z^3}{(r^2 + z^2)^{5/2}}$$

For example, let $Q = 100$ kN and $z = 2$ m

$$\sigma_z = \frac{3(100)}{2\pi} \frac{z^3}{(r^2 + z^2)^{5/2}} = \frac{381.97}{(r^2 + 4)^{5/2}} \quad \sigma_z \text{ (kN/m}^2)$$

Using the last equation σ_z is calculated for different values of r and plotted as shown in Fig 9.4.

r (m)	σ_z (kN/m²)
0	11.94
1	6.83
2	2.11
3	0.63

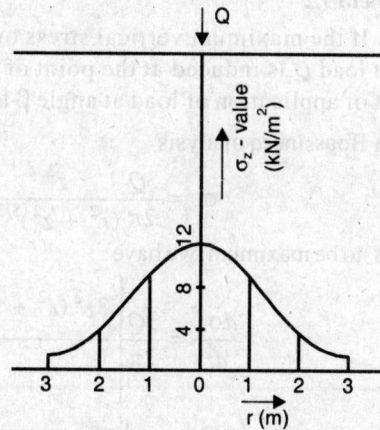

Fig. 9.4. Vertical stress distribution on horizontal plane

9.5 Vertical Stress Distribution on a Vertical Line

The vertical stress distribution on a vertical line due to a point load acting on the surface of soil mass can be determined using Boussinesq equation for vertical stress due to point load. Keeping r constant σ_z is evaluated for different values of z. The Boussinesq equation for vertical stress due to point load is

$$\sigma_z = \frac{3Q}{2\pi} \frac{z^3}{(r^2 + z^2)^{5/2}}$$

z (m)	σ_z (kN/m²)
0.5	0.16
1.0	0.85
2.0	2.11
3.0	2.12
4.0	1.71
5.0	1.32
6.0	1.02

For example, let $Q = 100$ kN and $r = 2$ m

Then

$$\sigma_z = \frac{150}{\pi} \frac{z^3}{(4 + z^2)^{5/2}}$$

σ_z is calculated for different values of z using the last equation and plotted as shown in Fig. 9.5

Problem 9.1

A concentrated load 10 kN acts on the surface of a soil mass. Using Boussinesq analysis find the vertical stress at points (i) 3 m below the surface on the axis of loading and (ii) at radial distance of 2 m from axis of loading but at same depth of 3m.

(i) $r = 0, z = 3m$

$$\sigma_z = \frac{3Q}{2\pi} \frac{z^3}{(r^2 + z^2)^{5/2}} = \frac{3Q}{2\pi z^2} = \frac{3(10)}{2\pi(9)} = 0.53 \text{ kN/m}^2$$

(ii) $r = 2 \text{ m}, z = 3 \text{ m}$

$$\sigma_z = \frac{3(10)}{2\pi} \frac{27}{(4+9)^{5/2}} = 0.04 \text{ kN/m}^2$$

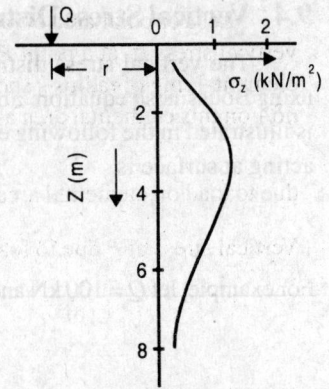

Fig. 9.5. Vertical stress distribution on vertical line

Problem 9.2

If the maximum vertical stress on a vertical line at radial distance r from the axis of a point load Q is induced at the point of intersection of the vertical line with line drawn from point of application of load at angle β to the vertical, find angle β using Boussinesq analysis.

From Boussinesq analysis

$$\sigma_z = \frac{3Q}{2\pi} \frac{z^3}{(r^2 + z^2)^{5/2}}$$

For σ_z to be maximum, we have

$$\frac{d\sigma_z}{dz} = \frac{3Q}{2\pi} \left[\frac{3z^2(r^2 + z^2)^{5/2} - z^3 \frac{5}{2}(r^2 + z^2)^{3/2} \cdot 2z}{(r^2 + z^2)^5} \right] = 0$$

$$3z^2(r^2 + z^2)^{5/2} - 5z^4(r^2 + z^2)^{3/2} = 0$$

$$[3(r^2 + z^2) - 5z^2](r^2 + z^2)^{3/2} z^2 = 0$$

$$(r^2 + z^2) = \frac{5}{3} z^2 = 1.67 z^2$$

$$0.67z^2 = r^2$$

$$z = \sqrt{\frac{r^2}{0.67}} = 1.222r$$

From Fig 9.6

$$\tan \beta = \frac{r}{z} = \frac{1}{1.222} = 0.8183$$

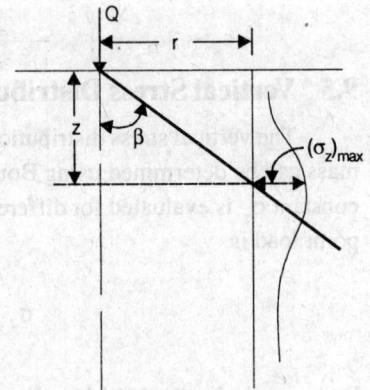

Fig. 9.6. Maximum vertical stress on a vertical line

$$\therefore \qquad \beta = \tan^{-1} 0.8183 = 39°$$

9.6   Vertical Stress under Uniformly Loaded Cirular Area

The vertical stress in a homogeneous soil mass under a uniformly loaded circular area can be found using Boussinesq analysis.

Let a circular area of radius a carry a uniformly distributed load of intensity q. To determine the vertical stress σ_z at a point P located at depth z below the center O of circular area, we consider an elemental ring of radius r and thickness dr. Let δA be an elemental area of this ring. Treating the load $q\delta A$ on this elemental area as a point load and using Boussinesq theory, we have vertical stress at P

due to load on elemental area δA, $\delta\sigma_z = \dfrac{3(q.\delta A)}{2\pi}\dfrac{z^3}{(r^2 + z^2)^{5/2}}$

Vertical stress at P due to load on elemental ring,

$$\Delta\sigma_z = \sum \delta\sigma_z = \frac{3q}{2\pi}\frac{z^3}{(r^2 + z^2)^{5/2}} \sum \delta A$$

$$= \frac{3q}{2\pi}\frac{z^3}{(r^2 + z^2)^{5/2}}(2\pi r dr)$$

Vertical stress at P due to load on entire circular area,

$$\sigma_z = 3qz^3 \int_0^a \frac{rdr}{(r^2 + z^2)^{5/2}}$$

We put $r^2 + z^2 = t^2$

Then $\quad 2\,rdr = 2\,tdt$

$\quad\quad\quad rdr = tdt$

Also, when $r = 0, t = z$

When $\quad r = a,\ t = \sqrt{a^2 + z^2}$

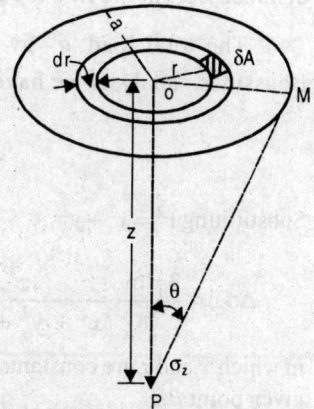

Fig. 9.7. Vertical stress under loaded circular area

$$\sigma_z = 3qz^3 \int_z^{\sqrt{a^2 + z^2}} \frac{rdr}{(r^2 + z^2)^{5/2}} = 3qz^3 \int_z^{\sqrt{a^2 + z^2}} \frac{dt}{t^4}$$

$$= 3qz^3 \left[\frac{-1}{3t^3}\right]_z^{\sqrt{a^2 + z^2}}$$

$$= qz^3 \left[\frac{1}{z^3} - \frac{1}{(a^2 + z^2)^{3/2}}\right]$$

i.e.,
$$\sigma_z = q\left[1 - \left\{\frac{1}{1 + \left(\dfrac{a}{z}\right)^2}\right\}^{3/2}\right] \qquad \text{...9.6 } (i)$$

Eq 9.6 (i) is also written as

$$\sigma_z = K_B q \qquad \text{...9.6 } (ii)$$

where K_B is Boussinesq influence factor for vertical stress below centre of uniformly distributed circular load and is given by

$$K_B = 1 - \left\{\frac{1}{1 + \left(\dfrac{a}{z}\right)^2}\right\}^{3/2} \qquad \text{...9.6 } (iii)$$

Eq. 9.6 (*ii*) is useful when Table 9.2 is provided. If θ is the angle which a line joining P with the outer edge of the circular loaded area makes with the central axis, as shown in Fig 9.7, we have

$$1 + \left(\frac{a}{z}\right)^2 = \frac{z^2 + a^2}{z^2} = \left(\frac{PM}{PO}\right)^2 = \sec^2 \theta$$

Substituting in Eq. 9.6 (*i*),

$$\sigma_z = q\left[1 - \left\{\frac{1}{\sec^2 \theta}\right\}^{3/2}\right] = q[1 - \cos^3 \theta] \qquad \qquad ...9.6 \,(iv)$$

9.7   Vertical Stress Due to Line Load

Let an infinitely long line load of intensity q' per unit length acting on the surface be directed along Y-axis. To find the vertical stress at point P located at depth z and at perpendicular distance x from the line load, we consider load on an elemental length δy as shown in Fig 9.8.

The total load q' δy on this elemental length can be treated as a point load. Using Boussinesq analysis we have vertical stress at P due to elemental load as

$$\Delta \sigma_z = \frac{3(q' \delta y)}{2\pi} \frac{z^3}{(r^2 + z^2)^{5/2}}$$

Substituting $r^2 = x^2 + y^2$

$$\Delta \sigma_z = \frac{3q' \delta y}{2\pi} \frac{z^3}{(x^2 + y^2 + z^2)^{5/2}}$$

in which x and z are constants for the given point P.

Vertical stress at P due to entire line load,

$$\sigma_z = \frac{3q' z^3}{2\pi} \int_{-\infty}^{\infty} \frac{1}{(x^2 + y^2 + z^2)^{5/2}} \, dy$$

$$= \frac{3q' z^3}{\pi} \int_{o}^{\infty} \frac{1}{(x^2 + y^2 + z^2)^{5/2}} \, dy$$

$$= \frac{2q' z^3}{\pi (x^2 + z^2)^2}$$

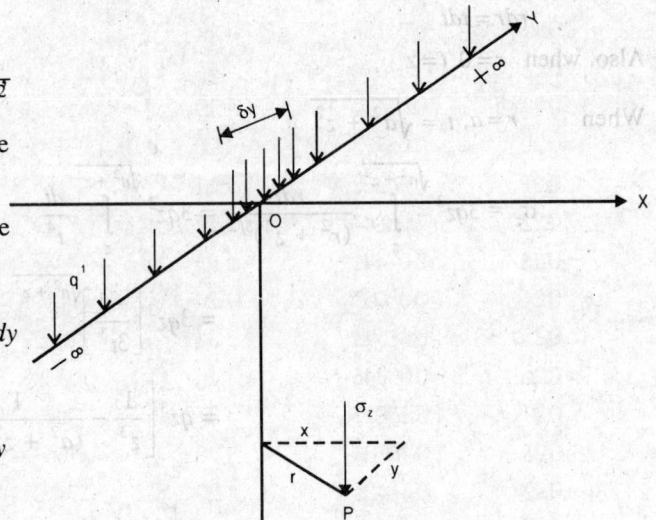

Fig. 9.8. Vertical stress Due to Line Load

or

$$\sigma_z = \frac{2q'}{\pi z}\left[\frac{1}{1 + \left(\frac{x}{z}\right)^2}\right]^2 \qquad \qquad ...9.7 \,(i)$$

If the point P is situated directly below the line load then $x = 0$, and

$$\sigma_z = \frac{2q'}{\pi z} \qquad \qquad ...9.7 \,(ii)$$

Table 9.2 Boussinesq Influence Values K_B for Vertical Stress Under Centre of Loaded Circular Area

r/z	K_B	r/z	K_B	r/z	K_B
0.00	0.00000	1.00	0.64645	2.00	0.91056
0.01	0.00015	1.01	0.65171	2.02	0.91267
0.02	0.00060	1.02	0.65689	2.04	0.91472
0.03	0.00135	1.03	0.66200	2.06	0.91672
0.04	0.00239	1.04	0.66703	2.08	0.91865
0.05	0.00374	1.05	0.67198	2.10	0.92053
0.06	0.00537	1.06	0.67686	2.15	0.92499
0.07	0.00731	1.07	0.68166	2.20	0.92914
0.08	0.00952	.08	0.68639	2.25	0.93301
0.09	0.01481	1.10	0.69562	2.30	0.93661
0.10	0.01788	1.11	0.70013	2.35	0.93997
0.12	0.02122	1.12	0.70457	2.40	0.94310
0.13	0.02482	1.13	0.70894	2.45	0.94603
0.14	0.02870	1.14	0.71384	2.50	0.94877
0.15	0.03282	1.15	0.71747	2.55	0.95134
0.16	0.03721	1.16	0.72163	2.60	0.95374
0.17	0.04183	1.17	0.72572	2.65	0.95599
0.18	0.04670	1.18	0.72975	2.70	0.95810
0.19	0.05180	1.19	0.73373	2.75	0.96008
0.20	0.05713	1.20	0.73763	2.80	0.96195
0.21	0.06268	1.21	0.74147	2.85	0.96371
0.22	0.06844	1.22	0.74525	2.90	0.96536
0.23	0.07441	1.23	0.74896	2.95	0.96691
0.24	0.08057	1.24	0.75262	3.00	0.96838
0.25	0.08692	1.25	0.75622	3.10	0.97106
0.26	0.09346	1.26	0.75076	3.20	0.97346
0.27	0.10017	1.27	0.76323	3.30	0.97561
0.28	0.10704	1.28	0.76666	3.40	0.97753
0.29	0.11408	1.29	0.77003	3.50	0.97927
0.30	0.12126	1.30	0.77334	3.60	0.98083
0.31	0.12859	1.31	0.77660	3.70	0.98224
0.32	0.13604	1.32	0.77981	3.80	0.98352
0.33	0.14363	1.33	0.78296	3.90	0.98464
0.34	0.15133	1.34	0.78606	4.00	0.98573
0.35	0.15915	1.35	0.78911	4.20	0.98757
0.36	0.16706	1.36	0.79211	4.40	0.98911
0.37	0.17507	1.37	0.79507	4.60	0.99041
0.38	0.18317	1.38	0.79797	4.80	0.99152

r/z	K_B	r/z	K_B	r/z	K_B
0.39	0.19134	1.39	0.80083	5.00	0.99246
0.40	0.19959	1.40	0.80364	5.20	0.99327
0.41	0.20790	1.41	0.80640	5.40	0.99396
0.42	0.21627	1.42	0.80912	5.60	0.99457
0.43	0.22469	1.43	0.81179	5.80	0.99510
0.44	0.23315	1.44	0.81442	6.00	0.99556
0.45	0.24164	1.45	0.81701	6.20	0.99596
0.46	0.25017	1.46	0.81955	6.40	0.99632
0.47	0.25872	1.47	0.82206	6.60	0.99664
0.48	0.26729	1.48	0.82452	6.80	0.99692
0.49	0.27587	1.49	0.82694	7.00	0.99717
0.50	0.28446	1.50	0.82932	7.20	0.99740
0.51	0.29304	1.51	0.83167	7.40	0.99760
0.52	0.30162	1.52	0.83397	7.60	0.99778
0.53	0.31019	1.53	0.83624	7.80	0.99794
0.54	0.31875	1.54	0.83847	8.00	0.99809
0.55	0.32728	1.55	0.84067	8.20	0.99823
0.56	0.33579	1.56	0.84283	8.40	0.99835
0.57	0.34427	1.57	0.84495	8.60	0.99846
0.58	0.35271	1.58	0.84704	8.80	0.99856
0.59	0.36112	9.00	0.99865	1.59	0.84910
0.60	0.36949	1.60	0.85112	9.20	0.99874
0.61	0.37781	1.61	0.85312	9.40	0.99882
0.62	0.38608	1.62	0.85507	9.60	0.99889
0.63	0.39431	1.63	0.85700	9.80	0.99898
0.64	0.40247	1.64	0.85890	10.00	0.99901
0.65	0.41085	1.65	0.86077	10.20	0.99907
0.66	0.41863	1.66	0.86260	10.40	0.99912
0.67	0.42662	1.67	0.86441	10.60	0.99917
0.68	0.43454	1.68	0.86619	10.80	0.99922
0.69	0.44239	1.69	0.86794	11.00	0.99926
0.70	0.45018	1.70	0.86966	11.20	0.99930
0.71	0.45789	1.71	0.97136	11.40	0.99933
0.72	0.46553	1.72	0.97302	11.60	0.99937
0.73	0.47310	1.73	0.87467	11.80	0.99940
0.74	0.48059	1.74	0.87628	12.00	0.99943
0.75	0.48800	1.75	0.87787	12.25	0.99946
0.76	0.49533	1.76	0.87944	12.50	0.99949
0.77	0.50259	1.77	0.88098	12.75	0.99952
0.78	0.50976	1.78	0.88250	13.00	0.99955
0.79	0.51685	1.79	0.88399	13.40	0.99959
0.80	0.52386	1.80	0.88546	13.80	0.99962
0.81	0.53079	1.81	0.88691	14.00	0.99964
0.82	0.53763	1.82	0.88833	14.50	0.99967

r/z	K_B	r/z	K_B	r/z	K_B
0.83	0.54439	1.83	0.88974	15.00	0.99971
0.84	0.55106	1.84	0.89112	15.40	0.99973
0.85	0.55765	1.85	0.89248	15.80	0.99975
0.86	0.56416	1.86	0.89382	16.00	0.99976
0.87	0.57058	1.87	0.89514	16.40	0.99977
0.88	0.57692	1.88	0.89643	16.80	0.99979
0.89	0.58317	1.89	0.89771	17.00	0.99980
0.90	0.58934	1.90	0.89897	17.40	0.99981
0.91	0.59542	1.91	0.90021	17.80	0.99982
0.92	0.60142	1.92	0.90143	18.00	0.99983
0.93	0.60734	1.93	0.90263	20.00	0.99988
0.94	0.61317	1.94	0.90382	25.00	0.99994
0.95	0.61892	1.95	0.90498	30.00	0.99996
0.96	0.62549	1.96	0.90613	40.00	0.99998
0.97	0.63018	1.97	0.90726	50.00	0.99999
0.98	0.65568	1.98	0.90838	100.00	1.00000
0.99	0.64110	1.99	0.90948	∞	1.00000

9.8   Vertical Stress under Strip Load

Let an infinitely long strip of width B be subjected to a uniformly distributed load of intensity q per unit area and P be a point at depth z below the centre line of the strip as shown in Fig. 9.9. We consider an elemental strip of width dx at distance x from the centre line of strip. The load $q.\,dx$ per unit length on this elemental strip can be treated approximately as a line load. Then using the equation for vertical stress due to line load [Eq. 9.7 (*i*)] obtained from Boussinesq analysis.

Vertical stress at P due to load on elemental strip is given by

$$\Delta\sigma_z = \frac{2\,(qdx)}{\pi z}\left[\frac{1}{1+\left(\dfrac{x}{z}\right)^2}\right]^2$$

Vertical stress at P due to entire strip load is given by

$$\sigma_z = \frac{2q}{\pi z}\int_{-B/2}^{B/2}\frac{dx}{\left[1+\left(\dfrac{x}{z}\right)^2\right]^2}$$

$$= \frac{4q}{\pi z}\int_{0}^{B/2}\frac{dx}{\left[1+\left(\dfrac{x}{z}\right)^2\right]^2}$$

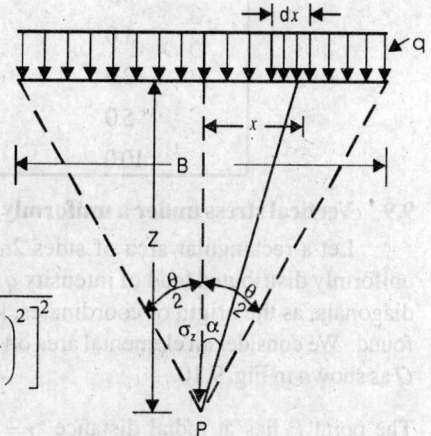

Fig. 9.9. Vertical stress under strip load

Referring to Fig. 9.9, we can write $\dfrac{x}{z} = \tan\alpha$

So that $dx = z\sec^2\alpha.d\alpha$

Also, when $x = 0,\ \alpha = 0$

$$x = \frac{B}{2}, \quad \alpha = \frac{\theta}{2}$$

$$\therefore \quad \sigma_z = \frac{4q}{\pi z} \int_o^{\theta/2} \frac{z \sec^2 \alpha . d\alpha}{\left[1 + \tan^2 \alpha\right]^2} = \frac{4q}{\pi} \int_o^{\theta/2} \cos^2 \alpha \, d\alpha$$

After integration and substitution of limits we get

$$\sigma_z = \frac{q}{\pi}(\theta + \sin \theta) \qquad \qquad \qquad ...9.8\,(i)$$

where θ is in radians.

As an example, let us find the vertical stress at depth $z = 0.1B$ below the centre line of strip.

When $$z = 0.1B, \tan \frac{\theta}{2} = \frac{B/2}{0.1\,B} = \frac{1}{0.2} = 5$$

$$\therefore \quad \theta = 2 \tan^{-1} 5 = 2.747 \text{ radians.}$$

$$\sigma_z = \frac{q}{\pi}\ (2.747 + \sin 2.747) = 0.997q = 99.7 \% \text{ of } q$$

In Table 9.3 is shown the values of vertical stress at different depths below the centre line of the strip load.

Table 9.3. Vertical stress under centre line of strip load

z/B ratio	Vertical stress σ_z as percentage of q
0.1	99.7
0.2	97.7
0.5	81.7
1.0	55.0
2.0	30.6
5.0	12.6
10.0	6.4

9.9   Vertical stress under a uniformly loaded rectangular area

Let a rectangular area of sides $2a$ and $2b$ on the surface of a homogeneous deposit carry a uniformly distributed load of intensity q over its entire area. Choosing O the point of intersection of diagonals, as the origin of coordinates, let $P\ (x, y, z)$ be the point at which vertical stress σ_z is to be found. We consider an elemental area $\delta A = \delta\xi.\delta\eta$ located at (ξ, η) with respect to origin of coordinates O as shown in Fig. 9.10.

The point P lies at radial distance $r = \sqrt{(x - \xi)^2 + (y - \eta)^2}$ from the vertical line through this elemental area. Treating the load on elemental area δA as concentrated load and using Boussinesq analysis we have vertical stress at P due to load on elemental area δA,

$$\Delta\sigma_z = \frac{3(q \cdot \delta\xi \cdot \delta\eta)}{2\pi} \frac{z^3}{\left[(x - \xi)^2 + (y - \eta)^2 + z^2\right]^{5/2}}$$

Hence the vertical stress at P due to entire loaded rectangular area,

$$\sigma_z = \frac{3qz^3}{2\pi} \int_{-a}^{a} \int_{-b}^{b} \frac{d\xi \cdot d\eta}{\left[(x-\xi)^2 + (y-\eta)^2 + z^2\right]^{5/2}}$$

The integral in the above equation has been evaluated and expression for σ_z has been obtained by Florin (1959, 61). The expression is too lengthy and not used in practice. The expression for vertical stress $(\sigma_z)_0$ under point O obtained by putting $x = y = 0$, is relatively simpler for use in practice.

$$(\sigma_z)_o = \frac{3qz^3}{2\pi} \int_{-a}^{a} \int_{-b}^{b} \frac{d\xi \cdot d\eta}{(\xi^2 + \eta^2 + z^2)^{5/2}}$$

After evaluating the integral, we obtain

$$(\sigma_z)_0 = \frac{2q}{\pi} \left[\frac{abz(a^2 + b^2 + 2z^2)}{(a^2 + z^2)(b^2 + z^2)\sqrt{a^2 + b^2 + z^2}} + \sin^{-1} \frac{ab}{\sqrt{a^2 + z^2}\sqrt{b^2 + z^2}} \right] \qquad \ldots 9.9\,(i)$$

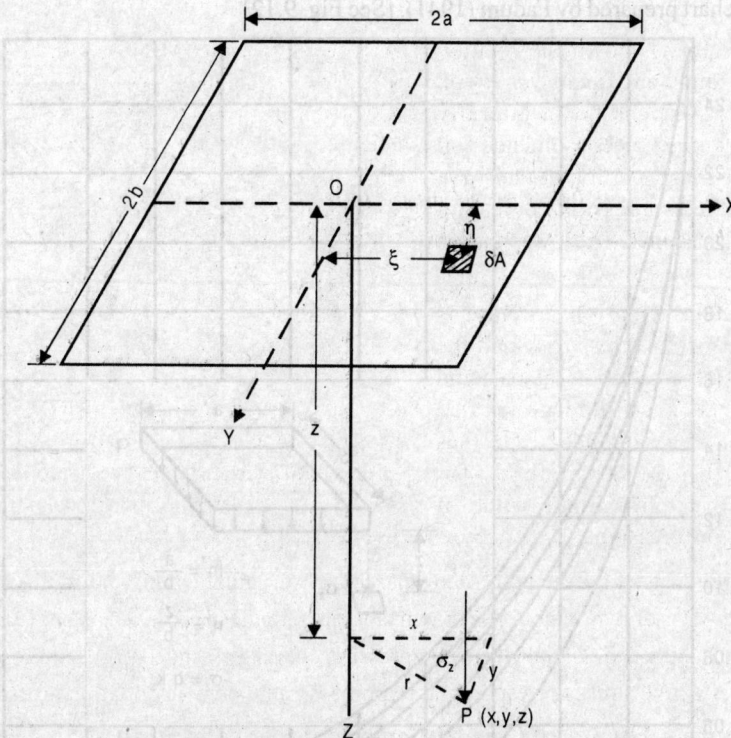

Fig. 9.10. Vertical stress due to rectangular loaded area

From the principle of superposition, we see that the vertical stress $(\sigma_z)_c$ under the corner of a uniformly loaded rectangular area of sides a and b is one-fourth of the vertical stress $(\sigma_z)_0$ given by

Eq. 9.9 (i) Based on this idea and introducing non-dimensional quantities $m' = \dfrac{a}{b}$ and $n' = \dfrac{z}{b}$,

Steinbrenner (1936) gave the following equation

$$(\sigma_z)_c = q.K_s \qquad \ldots 9.9\,(ii)$$

where $K_s = \dfrac{1}{2\pi}\left[\dfrac{m'n'}{\sqrt{1 + m'^2 + n'^2}}\dfrac{1 + m'^2 + 2n'^2}{(1 + n'^2)(m'^2 + n'^2)} + \sin^{-1}\dfrac{m'}{\sqrt{m'^2 + n'^2}\sqrt{1 + n'^2}}\right]$ ...9.9 (iii)

K_s is referred to as Steinbrenner influence factor, whose value can be conveniently obtained from curves in Fig. 9.11.

Newmark (1935) has given the following equation for $(\sigma_z)_c$ in which $m = \dfrac{a}{z}$ and $n = \dfrac{b}{z}$.

$$(\sigma_z)_c = q.K \qquad\qquad ...9.9\,(iv)$$

where influence factor K is given by

$$K = \dfrac{1}{4\pi}\left[\dfrac{2mn\sqrt{(m^2 + n^2 + 1)}}{m^2 + n^2 + m^2n^2 + 1}\dfrac{m^2 + n^2 + 2}{m^2 + n^2 + 1} + \sin^{-1}\dfrac{2mn\sqrt{(m^2 + n^2 + 1)}}{m^2 + n^2 + m^2n^2 + 1}\right] \qquad ...9.9\,(v)$$

In the above equation m and n are interchangeable. The values of influence factor K can be conveniently obtained from Table 9.4. Eq. 9.9 (iv) can also be written as $(\sigma_z)_c = q\,I_\sigma$. Influence value I_σ can be obtained from chart prepared by Fadum (1941). [See Fig. 9.12]

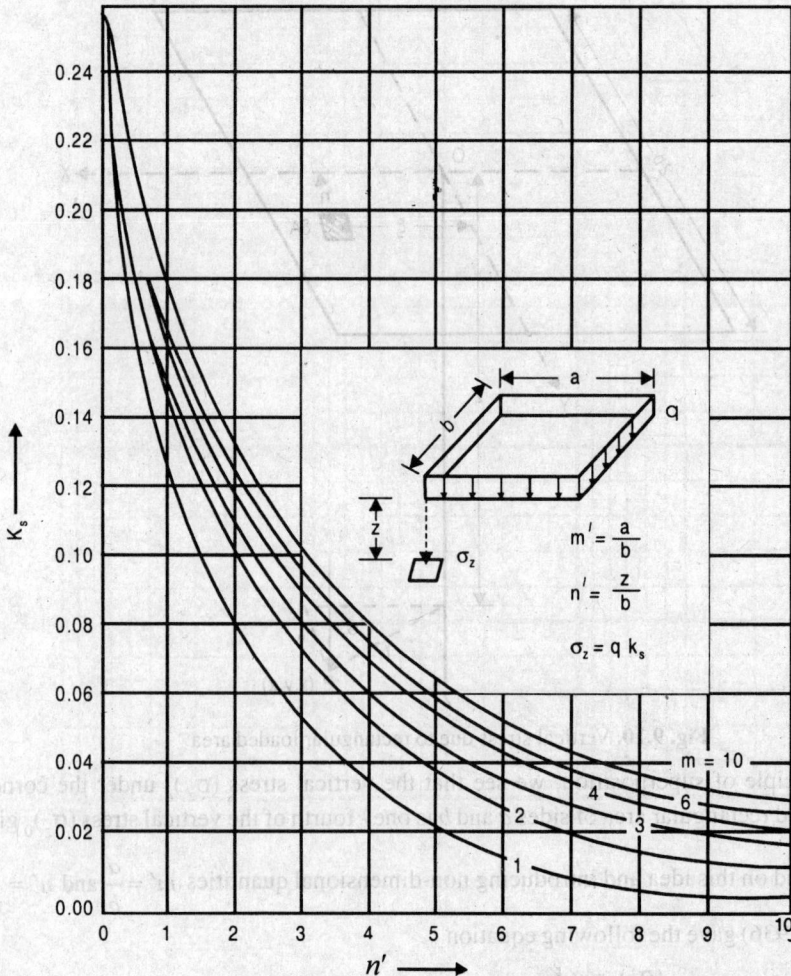

Fig. 9.11. Steinbrenner influence factor, K_s

Again by principle of superposition we can use expression for $(\sigma_z)_c$ in Eq. 9.9 (*ii*) or Eq. 9.9 (*iv*) to find vertical stress below any point inside or outside a loaded rectangle, as shown in the following examples.

Referring to Fig. 9.13, vertical stress at depth z below point P, due to uniformly distributed load of intensity q acting on rectangle abcd can be obtained as

$$\sigma_z = q(K_1 + K_2 + K_3 + K_4)$$

where K_1, K_2, K_3, and K_4 are influence factors corresponding to the four rectangles *Pgdf, Phcf*, Pgae and Pebh. respectively, with P as common corner.

Referring to Fig. 9.14, vertical stress at any depth z below point P due to uniformly distributed load of intensity q acting on rectangle abcd, can be obtained as

$$\sigma_z = q(K_1 - K_2 - K_3 + K_4)$$ where K_1, K_2, K_3, and K_4 are influence factors corresponding to the four rectangles Pfbh, Pfcg, Peah and Pedg respectively with P as common corner.

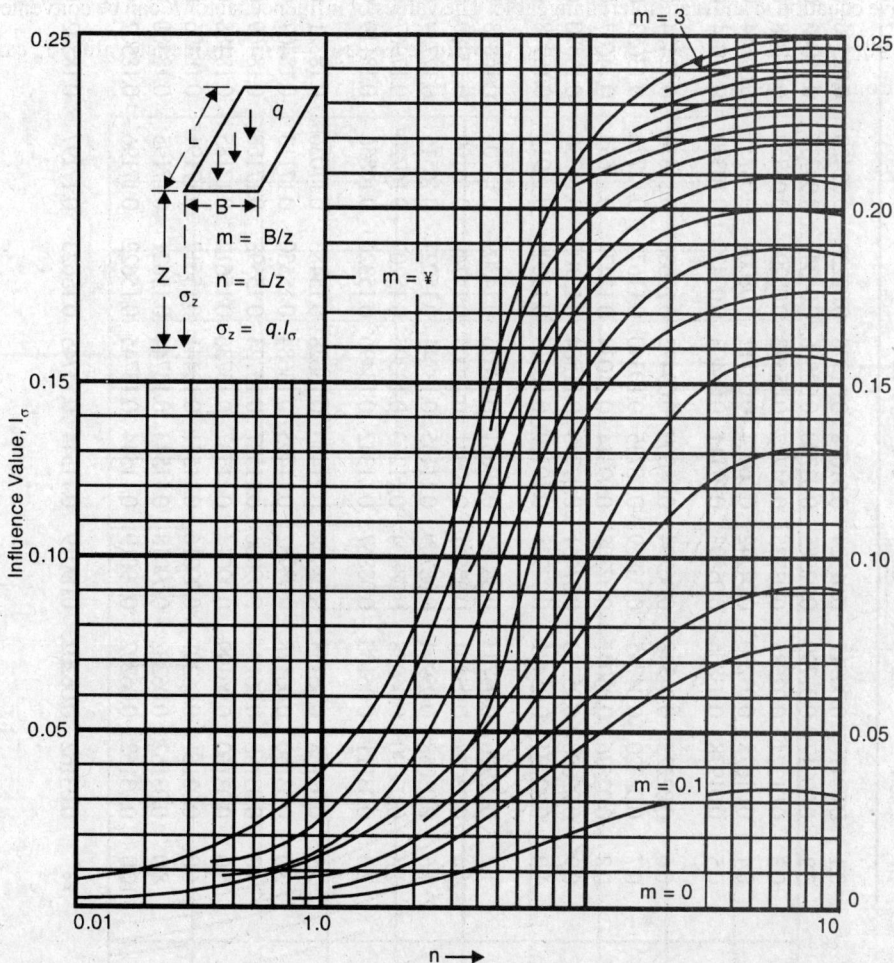

Fig. 9.12. Influence factors for vertical stress beneath a corner of a uniformly loaded rectangular area (After Fadum, 1941)

Note :
$$I_\sigma = \frac{1}{4\pi}\left[\frac{2mn\sqrt{(m^2+n^2+1)}}{m^2+n^2+m^2n^2+1}\frac{m^2+n^2+2}{m^2+n^2+1} + \tan^{-1}\frac{2mn\sqrt{(m^2+n^2+1)}}{m^2+n^2-m^2n^2+1}\right]$$

Table 9.4. Influence factor K for vertical stress under corner of loaded rectangular area based on Boussinessq analysis (After New mark, 1935)

m	0.1	0.2	0.3	0.4	0.5	0.6	0.7	0.8	0.9	1.0	1.2	1.4
0.1	0.00470	0.00917	0.01324	0.01678	0.01978	0.02223	0.02420	0.02576	0.02698	0.02794	0.02926	0.03007
0.2	0.00917	0.01790	0.02585	0.03280	0.03866	0.04348	0.04735	0.05042	0.05283	0.05471	0.05733	0.05894
0.3	0.01324	0.02585	0.03735	0.04742	0.05593	0.06294	0.06859	0.07308	0.07661	0.07938	0.08323	0.08561
0.4	0.01678	0.03280	0.04742	0.06024	0.07111	0.08009	0.08735	0.09314	0.09770	0.10129	0.10631	0.10941
0.5	0.01978	0.03866	0.05593	0.07111	0.08403	0.09472	0.10340	0.11034	0.11584	0.12018	0.12626	0.13003
0.6	0.02223	0.04348	0.06294	0.08009	0.09472	0.10688	0.11679	0.12474	0.13105	0.13605	0.14309	0.14749
0.7	0.02420	0.04735	0.06859	0.08735	0.10340	0.11679	0.12772	0.13653	0.14356	0.14914	0.15703	0.16199
0.8	0.02576	0.05042	0.07308	0.09314	0.11034	0.12474	0.13653	0.14607	0.15370	0.15978	0.16843	0.17389
0.9	0.02698	0.05283	0.07661	0.09770	0.11584	0.13105	0.14356	0.15370	0.16185	0.16835	0.17766	0.18357
1.0	0.02794	0.05471	0.07938	0.10129	0.12018	0.13605	0.14914	0.15978	0.16835	0.17522	0.18508	0.19139
1.2	0.02926	0.05733	0.08323	0.10631	0.12626	0.14309	0.15703	0.16843	0.17766	0.18508	0.19584	0.20278
1.4	0.03007	0.05894	0.08561	0.10941	0.13003	0.14749	0.16199	0.17389	0.18357	0.19139	0.20278	0.21020
1.6	0.03058	0.05994	0.08709	0.11135	0.13241	0.15027	0.16515	0.17739	0.18737	0.19546	0.20731	0.21509
1.8	0.03090	0.06058	0.08804	0.11260	0.13395	0.15207	0.16720	0.17967	0.18986	0.19814	0.21032	0.21836
2.0	0.03111	0.06100	0.08867	0.11342	0.13496	0.15326	0.16856	0.18119	0.19152	0.19994	0.21235	0.22058
2.5	0.03138	0.06155	0.08948	0.11450	0.13628	0.15483	0.17036	0.18321	0.19375	0.20236	0.21512	0.22364
3.0	0.03150	0.06178	0.08982	0.11495	0.13684	0.15550	0.17113	0.18407	0.19470	0.20341	0.21633	0.22499
4.0	0.03158	0.06194	0.09006	0.11527	0.13724	0.15598	0.17168	0.18469	0.19540	0.20417	0.21722	0.22600
5.0	0.03160	0.06199	0.09014	0.11537	0.13736	0.15612	0.17185	0.18488	0.19561	0.20440	0.21749	0.22632
6.0	0.03161	0.06201	0.09016	0.11541	0.13441	0.15617	0.17191	0.18496	0.19569	0.20449	0.21760	0.22644
8.0	0.03162	0.06202	0.09018	0.11543	0.13744	0.15621	0.17195	0.18500	0.19574	0.20455	0.21767	0.22652
10.0	0.03162	0.06202	0.09019	0.11544	0.13745	0.15622	0.17196	0.18502	0.19576	0.20457	0.21769	0.22654
∞	0.03162	0.06202	0.09019	0.11544	0.13745	0.15623	0.17197	0.18502	0.19577	0.20459	0.21770	0.22656

Table 9.4 (Continued)

m	1.6	1.8	2.0	2.5	3.0	4.0	5.0	6.0	8.0	10.0	∞
0.1	0.03058	0.03090	0.03111	0.03138	0.03150	0.03158	0.03160	0.03161	0.03162	0.03162	0.03162
0.2	0.05994	0.06058	0.06100	0.06155	0.06178	0.06194	0.06199	0.06201	0.06202	0.06202	0.06202
0.3	0.08709	0.08804	0.08867	0.08948	0.08982	0.09006	0.09014	0.09016	0.09018	0.09019	0.09019
0.4	0.11135	0.11260	0.11342	0.11450	0.11495	0.11527	0.11537	0.11541	0.11543	0.11544	0.11544
0.5	0.13241	0.13395	0.13496	0.13628	0.13684	0.13724	0.13736	0.13741	0.13744	0.13745	0.13747
0.6	0.15027	0.15207	0.15326	0.15483	0.15550	0.15598	0.15612	0.15617	0.15621	0.15622	0.15623
0.7	0.16515	0.16720	0.16856	0.17036	0.17113	0.17168	0.17185	0.17191	0.17195	0.17196	0.17197
0.8	0.17739	0.17967	0.18119	0.18321	0.18407	0.18469	0.18488	0.18496	0.18500	0.18502	0.18502
0.9	0.18737	0.18986	0.19152	0.19375	0.19470	0.19540	0.19561	0.19569	0.19574	0.19576	0.19577
1.0	0.19546	0.19814	0.19994	0.20236	0.20341	0.20417	0.20440	0.20449	0.20455	0.20457	0.20459
1.2	0.20731	0.21032	0.21235	0.21512	0.21633	0.21722	0.21749	0.21760	0.21767	0.21769	0.21770
1.4	0.21509	0.21836	0.22058	0.22364	0.22499	0.22600	0.22632	0.22644	0.22652	0.22654	0.22656
1.6	0.22025	0.22372	0.22610	0.22940	0.23088	0.23200	0.23235	0.23249	0.23258	0.23261	0.23263
1.8	0.22372	0.22736	0.22986	0.23336	0.23499	0.23617	0.23656	0.23671	0.23681	0.23684	0.23686
2.0	0.22610	0.22986	0.23247	0.23613	0.23782	0.23912	0.23954	0.23970	0.23981	0.23985	0.23987
2.5	0.22940	0.23336	0.23613	0.24010	0.24196	0.24344	0.24392	0.24412	0.24425	0.24429	0.24432
3.0	0.23088	0.23496	0.23782	0.24196	0.24394	0.24554	0.24608	0.24630	0.24646	0.24650	0.24654
4.0	0.23200	0.23617	0.23912	0.24344	0.24554	0.24729	0.24791	0.24817	0.24836	0.24841	0.24886
5.0	0.23255	0.23656	0.23954	0.24392	0.24608	0.24791	0.24857	0.24886	0.24907	0.24914	0.24919
6.0	0.23249	0.23671	0.23970	0.24412	0.24630	0.24817	0.24886	0.24916	0.24939	0.24946	0.24952
8.0	0.23258	0.23681	0.23981	0.24425	0.24646	0.24836	0.24907	0.24939	0.24964	0.24972	0.24980
10.0	0.23261	0.23684	0.23985	0.24429	0.24650	0.24841	0.24914	0.24946	0.24972	0.24981	0.24989
∞	0.23263	0.23686	0.23987	0.24432	0.24654	0.24846	0.24919	0.24952	0.24980	0.24989	0.25000

9.10 Equivalent Point Load Method

This is an approximate method to find the vertical stress at any point inside a soil mass due to a uniformly distributed load acting on an area of any shape. The loaded area is divided into a number of smaller areas, referred to as area units, and the total load on each area unit is considered to be a concentrated load acting at its centroid. Using Boussinesq equation for vertical stress due to point load, the vertical stress at a given point can be obtained by principle of superposition.

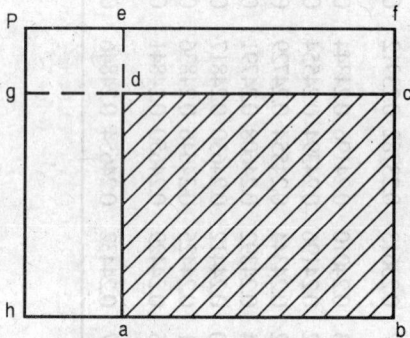

Fig. 9.13. Point P lies inside loaded rectangular area

To illustrate the method, let it be required to find the vertical stress at depth z below a point P, due to a uniformly distributed load of intensity q acting over a rectangle abcd as shown in Fig. 9.15. The rectangle *abcd* is divided into four rectangles and the total loads Q_1, Q_2, Q_3, Q_4 on each of the rectangles are assumed to act at their respective centroids. The vertical stress at depth z below point P can be obtained as

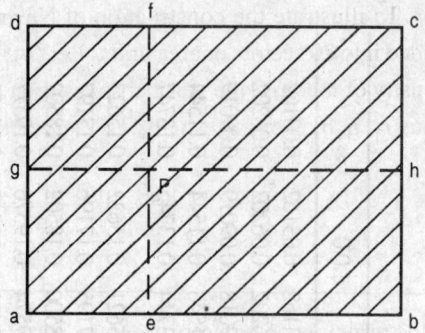

Fig. 9.14. Point P lies outside loaded rectangular area

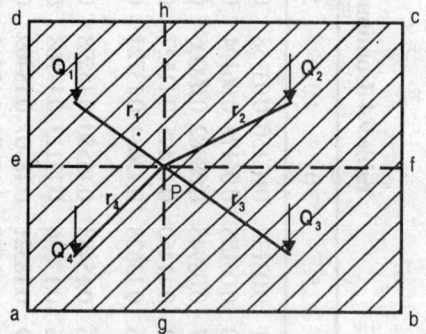

Fig. 9.15. Equivalent point loads

$$\sigma_z = \frac{3z^3}{2\pi}\left[\frac{Q_1}{(r_1^2 + z^2)^{5/2}} + \frac{Q_2}{(r_2^2 + z^2)^{5/2}} + \frac{Q_3}{(r_3^2 + z^2)^{5/2}} + \frac{Q_4}{(r_4^2 + z^2)^{5/2}}\right]$$

The accuracy of the result can be increased by increasing the number of area units and thus decreasing the size of each area unit. The error involved can be limited to 3 percent if the length of each area unit is kept less than one-third the depth at which vertical stress is required.

9.11 Newmark's Influence Chart

For finding vertical stress at any point under a uniformly loaded area of any shape, the use of Newmark's influence chart is more convenient compared to equivalent point load method. The influence chart or influence diagram originally suggested by Newmark (1942) consists of a number of concentric circles divided into area units by radiating lines such that vertical stress at a given depth below the centre of circles, due to load on each area unit is same.

To illustrate the construction of Newmark's chart, let a loaded circular area of radius r_1 be divided into 20 sectors or area units. Let OA_1B_1 be one such area unit as shown in Fig. 9.16. If q is the intensity of loading, the vertical stress σ_z at depth z below centre O, due to load on circular area of radius r_1, is by Boussinesq analysis, given by

$$\sigma_z = q\left[1 - \frac{1}{\left\{1 + \left(\dfrac{r_1}{z}\right)^2\right\}^{3/2}}\right] \qquad \text{...9.11 }(i)$$

By the principle of superposition, the vertical stress at the same point below O, due to load on area unit OA_1B_1 can be written as

$$\frac{\sigma_z}{20} = \frac{q}{20}\left[1 - \frac{1}{\left\{1 + \left(\dfrac{r_1}{z}\right)^2\right\}^{3/2}}\right] = I_f\, q \qquad \text{...9.11 }(ii)$$

where

$$I_f = \frac{1}{20}\left[1 - \frac{1}{\left\{1 + \left(\dfrac{r_1}{z}\right)^2\right\}^{3/2}}\right] \qquad \text{...9.11 }(iii)$$

I_f is called influence value and is given by

$$I_f = \frac{1}{(\text{number of circles})(\text{number of rays})}$$

In this illustration,

$$I_f = \frac{1}{(10)(20)} = 0.005$$

$$\left[1 - \frac{1}{\left\{1 + \left(\dfrac{r_1}{z}\right)^2\right\}^{3/2}}\right] = 0.005 \times 20 = 0.1 \quad \text{...9.11 }(iv)$$

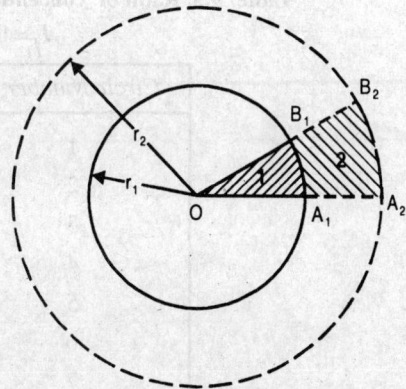

Fig. 9.16. Construction of Newmark's chart

We can choose suitable value for z. Suppose we take $z = 5$ cm and solve Eq. 9.11 (iv) for r_1, we get $r_1 = 1.36$ cm. Hence if we draw a circle of radius $r_1 = 1.36$ cm and divide it into 20 equal area units, each area unit will cause a vertical stress $\dfrac{\sigma_z}{20} = 0.005q$ at a depth of 5 cm below O.

Let r_2 be the radius of second concentric circle. If we extend the radial lines of the first circle, the space between the two concentric circles will also be divided into 20 area units. In Fig. 9.16, $A_1\,A_2\,B_2\,B_1$ is one such area unit and load on this should also cause vertical stress equal to $0.005q$ at 5 cm depth below O. Hence, vertical stress caused by loaded sector $OA_2\,B_2$ is, from Eq. 9.11 (ii), given by

$$\frac{\sigma_z}{20} = \frac{q}{20}\left[1 - \frac{1}{\left\{1+\left(\frac{r_2}{z}\right)^2\right\}^{3/2}}\right] = 2(0.005)q$$

$$1 - \frac{1}{\left\{1+\left(\frac{r_2}{z}\right)^2\right\}^{3/2}} = 2\cdot(0.005)\,20 = 2 \times 0.1 = 0.2 \qquad\qquad ...9.11\,(v)$$

Substituting $z = 5$ cm and solving Eq. 9.11(v) for r_2 we get $r_2 = 2.00$ cm. Similarly the radii of 3rd, 4th, 5th, 6th, 7th, 8th and 9th circles can be calculated. Clearly the radius of 10th circle is given by

$$1 - \frac{1}{\left\{1+\left(\frac{r_{10}}{5}\right)^2\right\}^{3/2}} = 10 \times 0.1 = 1$$

whence, $r_{10} = \infty$,

The radii of different concentric circles is calculated using the following final expression for r_i and tabulated as shown in Table. 9.5.

$$r_i = 5\sqrt{\frac{1}{(1-0.1i)^{2/3}} - 1}$$

Table: 9.5. Radii of concentric circles for construction of Newmark's chart
$I_f = 0.005$ and $z = 5$ cm.

Circle Number (i)	Radius of Circle r_i in cm
1	1.35
2	2.00
3	2.59
4	3.18
5	3.83
6	4.59
7	5.54
8	6.94
9	9.54
9 ½	12.62
10	∞

Use of Newmark's influence chart

To use the chart for finding vertical stress at a depth Z (metre) below a given point, due to a loaded area, the plan of loaded area is drawn on a tracing paper to a scale determined as shown below using z value of Newmark's chart.

$$z\,(cm) = Z\,(m)$$

$$\therefore \qquad 1\ cm = \frac{Z}{z}\ m$$

The plan of loaded area is then superimposed on the Newmark's chart in such a way that the point below which the vertical stress is required coincides with the centre of concentric circles. The number of area units enclosed, including fractions, within the plan of loaded area is found by counting. The vertical stress σ_z is then given by

$$\sigma_z = N_A \times I_f \times q \qquad \qquad ...9.11\,(vi)$$

where N_A = number of area units enclosed within the plan of loaded area

         I_f = influence value of Newmark's chart

         q = intensity of load on given area.

9.12 Westergaard Analysis

Boussinesq analysis is applicable to homogeneous soil deposits. In sedimentary soil deposits we commonly come across clay strata reinforced with lenses of coarse material between them. The problem of finding stresses in such soil deposits due to surface loading was solved by Westergaard (1938).

In Westergaard analysis the following assumptions are made.

1. The soil mass is an elastic medium.
2. Modulus of elasticity E of soil is constant.
3. The soil mass is semi-infinite in extent, *i.e.* it is bound by the ground surface and extends infinitely in all other directions.
4. The soil mass is assumed to be laterally reinforced by numerous closely spaced horizontal sheets of negligible thickness of an infinitely rigid material which prevent the mass as a whole from undergoing any lateral strain.

The Westergaard equation for vertical stress at a point P (Fig. 9.2) located at depth z below the surface and at radial distance r from the axis of loading, due to point load Q acting on the surface is

$$\sigma_z = \frac{Q}{2\pi\eta^2 z^2}\left[\frac{1}{1+\left(\dfrac{r}{\eta z}\right)^2}\right]^{3/2} \qquad \qquad ...9.12\,(i)$$

where $\eta = \sqrt{\dfrac{1-2\mu}{2(1-\mu)}}$

The use of Poisson's ratio μ of 0.2 to 0.3 in the above equation is quite realistic for many types of stratified soil deposits..

In many cases, for practical purposes, μ is taken as zero and then we will have

$\eta = \sqrt{\dfrac{1}{2}}$ so that

$$\sigma_z = \frac{Q}{\pi z^2}\left[\frac{1}{1+2\left(\dfrac{r}{z}\right)^2}\right]^{3/2} = \frac{Q}{z^2}\cdot K_w \qquad \qquad ...9.12\,(ii)$$

where $K_w = \dfrac{1}{\pi}\left[\dfrac{1}{1+2\left(\dfrac{r}{z}\right)^2}\right]^{3/2}$

K_w is called Westergaard influence factor for vertical stress due to surface point load. Table 9.6 gives I_w for different values of $\left(\dfrac{r}{z}\right)$ ratio.

Using Eq. 9.12 (*i*) the vertical stress at a depth z under the centre of a uniformly distributed load of intensity q acting on circular area of radius 'a' at surface, is obtained as

$$\sigma_z = q\left[1 - \left\{\frac{1}{1 + \left(\dfrac{a}{\eta z}\right)^2}\right\}^{1/2}\right]$$ ...9.12 (*iii*)

An influence chart based on Westergaard analysis was first obtained by Fenske (1951) using Eq. 9.12(*iii*) and method adopted by Newmark earlier.

With the data given in Table 9.7 an influence chart has been constructed. For using the chart the scale for drawing the plan of loaded area is obtained by equating the value of ηz in cm to the modified depth ηZ in m where Z is the depth of point at which the vertical stress is required in the given problem.

Table 9.6. Westergaard influence factor

r/z	K_w	r/z	K_w	r/z	K_w	r/z	K_w	r/z	K_w
0.00	0.3183	0.44	0.1949	0.88	0.0783	1.32	0.0335	1.76	0.0165
0.02	0.3178	0.46	0.1875	0.90	0.0751	1.34	0.0324	1.78	0.0160
0.04	0.3168	0.48	0.1803	0.92	0.0721	1.36	0.0312	1.80	0.0156
0.06	0.3149	0.50	0.1733	0.94	0.0692	1.38	0.0302	1.82	0.0151
0.08	0.3123	0.52	0.1664	0.96	0.0664	1.40	0.0292	1.84	0.0147
0.10	0.3090	0.54	0.1598	0.98	0.0638	1.42	0.0282	1.86	0.0143
0.12	0.3050	0.56	0.1534	1.00	0.0613	1.44	0.0273	1.88	0.0139
0.14	0.3005	0.58	0.1471	1.02	0.0589	1.46	0.0264	1.90	0.0135
0.16	0.2953	0.60	0.1411	1.04	0.0566	1.48	0.0255	1.92	0.9131
0.18	0.2897	0.62	0.1353	1.06	0.0544	1.50	0.0247	1.94	0.0128
0.20	0.2836	0.64	0.1298	1.08	0.0523	1.52	0.0239	1.96	0.0124
0.22	0.2771	0.66	0.1244	1.10	0.0503	1.54	0.0231	1.98	0.0121
0.24	0.2703	0.68	0.1192	1.12	0.0484	1.56	0.0223	2.00	0.0118
0.26	0.2632	0.70	0.1142	1.14	0.0466	1.58	0.0217	2.10	0.0103
0.28	0.2558	0.72	0.1035	1.16	0.0499	1.60	0.0210	2.20	0.0091
0.30	0.2483	0.74	0.1050	1.18	0.0432	1.62	0.0204	2.30	0.0081
0.32	0.2407	0.76	0.1006	1.20	0.0416	1.64	0.0198	2.40	0.0072
0.34	0.2331	0.78	0.0964	1.22	0.0401	1.66	0.0192	2.50	0.0064
0.36	0.2254	0.80	0.0925	1.24	0.0386	1.68	0.0186	2.60	0.0058
0.38	0.2175	0.82	0.0887	1.26	0.0373	1.70	0.0180	2.70	0.0052
0.40	0.2099	0.84	0.0850	1.28	0.0300	1.72	0.0175	2.80	0.0047
0.42	0.2023	0.86	0.0815	1.30	0.0347	1.74	0.0160	3.00	0.0038

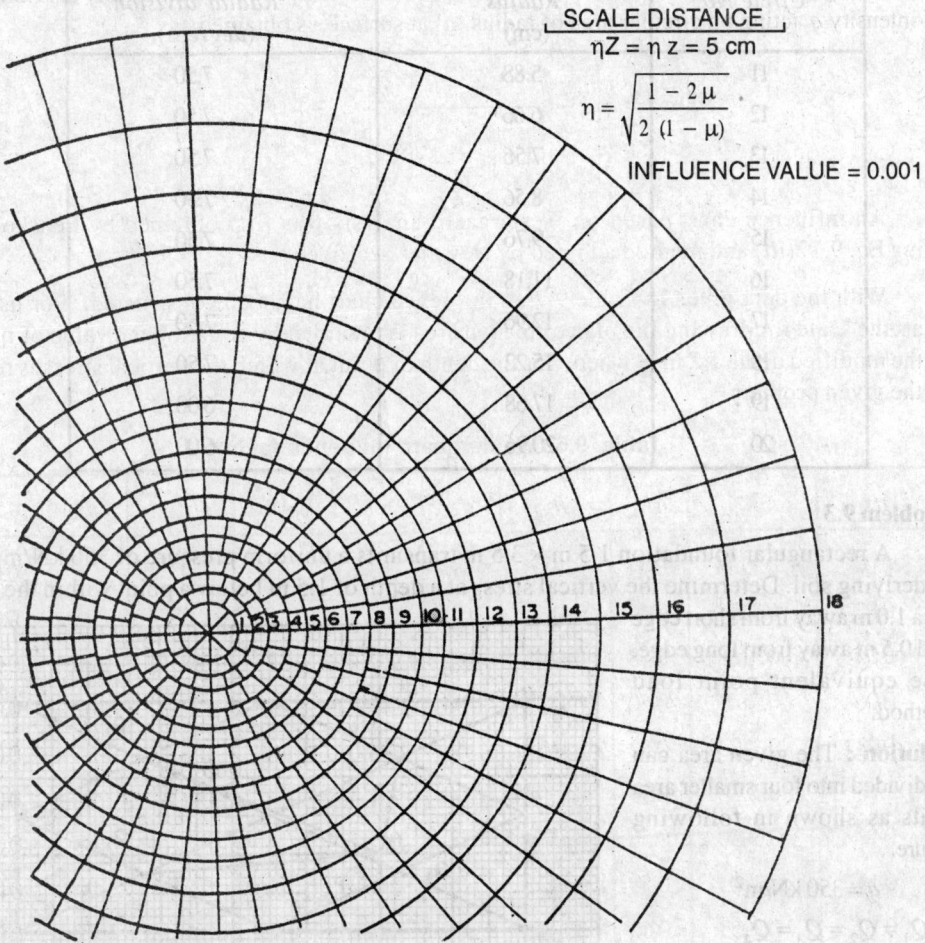

SCALE DISTANCE
$$\eta Z = \eta\, z = 5 \text{ cm}$$

$$\eta = \sqrt{\frac{1 - 2\mu}{2\,(1 - \mu)}}$$

INFLUENCE VALUE = 0.001

Table 9.7. Data for Newmark's chart based on Westergaard Analysis

Circle No.	Radius (cm)	Radial division (degrees)
1	0.64	45.00
2	1.02	30.00
3	1.46	18.00
4	1.88	15.00
5	2.36	11.25
6	2.80	11.25
7	3.32	9.00
8	3.86	9.00
9	4.50	7.50
10	5.16	7.50

Contd.

Circle No.	Radius (cm)	Radial division (degrees)
11	5.88	7.50
12	6.66	7.50
13	7.56	7.50
14	8.56	7.50
15	9.76	7.50
16	11.18	7.50
17	12.96	7.50
18	15.22	7.50
19	17.68	6.00
20	20.96	6.00

Problem 9.3

A rectangular foundation 1.5 m × 3.5 m transmits a uniform pressure of 350 kN/m² to the underlying soil. Determine the vertical stress at a depth of 1.5 m below a point within the loaded area 1.0 m away from short edge and 0.5 m away from long edge. Use equivalent point load method.

Solution : The given area can be divided into four smaller area units as shown in following figure.

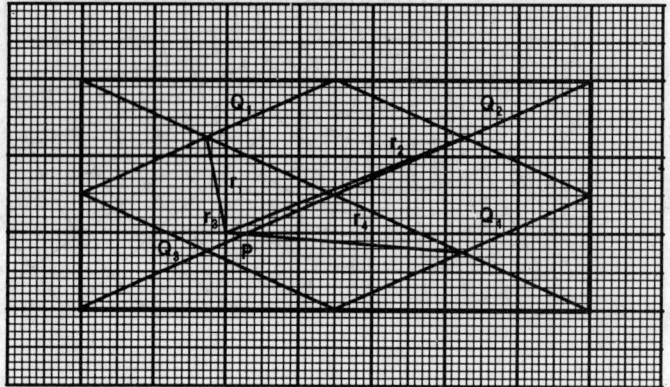

$q = 350$ kN/m²

$Q_1 = Q_2 = Q_3 = Q_4$

$= (350)(0.75 \times 1.75) = 459.4$ kN

Vertical stress at depth 1.5 m below point P,

$$\sigma_z = \frac{3(459.4)(1.5)^3}{2\pi} \left[\frac{1}{\left\{(0.65)^2 + (1.5)^2\right\}^{5/2}} + \frac{1}{\left\{(1.75)^2 + (1.5)^2\right\}^{5/2}} + \frac{1}{\left\{(0.2)^2 + (1.5)^2\right\}^{5/2}} + \frac{1}{\left\{(0.2)^2 + (1.5)^2\right\}^{5/2}} \right]$$

$= 637.25$ kN/m²

Problem : 9.4

Three parallel strip foundations each 3 m wide 5 m apart centre to centre transmit contact pressures of 200 kN/m², 150 kN/m² and 100 kN/m² respectively. Calculate the intensity of vertical stress due to the combined loads beneath the centre of each footing at 3m depth. Use Boussinesq line load approximation.

$$200 \text{ kN/m}^2 \qquad 150 \text{ kN/m}^2 \qquad 100 \text{ kN/m}^2$$

|←— 3m —→| |←— 3m —→| |←— 3m —→|

3m

$P_1 \qquad\qquad\qquad P_2 \qquad\qquad\qquad P_3$

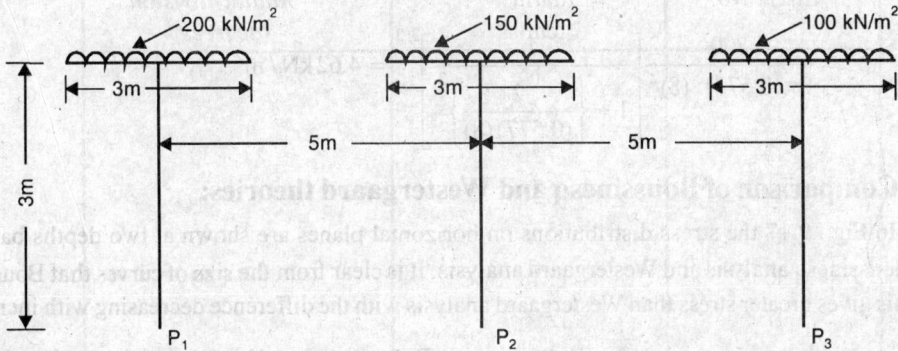

Solution : Assumed equivalent line loads are

$$q'_1 = 200 \times 3 = 600 \text{ kN/m}$$
$$q'_2 = 150 \times 3 = 450 \text{ kN/m}$$
$$q'_3 = 100 \times 3 = 300 \text{ kN/m}$$

$$(\sigma_z)_{atP_1} = \left[\frac{2(600)}{\pi(3)} \left[\frac{1}{1+0} \right]^2 + \frac{2(450)}{\pi(3)} \left[\frac{1}{1+\left(\frac{5}{3}\right)^2} \right]^2 + \frac{2(300)}{\pi(3)} \left[\frac{1}{1+\left(\frac{10}{3}\right)^2} \right]^2 \right] = 134.45 \text{ kN/m}^2$$

$$(\sigma_z)_{atP_2} = \left[\frac{2(600)}{\pi(3)} \left[\frac{1}{1+\left(\frac{5}{3}\right)^2} \right]^2 + \frac{2(450)}{\pi(3)} \left[\frac{1}{1+0} \right]^2 + \frac{2(300)}{\pi(3)} \left[\frac{1}{1+\left(\frac{5}{3}\right)^2} \right]^2 \right] = 108.87 \text{ kN/m}^2$$

$$(\sigma_z)_{atP_3} = \left[\frac{2(600)}{\pi(3)} \left[\frac{1}{1+\left(\frac{10}{3}\right)^2} \right]^2 + \frac{2(450)}{\pi(3)} \left[\frac{1}{1+\left(\frac{5}{3}\right)^2} \right]^2 + \frac{2(300)}{\pi(3)} \left[\frac{1}{1+0} \right]^2 \right] = 71.22 \text{ kN/m}^2$$

Problem 9.5

A concentrated load of 800 kN acts at the ground surface. Compute the vertical stresses at 8m depth (*i*) on the axis of load and (*ii*) 2 m away from the axis. Use Westergaard analysis Take μ = 0.25

Solution :

$$\eta = \sqrt{\frac{1-2\mu}{2(1-\mu)}} = 0.577$$

(*i*) $z = 8 \, m, r = 0$

$$\sigma_z = \frac{Q}{2\pi\eta^2 z^2} \left[\frac{1}{1+\left(\frac{r}{\eta z}\right)^2} \right]^{3/2} = \frac{800}{2\pi(0.577)^2 (8)^2} \left[\frac{1}{1+0} \right]^{3/2} = 5.97 \text{ kN/m}^2$$

(*ii*) $z = 8\,m, r = 2m$

$$\sigma_z = \frac{800}{2\pi(0.577)^2(8)^2}\left[\frac{1}{1 + \left\{\frac{2}{(0.577)(8)}\right\}^2}\right]^{3/2} = 4.62\,kN/m^2$$

9.13 Comparison of Boussinesq and Westergaard theories:

In Fig . 9.17 the stress distributions on horizontal planes are shown at two depths based on both Boussinesq analysis and Westergaard analysis. It is clear from the size of curves that Boussinesc analysis gives greater stress than Westergaard analysis with the difference decreasing with increasing $\left(\frac{r}{z}\right)$ ratio and also increasing depth. In practice Boussinesq analysis is widely used as it gives conservative values. However homogeneous soil deposits of large depth are near to assumptions made in Boussinesq analysis and sedimentary deposits or stratified soil deposits are approximately nearer to assumptions made in Westergaard analysis.

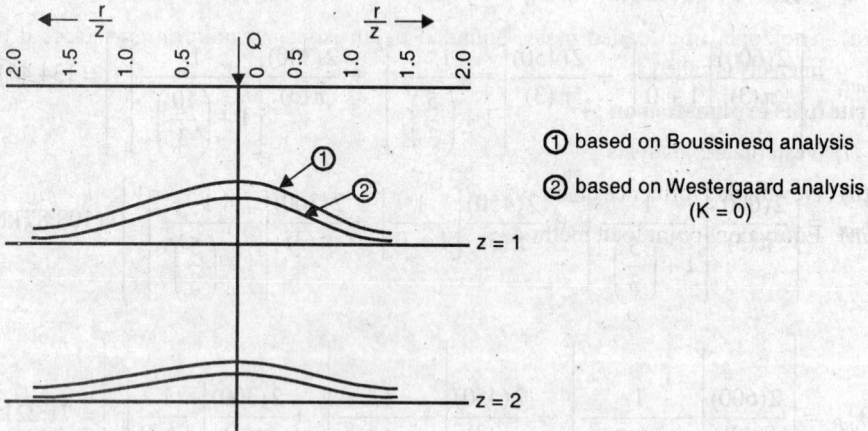

Fig. 9.17. Comparison of Boussinesq and Westergaard theories

EXERCISE–9

9.1   An elevated structure with a total load of 1200 kN is supported on a tower with 4 legs. The legs rest on piers located at the corners of a square of side 7.5 m. Compute the vertical stress increment at a point 6 m below the centre of the structure.

9.2   An annular ring footing of external and internal radii of 8 m and 4 m respectively transmits a pressure of 100 kN/m^2.Compute the vertical stresses at depths 0.5m, 1m, 2m, 4m, 8m below the centre. Draw the stress distribution curve with depth.

9.3   The plan of foundation is given in the adjacent figure Determine the vertical stress increment if the contact pressure is 40 kN/m^2 uniformly distributed over the area, at a depth of 5 m below the point *P* marked in the figure, (*i*) by equivalent point load method and (*ii*) by use of Newmark's chart

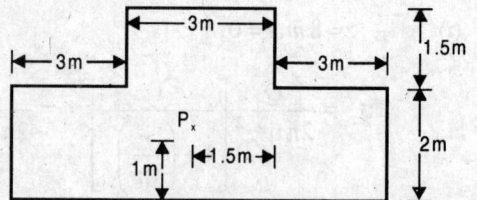

9.4   An elastic medium carries at its surface a uniform load of 250 kN/m², covering a rectangular area 4 m × 3 m. Find the vertical stress, at a depth of 5 m below the centroid of area (assume Boussinesq influence factor for one quadrant = 0.0474.).

9.5   Fill in the blanks:

(*i*)   Theoretically, the vertical stress at surface under a point load is ....

(*ii*) The curve connecting all the points below the ground surface of equal vertical stress is called................

(*iii*) The distribution of vertical stress under a uniformly loaded area of any shape can be calculated with the help of..................

(*iv*) In the Newmark's chart based on Boussineq theory, influence value I_f is equal to the reciprocal of the product of....... and ....

(*v*)   The Westergaard analysis is used for............soil deposits.

(*vi*) In Westergaard theory, lateral pressure is assumed to be ....

(*vii*) A material that has identical properties, in all directions, through any point of it, is called ....

(*viii*) The zone, in a loaded mass, bounded by an isobar of an arbitrary selected pressure-intensity is called a ....

9.6   Write brief explanation on :–

(*i*) Westergaard analysis.

(*ii*) Newmark's influence chart

(*iii*) Equivalent point load method

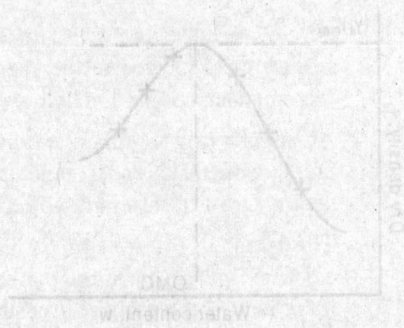

CHAPTER 10

COMPACTION

10.1 Compressibility

Compressibility of soil mass is an engineering property by virtue of which the soil mass is capable of undergoing compression or decrease in volume when subjected to compressive loads. The two processes, namely compaction and consolidation, both involve reduction in volume but in practice only consolidation is associated with compressibility.

Compaction is the process in which rapid reduction in volume takes place due to sudden application of loads as caused by ramming, tamping, rolling and vibration.

Consolidation is the process in which gradual reduction in volume takes place due to sustained loading.

10.2 Compaction

During compaction the reduction in volume is mainly due to expulsion of pore air and rearrangement of particles resulting in their closer packing. Compaction of a soil mass results in increase in dry density. The dry density attained depends on water content, amount and type of compaction. The amount and type of compaction determine the compactive effort. For a specific amount of compactive energy applied on soil, the mass attains maximum dry density at a particular water content. This water content is referred to as optimum water content.

(see Fig. 10.1). It is of interest to note the difference between the two processes. Whereas consolidation is the process in which gradual reduction in volume of soil mass occurs under sustained loading and is mainly due to expulsion of pore water, compaction is the process in which sudden reduction in volume of soil mass occurs under instant application of loads and is mainly due to the expulsion of pore air.

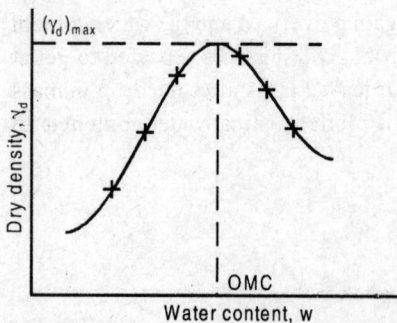

Fig. 10.1. γ_d-w relation from compaction test

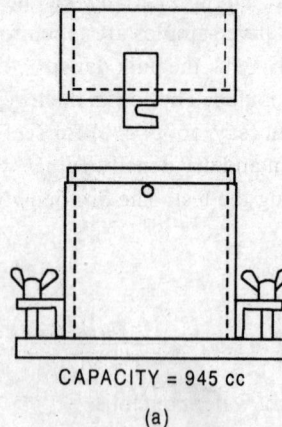

CAPACITY = 945 cc

(a)

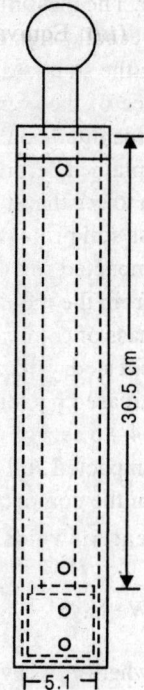

(b)

Fig 10.2. (a) Proctor mould and (b) rammer for standard Proctor test

10.3 Standard Proctor Test

To study the compaction characteristics laboratory compaction tests have been developed. The standard Proctor test developed by R.R. Proctor (1933) is widely used as a guide and a basis of comparison for compaction of earthen embankments and subgrade of highway pavements. The equipment used in the test consists of (i) cylindrical metal mould with detachable base plate having an internal diameter of 4 inches (10.16 cm), internal height of 4.6 inches (11.68 cm) and internal volume of 1/30 cu.ft.(945 cc), (ii) collar of 2 inches (5 cm) effective height and (iii) rammer of mass 5.5 lb (2.5 kg) with a height of fall of 1 foot (30.48 cm). The procedure essentially consists of compacting the soil at different water contents and finding the corresponding dry densities. At each water content the mould with base plate is filled in three layers, each layer being given 25 blows from the standard rammer. Dry density is plotted against water content to obtain the compaction curve. The compaction curve represents the relation between dry density and water content. As it is clear from Fig. 10.1, the dry density increases with the water content, reaches a maximum value and thereafter decreases. The values of maximum dry density and optimum water content are obtained corresponding to the peak of the compaction curve. IS: 2720 (Part VII) 1965 recommends mould with a 100 mm internal diameter and 127.5 mm effective height, whose internal volume is 1000 ml. The rammer has modified mass of 2.6 kg with a drop of 310 mm.

Brief experimental procedure: About 3 kg of dry soil with all lumps pulverized and passing through 4.75 mm sieve is taken in tray. The quantity of water to be added for the first trial is computed. For this purpose it is usually convenient to take the water content for the first trial as about 4% for coarse grained soil and 8% for fine grained soil. The computed quantity of water is added to the soil in the tray and mixed thoroughly with hand to ensure uniform distribution of water. The mass of mould with base plate (M_1) is found. The mould is filled with some quantity of the wet soil taken from the tray and compacted with 25 uniformly distributed blows on the surface, using the standard rammer. The compacted soil should be about 1/3 of the height of mould. The surface of the compacted soil is scratched with a knife to ensure bond with the next layer. The collar is fitted on the mould and the soil for the second layer is put inside the mould and compacted as explained before. In similar manner the third layer of compacted soil is obtained with care being taken to see that it does not protrude more than 6mm into the collar. The collar is removed and the excess soil projecting above the top of the mould is trimmed off. The mass of mould + base plate + compacted soil (M_2) is found. The soil is removed and put back in the tray. While removing the soil from the mould, representative samples are taken for water content determination. Knowing the mass of compacted soil (M_2-M_1), the bulk density γ is calculated. After determining the water content ω, dry density γ_d is computed. The soil in the tray is again pulverized and the water content is increased by suitable amount (say, about 4%) for second trial. The steps are repeated to get at least 4 – 5 sets of water content and dry density values with at least 2 trials after the drop in mass of compacted soil occurs during the test. The dry density γ_d is plotted against water content ω to obtain the compaction curve.

Percent Air Voids Lines

We have
$$\gamma_d = \frac{(1 - n_a)G\gamma_w}{1 + \omega G}$$

where γ_d = dry unit weight at water content ω.

n_a = percentage air voids

G = specific gravity of soil particles

γ_w = unit weight of water

For a particular value of n_a, γ_d can be evaluated for different values of ω and plotted to obtain the curve referred to as percent air voids line. The following example illustrates 30 per cent air voids line.

Given $G = 2.7$, $\gamma_w = 9.81$ kN/m³, for $n_a = 30\%$

we have
$$\gamma_d = \frac{(1-0.3)(2.7)(9.81)}{1+\omega(2.7)} = \frac{18.54}{1+2.7\omega}$$

ω (%)	γ_d (kN/m³)
10	20.86
20	17.20
40	12.73
60	10.11

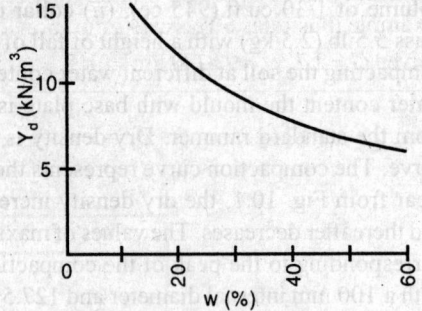

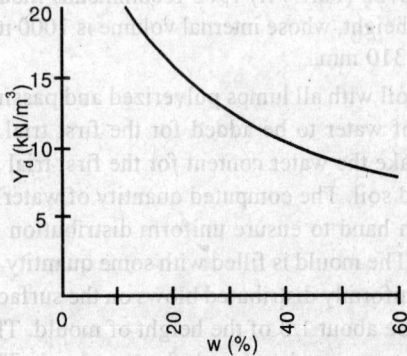

30 percent air voids line

Zero air voids line

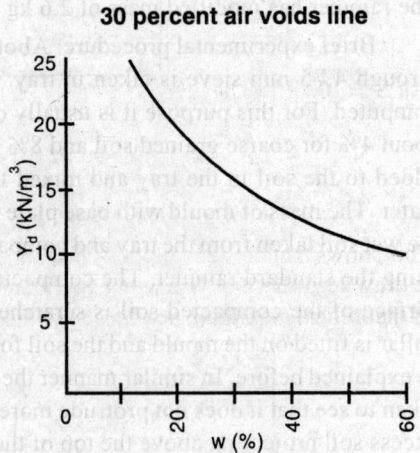

60 per cent saturation line

Zero Air Voids Line

For any water content ω the theoretical maximum dry density is obtained when $n_a = 0$

For $\quad n_a = 0$, $\gamma_d = \dfrac{(1-n_a)G\gamma_\omega}{1+\omega G} = \dfrac{G\gamma_\omega}{1+\omega G}$

This equation is used to plot the zero air voids line as shown in the following example.
Given $G = 2.7$, $\gamma_w = 9.81$ kN/m³

we have
$$\gamma_d = \frac{(2.7)(9.81)}{1+\omega(2.7)} = \frac{26.49}{1+2.7\omega}$$

ω (%)	γ_d (kN/m³)
10	20.86
20	17.20
40	12.73
60	10.11

Percent Saturation Lines

We have $$\gamma_d = \frac{G\gamma_\omega}{1 + \dfrac{\omega G}{S_r}}$$

Using the above equation, for a particular value of degree of saturation S_r, γ_d can be computed for different values of ω and plotted to obtain a curve called per cent saturation line. The following example illustrates 60 percent saturation line

Given $G = 2.7$, $\gamma_\omega = 9.81$ kN/m^3, for $S_r = 60\%$

we have $$\gamma_d = \frac{(2.7)(9.81)}{1 + \dfrac{\omega(2.7)}{0.6}} = \frac{26.49}{1 + 4.5\omega}$$

ω (%)	γ_d (kN/m^3)
10	18.27
20	13.94
40	9.46
60	7.16

For $S_r = 100\%$, we have

$$\gamma_d = \frac{G\gamma_\omega}{1 + \dfrac{\omega G}{S_r}} = \frac{G\gamma_w}{1 + \omega G}$$

This shows that the zero air voids line and 100 percent saturation line coincide. It is the usual practice to draw zero air voids line with experimental compaction curve and serves as a means of comparison of theoretical maximum dry density with its experimental value for any water content (see Fig. 10.3).

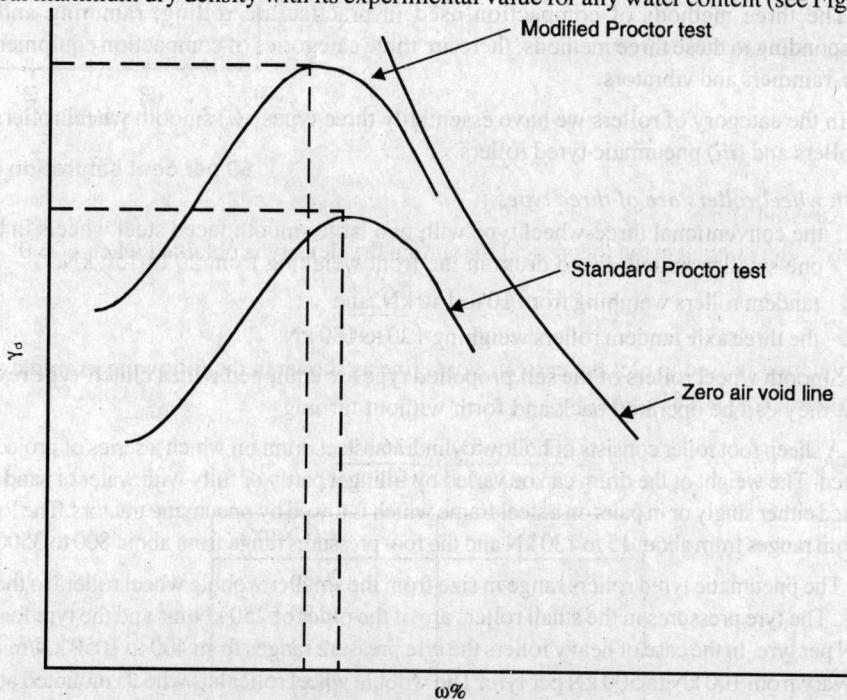

Fig 10.3. Comparison of compaction curves

10.4 Modified Proctor test

With the advent of heavy vehicles and the need for higher compaction, the modified Proctor test was developed to give a higher standard of compaction. As this test was standardized by American Association of State Highway Officials, it is also referred to as modified AASHO test. The test procedure is similar to that of standard Proctor test, except for the application of higher compactive effort. The mould used is the same as in standard Proctor test (Proctor mould of capacity 1/30 Cu.ft or 945 ml). But the soil is compacted in 5 layers giving 25 blows to each layer, with rammer of mass 10 lb (4.54 kg) and height of fall of 18 in (45.72 cm). IS: 2720 (Part VIII) - 1965 recommends the use of 4.89 kg rammer with a drop of 45 cm for heavy compaction.

Typical compaction curves obtained for same soil from standard Proctor test and modified Proctor test are shown in Fig. 10.3 It is clear from the figure that the effect of increasing compactive effort is to cause increase in maximum dry density and decrease in optimum moisture content.

Comparison of Compactive Energies

Compactive energy used in standard Proctor test

$$= \frac{2.5 \times 30.48 \times 25 \times 3}{945} = 6.05 \, \text{kg} - \text{cm} / \text{ml}$$

Compactive energy used in modified Proctor test

$$= \frac{4.54 \times 45.72 \times 25 \times 5}{945} = 27.46 \, \text{kg} - \text{cm} / \text{ml}$$

Thus the compactive energy used in modified Proctor test is (27.46/6.05 = 4.54), about $4\frac{1}{2}$ times that used in standard Proctor test.

10.5 Field Compaction Methods

The three methods of compaction used in practice are rolling, ramming and vibration. Corresponding to these three methods, there are three categories of compaction equipments, namely, rollers, rammers and vibrators.

In the category of rollers we have essentially three types : (*i*) smooth wheel rollers, (*ii*) sheep foot rollers and (*iii*) pneumatic tyred rollers.

Smooth wheel rollers are of three types :
   .   the conventional three-wheel type with two large smooth faced steel wheels in the rear and one smaller smooth faced drum in the front weighing from 20 to 150 kN,
   2.   tandem rollers weighing from 10 to 140 kN, and
   3.   the three axle tandem rollers weighing 120 to 180 kN.

Smooth wheel rollers of the self propelled type are equipped with a clutch-type reversing gear so that they can be operated back and forth without turning.

A sheep foot roller consists of hollow cylindrical steel drum on which a series of projecting feet are mounted. The weight of the drum can be varied by filling it partly or fully with water or sand and they are mounted either singly or in pairs on a steel frame which is towed by pneumatic tractors. The loaded weight per drum ranges from about 15 to 130 kN and the foot pressures range from about 800 to 3500 kN/m².

The pneumatic tyred rollers range in size from the smaller wobble wheel rollers to the very heavy rollers. The tyre pressures in the small rollers are of the order of 250 kN/m² and the tyre loads are about 7.5 kN per tyre. In the case of heavy rollers the tyre pressure ranges from 400 to 1050 kN/m² and the tyre loads vary from 100 kN to 500 kN per tyre. The wobble wheel roller has wheels mounted at slight angle with respect to the axle to induce kneading of the soil during rolling.

Rammers for compacting soil are used in places where use of rollers are not feasible. The mechanically operated type comprise of pneumatic and internal combustion type, weighing from 300 to 1500 N. Internal combustion type jumping rammers, known as frog rammers, weigh upto 10 kN. Vibrators consist of a vibrating unit of either the out of balance weight type or a pulsating hydraulic type mouned on a screed, plate or roller.

Suitability of Compaction Equipments

The suitability of compaction equipments for compacting different types of soils is made clear in the following brief discussion.

Smooth wheel rollers are most suited for compacting coarse grained soils but can be used satisfactorily on moderately cohesive soils.

Sheep foot rollers are most suitable for compacting cohesive soils. The kneading action produced results in a better bond between compacted layers. They are not effective on coarse-grained cohesionless soils.

Pneumatic tyred rollers are suitable for both cohesionless soils and cohesive soils. The action of pneumatic tyred rollers is a combination of pressure and kneading. Vibratory rollers are effective in compacting cohesionless soils.

Rammers are used for compacting soils in confined places where the use of rollers is not feasible

10.6 Placement Water Content

Placement water content is the water content at which the soil is compacted in the field. Depending on soil type and other requirements, the placement water content may be equal to, less than or greater than the optimum water content determined in laboratory. For example cohesive subgrades under pavements should preferably be compacted wet of optimum so that they may not exhibit large expansions and swelling pressure on submergence. Highway embankments on cohesive soils should be compacted somewhat dry of optimum in order to achieve high strength and resistance to deformation and low volume compressibility. High earth dams should be compacted at a placement water content 1 to 2.5 per cent less than the optimum to reduce the probability of the development of high pore pressure. However, the impervious cores of earth dams should desirably be compacted on the wet side of optimum in order to achieve low permeability and greater safety against cracking due to differential settlements or other causes.

10.7 Field Compaction Control

The field compaction control consists of the determination of (*i*) water content at which the soil has ben compacted and (*ii*) the bulk density, in order to find out the degree of compaction achieved. As the work is in progress, rapid methods of testing are to be adopted.

For rapid determination of water content the Proctor needle method has been in use for long time. The Proctor needle consists of a needle point attached to graduated needle shank which in turn is attached to a spring loaded plunger. Needle points of varying cross-sectional areas are available to suit the measurement of a wide range of penetration resistance. The penetration resistance is read on the calibrated stem [See Fig. 10.4]

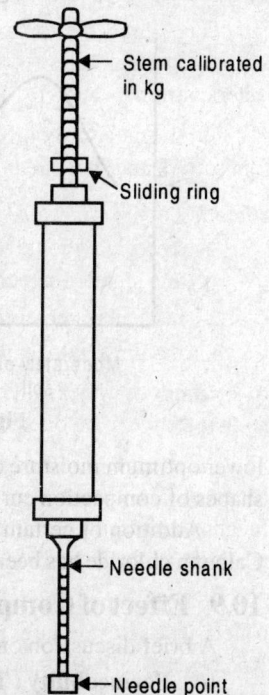

Fig. 10.4. Proctor needle

A calibration curve is first obtained in the laboratory by plotting penetration resistance against water content. For this purpose, the penetration resistance is measured by inserting the Proctor needle in the soil compacted in the Proctor mould. The penetration resistance corresponding to different water contents is obtained by compacting the soil in the Proctor mould at different water contents in each trial. The calibration curve is used to obtain the water content of the compacted soil in the field. The penetration resistance for the compacted soil in the field is measured with the Proctor needle and water content corresponding to it is read off from the calibration curve.

The bulk density of the compacted soil in the field can be determined by core cutter method or sand replacement method.

The degree of compaction achieved in the field is measured by the ratio of field dry density to the maximum dry density obtained from laboratory compaction test. It is expressed as a percentage and also referred to as relative compaction or per cent compaction.

10.8 Factors affecting Compaction

The factors which affect the dry density of compacted soil are (i) water content (ii) amount and type of compaction, (iii) type of soil and (iv) addition of admixtures.

As is evident from laboratory compaction tests the dry density increases with water content, attains a maximum at optimum moisture content and thereafter decreases with further increase in water content.

The amount of compaction affects both maximum dry density and optimum water content. Increase of compactive effort, for a given soil, causes increase in maximum dry density and decrease in optimum water content. [see Fig. 10.3]

For a given compactive effort, the maximum dry density achieved depends to a large extent upon the soil type. Well graded coarse-grained soils attain much higher maximum dry density at

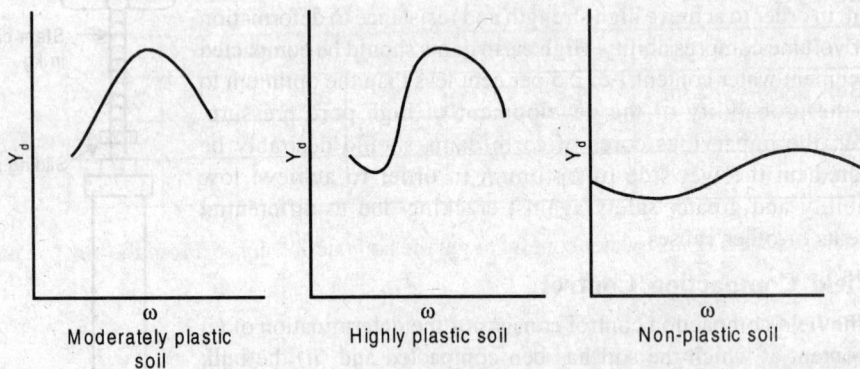

Fig 10.5. Typical shapes of compaction curves

lower optimum moisture content when compared with fine grained soils. It is of interest to note the shapes of compaction curves for different types of soils. [see Fig. 10.5]

Addition of certain admixtures to soils results in modification of its compaction properties. Calcium chloride has been a widely used chemical additive.

10.9   Effect of Compaction on some Soil properties

A brief discussion on effect of compaction on some soil properties is presented here.

(i)   **Permeability :** The effect of compaction is to decrease the permeability. In the case of fine grained soils it has been found that for the same dry density soil compacted wet of optimum will be less permeable than that compacted dry of optimum.

(ii) **Compressibility :** In case of soil samples initially saturated and having same void ratio, it has been found that in low pressure range a wet side compacted soil is more compressible than a dry side compacted soil, and vice versa in high pressure range.

(iii) **Pore pressure :** In undrained shear tests conducted on saturated samples of clay it has been found that lower pore pressures develop at low strains when the sample is compacted dry of optimum, compared to the case when the sample is compacted wet of optimum. But at high strains in both types of samples the development of pore pressure is same for same density and water content.

(iv) **Stress-strain relation :** Samples compacted dry of optimum produce much steeper stress-strain curves with peaks at low strains, whereas samples compacted wet of optimum, having the same density, produce much flatter stress-strain curves with increase in stress even at high strains.

(v) **Shrinkage and swelling :** At same density a soil compacted dry of optimum shrinks appreciably less than that compacted wet of optimum. Also the soil compacted dry of optimum exhibits greater swelling characteristics than samples of the same density compacted wet of optimum.

Problem:10.1

The following observations were made in a Standard Proctor Test.

Trial No.	1	2	3	4	5	6
Mass of wet soil (Kg)	1.70	1.89	2.03	1.99	1.96	1.92
Water content (%)	7.7	11.5	14.6	17.5	19.7	21.2

Volume of mould = 945 cc. $G = 2.67$

Determine maximum dry density and optimum moisture content. Also plot zero air voids line

Solution:

Trial No.	1	2	3	4	5	6
Bulk density (Kg/m³)	1798.9	2000	2148.2	2105.8	2074.1	2031.7
Water content (%)	7.7	11.5	14.6	17.5	19.7	21.2
Dry density (Kg/m³)	1670.3	1793.7	1874.5	1792.2	1732.7	1676.3

Dry density is plotted against water content to get the compaction curve. From the plot (on page 172) we obtain,

Maximum dry density, $(\gamma_d)_{max} = 1875$ Kg/m³

Optimum moisture content = 15%

To plot zero air voids line we use the following equation

$$\gamma_d = \frac{G\gamma_w}{1+wG} = \frac{(2.67)(1000)}{1+w(2.67)} = \frac{2670}{1+2.67w}$$

Data for plotting zero air voids line

w (%)	15	20	25
γ_d (Kg/m³)	1906.5	1740.5	1601.2

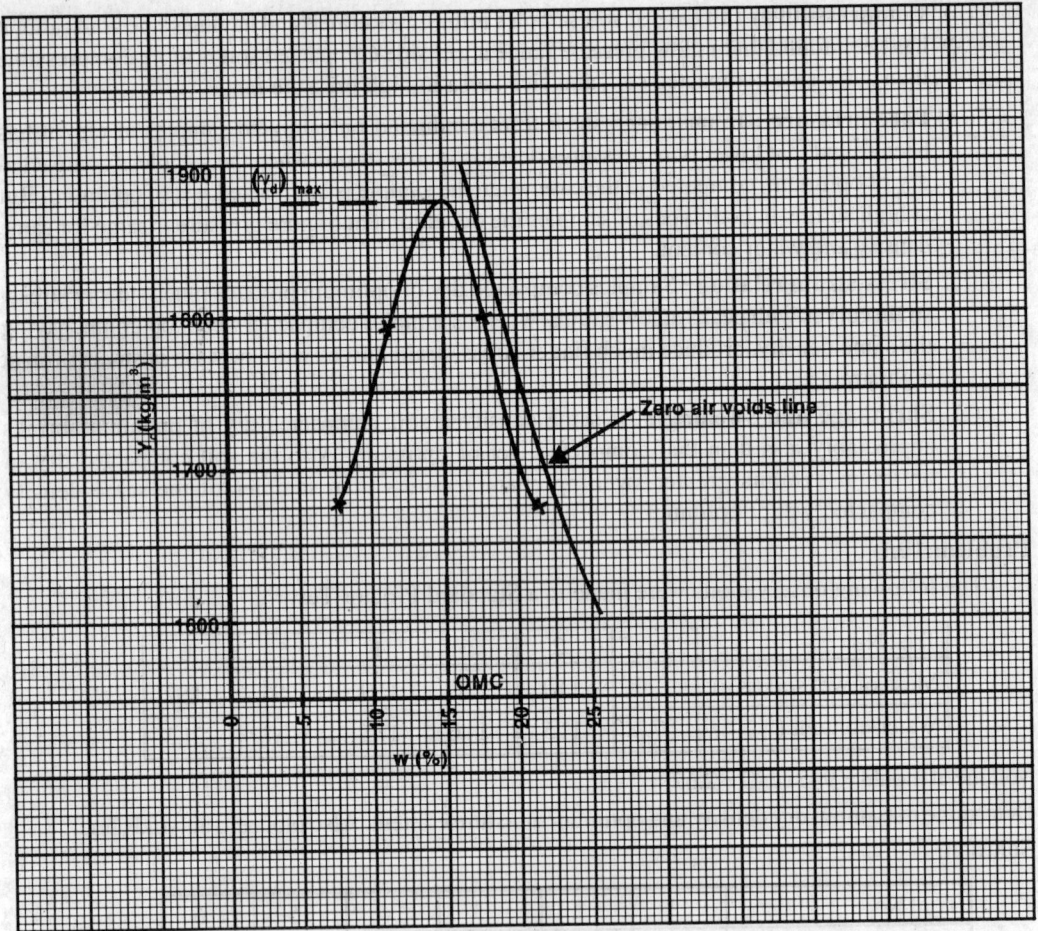

Figure for problem 10.1

EXERCISE–10

10.1   In a Standard Proctor Test the mould of 1 litre capacity weighs 12.5 N when empty. Successive trials gave the following results.

Weight of mould + wet soil (N)	29.6	30.1	31.5	31.2	30.8
Water content (%)	16.7	18.6	21.0	21.7	23.5

Determine maximum dry density and optimum water content. If $G = 2.7$ calculate degree of saturation and percentage air voids at maximum dry density.

10.2   List and explain the factors affecting compaction.

10.3   An earth embankment is to be compacted to a dry unit weight of 1.84 g/cc at a placement water content of 15%. The in-situ dry density and water content in the borrow pit are 1.77 g/cc and 8% respectively. How much excavation should be carried out in the borrow pit for each cu-m of embankment?

10.4 What is compaction of soil? How does it differ from consolidation? Describe briefly the different methods of compaction. State their relative merits and demerits.

10.5 Fill up the Blanks

(i) While consolidation results in _____ reduction in volume, under _____ loading, compaction is rapid _____, under a load of _____ duration.

(ii) Compaction increases _____ of the soil.

(iii) In Proctor's test, the weight of the hammer is _____ kg and its height of fall is _____ cm.

(iv) Fine grained soils have _____ OMC values than coarse grained soils.

(v) At water content much higher than OMC, the soil attains a _____ structure.

10.6 Name the following.

(i) When soil is compacted under the blows of hammer.

(ii) Water Content at which dry density attains maximum value.

(iii) A line showing the dry density-water content relation at fully saturated condition (i.e. theoretically maximum compaction)

(iv) Compaction of a soil, in a mould, under pressure of a rammer, which covers entire area of the mould.

(v) The line forming the peak of the water-density curves.

10.7 Write short notes on

(i) Proctor test (standard and modified)

(ii) Field compaction control

(iii) Effect of compaction on soil properties

CHAPTER 11

CONSOLIDATION

11.1 Introduction

As already discussed in the previous chapter (Art 10.1) consolidation is a process associated with the property of compressibility of soil mass. To recapitulate, compressibility is an engineering property by virtue of which a soil mass is capable of undergoing compression or reduction in volume under pressure. Consolidation is defined as the process in which gradual reduction in volume of soil mass occurs under sustained loading and is primarily due to expulsion of pore water. In the analysis of this process both water and soil particles are assumed to be relatively incompressible so that the decrease in volume is due entirely to the change in relative positions of soil particles with the particles coming closer to each other.

11.2 Primary Consolidation and Secondary Consolidation

When a saturated clay layer is subjected to compressive load, excess pore pressure develops and the water drains out of the clay layer into an adjacent layer of relatively more permeable soil like sand. The drainage of pore water takes place as long as there is excess pore water pressure and is accompanied by compression of soil. This process is known as primary consolidation, primary compression or primary time effect. Once the excess pore pressure becomes zero, that is the excess pore pressure gets fully dissipated, the primary compression under the applied stress ends. It is found in practice that some compression takes place even after the primary compression has ceased. This is referred to as secondary consolidation, secondary compression or secondary time effect and is due to highly viscous water between points of contact of soil particles being forced out and readjustment of particles due to creep effect. In most soil deposits, the secondary time effect is very small compared to primary consolidation and is often neglected. The remaining discussion in this chapter is focussed on primary consolidation with a brief reference to secondary consolidation at the end of this chapter.

11.3 The Spring Analogy

Terzaghi demonstrated the mechanics of consolidation by piston and spring analogy.

A saturated soil mass taken in a container consists of soil particles forming the skeleton of soil mass and voids filled with water. As shown in Fig. 11.1(c) the skeleton formed of soil particles can be assumed to be replaced by a number of springs and the water filling voids in soil mass by the water filling the cylinder. The compressive stress is caused by load applied on piston placed on top of the springs. An outlet with valve is provided to control drainage of water from out of the cylinder.

Let z_0 be the length of springs under a pressure of say 10 units as shown in Fig. 11.1(a). Let the length decrease to z_1 when the pressure is increased by say 2 units as shown in Fig. 11.1(b). In Figs. 11.1(c), (d), (e) and (f) springs with piston is shown placed in a container filled with water. In Fig. 11.1(c) the valve is open but no drainage takes place as the entire pressure of 10 units is borne by the springs and the pressure in water is zero. For soil mass, by analogy

$$\sigma = \sigma' + \bar{u}$$

where σ = total stress

                    σ' = effective stress

$$\bar{u} = \text{excess pore pressure}$$

In Fig. 11.1(d) additional pressure of 2 units acts and the valve is closed. Because water is incompressible the springs are prevented from undergoing any further compression and therefore the additional pressure will have to be borne by water.

By analogy $\sigma = \sigma' + \bar{u}$

$$12 = 10 + 2$$

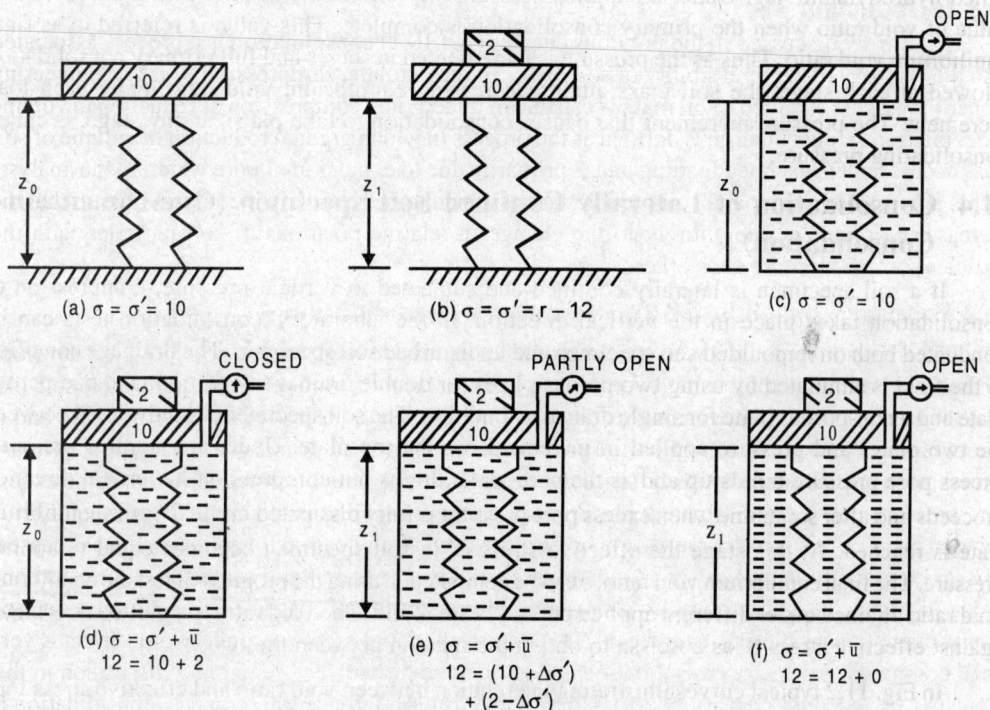

(a) $\sigma = \sigma' = 10$ (b) $\sigma = \sigma' = r - 12$ (c) $\sigma = \sigma' = 10$

(d) $\sigma = \sigma' + \bar{u}$ (e) $\sigma = \sigma' + \bar{u}$ (f) $\sigma = \sigma' + \bar{u}$
    $12 = 10 + 2$ $12 = (10 + \Delta\sigma')$ $12 = 12 + 0$
                              $+ (2 - \Delta\sigma')$

Fig. 11.1. Spring analogy

In Fig. 11.1(e) the valve is partly open and as the water starts flowing out transfer of additional pressure from water to spring commences and at any intermediate stage we have by analogy

$$\sigma = \sigma' + \bar{u}$$
$$12 = (10 + \nabla\sigma') + (2 - \nabla\sigma')$$

where $\nabla\sigma'$ is part of the additional pressure transferred to springs at that stage. In Fig. 11.1(f) the valve is shown fully open and the rate of drainage of water increases and finally the drainage stops when all the additional pressure is transferred from water to the springs. This is similar to the condition when the excess pore pressure has fully dissipated in the case of soil mass.

By analogy $\sigma = \sigma' + \bar{u}$

(b) $\sigma = \sigma' = 12$

The piston and spring analogy helps a beginner in understanding the process of primary consolidation. It is clear from the analogy model that in the case of a saturated soil mass subjected to an initial pressure σ and when no drainage is occurring

$$\sigma = \sigma' + u$$

where u is the pore pressure under static condition (non-flow condition). When the soil mass is subjected to additional compressive load, the pressure increment is first borne by water. The pressure that builds up in pore water due to load increment is referred to as excess pore pressure, excess hydrostatic pressure or hydrodynamic pressure. It is denoted by $\bar{u}$. When the drainage of water is permitted the excess pore pressure dissipates as the water drains out and compression of soil mass takes place with decrease in void ratio. The rate of compression depends on the permeability of soil and the boundary conditions for drainage. In soils of low permeability the drainage is very slow. The delay caused in consolidation by the slow drainage of water out of the saturated soil mass is called hydrodynamic lag. Under an applied pressure the soil mass will have reached a particular value of void ratio when the primary consolidation is complete. This value is referred to as final equilibrium void ratio. Thus as the pressure is incremented in stages and full primary consolidation allowed at each stage, the soil mass attains decreasing equilibrium void ratio under each load increment. The pressure increment that causes consolidation to take place, at any stage, is called consolidating pressure.

11.4 Consolidation of Laterally Confined Soil Specimen (One-Dimensional Consolidation)

If a soil specimen is laterally confined and subjected to vertical pressure, compression or consolidation takes place in the vertical direction. In the laboratory, consolidation tests can be conducted both on remoulded soil specimen and undisturbed soil specimen. The drainage condition in the field is simulated by using two porous plates for double drainage condition, and one porous plate and a non-porous plate for single drainage condition. The soil specimen is sandwiched between the two plates and pressure applied in increments on the top plate. Under any applied pressure, excess pore pressure builds up and as the pore water drains out compression in vertical direction proceeds and after sometime when excess pore pressure is fully dissipated *i.e.* $\bar{u} = 0$, the equilibrium state is reached. At this stage the effective stress σ' in soil specimen becomes equal to applied pressure. The final equilibrium void ratio, e can be computed. During the progress of test the equilibrium void ratio attained under different applied pressures are found. The void ratio e is plotted as ordinate against effective stress σ' as abscissa to obtain the relation between the two.

In Fig. 11.2 typical curves illustrating the relation between void ratio and effective stress for a laterally confined remoulded soil specimen are shown. The curve AB is obtained by increasing the applied pressure in increments allowing equilibrium stage to be reached under each pressure. If at stage corresponding to point B, the applied pressure is completely removed, the soil specimen expands as indicated by curve BC. However, the soil specimen will not attain again the original void ratio corresponding to beginning of test because it will have undergone some permanent compression which can be attributed to irreversible orientation undergone by soil particles. If the specimen is recompressed and the test continued the curves CD and DE are obtained. The curves AB and DE

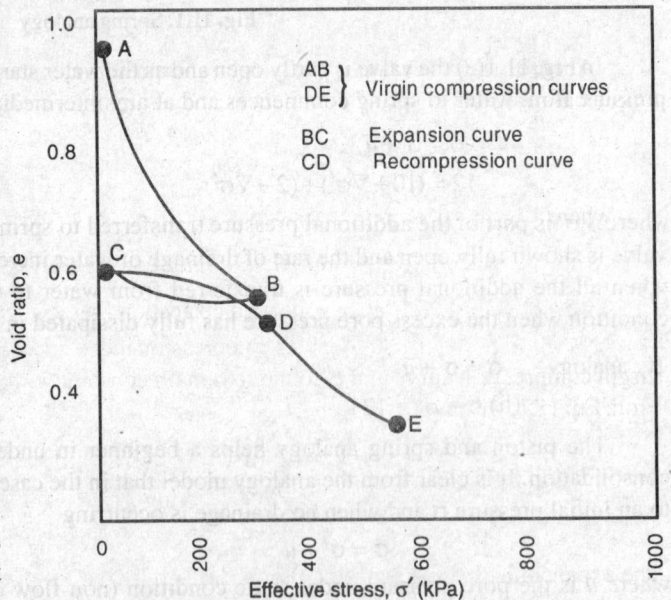

Fig. 11.2. $e - \sigma'$ curves for a remoulded specimen

correspond to consolidation of soil specimen during which at any stage, the applied pressure is greater than any pressure to which the soil specimen has been subjected to in the past. They are referred to as virgin compression curves. The curve BC is called expansion curve and the curve CD, the recompression curve. It is to be observed that point D of recompression curve does not coincide with B even though both correspond to same effective stress. Clearly D lies below B indicating that void ratio attained during recompression is less than that attained during virgin compression under the same applied pressure.

If void ratio e is plotted as ordinate on natural scale against effective stress σ' as abscissa on logarithmic scale, the virgin compression curves and the expansion curve become nearly straight lines as shown in Fig. 11.3.

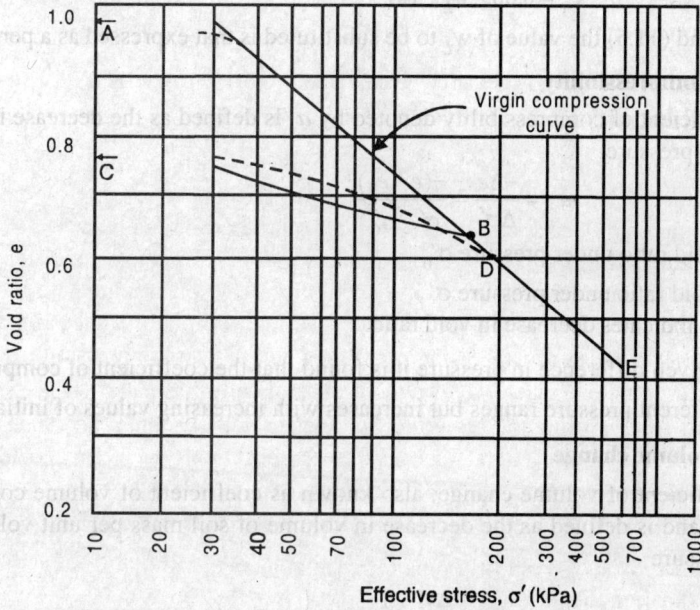

Fig.11.3. $e - \log \sigma'$ curves for a remoulded soil specimen

According to Terzaghi, the virgin compression curve can be defined by the following empirical relation.

$$e = e_0 - C_c \log_{10} \frac{\sigma'}{\sigma_0'} \qquad \qquad ...(11.1)$$

where

e_0 = initial void ratio corresponding to initial effective stress σ_0'

e = void ratio corresponding to increased effective stress σ'

C_c denotes compression index and it is the slope of straight line portion of virgin compression curve and is found to remain constant within a fairly large range of pressure. From Eq. 11.1, we have

$$C_c = \frac{e_0 - e}{\log_{10} \dfrac{\sigma'}{\sigma_0'}} = \frac{\Delta e}{\Delta \log_{10} \sigma'} \qquad \qquad ...(11.2)$$

The expansion curve on semi-log plot is defined by the following relation.

$$e_0 = e + C_s \log_{10} \frac{\sigma'}{\sigma'_0} \qquad\qquad ...(11.3)$$

C_s denotes expansion index or swelling index. It is the slope of straight line portion of expansion curve and is a measure of the increase in volume that occurs on removal of pressure.

Skempton (1944) has given the following equation for estimating C_c for remoulded clay sample

$$C_c = 0.007(w_L - 10) \qquad\qquad ...(11.4)$$

For undistuded clay of medium to low sensitivity the value of C_c is roughly equal to 1.3 times that corresponding to remoulded sample and therefore can be estimated by

$$C_c = 0.009(w_L - 10) \qquad\qquad ...(11.5)$$

In Eqs. (11.4) and (11.5) the value of w_L to be substituted is that expressed as a percentage.

Coefficient of Compressibility

The coefficient of compressibility denoted by a_v is defined as the decrease in void ratio per unit increase in pressure

$$a_v = \frac{-\Delta e}{\Delta \sigma'} = \frac{-(e - e_0)}{\sigma' - \sigma'_0} \qquad\qquad ...(11.6)$$

where e_0 = void ratio under pressure σ'_0

e = void ratio under pressure σ'

The minus sign indicates decrease in void ratio.

For any given difference in pressure it is found that the coefficient of compressibility is not constant for different pressure ranges but increases with increasing values of initial pressure σ'_0.

Coefficient of volume change

The coefficient of volume change, also known as coefficient of volume compressibility, is denoted by m_v and is defined as the decrease in volume of soil mass per unit volume due to unit increase in pressure

$$m_v = \frac{-\Delta V}{V_0} \cdot \frac{1}{\Delta \sigma'} \qquad\qquad ...(11.7)$$

Further we have $$e = \frac{V_v}{V_s}$$

Adding 1 to both sides, $1 + e = \dfrac{V_v}{V_s} + 1 = \dfrac{V_v + V_s}{V_s} = \dfrac{V}{V_s}$

i.e. $$V = (1 + e) V_s$$

$\therefore$ $$V_0 = (1 + e_0) V_s \qquad\qquad ...(11.7a)$$

$$V_1 = (1 + e_1) V_s$$

$$\Delta V = (V_0 - V_1) = (1 + e_0) V_s - (1 + e_1) V_s = (e_0 - e_1) V_s$$

i.e., $$\Delta V = \Delta e\, V_s \qquad\qquad ...(11.7b)$$

By substitution for V_0 and ΔV from Eq. 11.7(a) and 11.7(b) in Eq. 11.7, we get

$$m_v = \frac{-\Delta e}{1 + e_0} \cdot \frac{1}{\Delta \sigma'} = \frac{1}{1 + e_0}\left(\frac{-\Delta e}{\Delta \sigma'}\right)$$

$\therefore$ $$m_v = \frac{a_v}{1 + e_0} \qquad\qquad ...(11.8)$$

When the soil mass is laterally confined, the decrease in volume ΔV is proportional to decrease in thickness ΔH and the initial volume V_0 is proportional to initial thickness H_0. Therefore we can write

$$m_v = \frac{\Delta H}{H_0} \cdot \frac{1}{\Delta \sigma'} \qquad \qquad ...(11.9)$$

The compression ΔH due to pressure increment $\Delta \sigma'$ is given by

$$\Delta H = m_v\, H_0 . \Delta \sigma' \qquad \qquad ...(11.10)$$

Computation of consolidation settlement

When a soil stratum of initial thickness H_0 has fully consolidated under a pressure increment $\Delta \sigma'$, the final consolidation settlement ρ_f can be computed as

$$\rho_f = m_v \cdot H_0 \cdot \Delta \sigma' \qquad \qquad ...(11.11)$$

assuming that the pressure increment $\Delta \sigma'$ is transmitted uniformly over the thickness H_0. However, in practice, one should note that the vertical stress due to finite surface loading decreases non-linearly with depth. Hence we should apply Eq. 11.11 for computing the settlement of a thin layer dz and then integrate over total thickness H.

$$\Delta \rho_f = m_v \Delta \sigma' dz$$

$$\therefore \qquad \rho_f = \int_0^H m_v . \Delta \sigma' . dz \qquad \qquad ...(11.12)$$

In Eq. (11.12) both m_v and $\Delta \sigma'$ are variables. The integration may be performed numerically by dividing the total thickness H into a convenient number of layers and computing the settlement $\Delta \rho_f$ for each layer taking $\Delta \sigma'$ at the middle of a layer as representing the average pressure increment for that layer.

Another equation to compute the final settlement ρ_f is derived below.

$$\frac{\Delta H}{H} = \frac{\Delta V}{V} = \frac{e_0 - e}{1 + e_0}$$

$$\rho_f = \Delta H = \frac{e_0 - e}{1 + e_0} . H \qquad \qquad ...(11.13)$$

We have

$$C_c = \frac{e_0 - e}{\log_{10} \dfrac{\sigma'}{\sigma_0'}}$$

$$\therefore \qquad e_0 - e = C_c \log_{10} \frac{\sigma'}{\sigma_0'} \quad \text{where } \sigma' = \sigma_0' + \Delta \sigma'$$

Substituting in Eq. 11.13, we get

$$\rho_f = \frac{H}{1 + e_0} C_c \log_{10} \frac{\sigma_0' + \Delta \sigma'}{\sigma_0'} \qquad \qquad ...(11.14)$$

where H = initial thickness of consolidating layer

e_0 = initial void ratio of the layer

C_c = compression index

σ_0' = initial effective pressure at the middle of layer

$\Delta \sigma'$ = effective pressure increment at the middle of layer

Depending on state of consolidation soil deposits are divided into three types :

(*i*)   Preconsolidated deposit

(*ii*) Normally consolidated deposit and

(*iii*) Underconsolidated deposit

A soil deposit is said to be preconsolidated, precompressed or overconsolidated if it has in the past been fully consolidated under a pressure greater than the present overburden pressure acting on the soil. The preconsolidation may have been caused by a geologic overburden in the past or structural load which has been subsequently removed.

A soil deposit is said to be normally consolidated if it has never been subjected to a pressure greater than the present overburden pressure and has been fully consolidated under the presently acting pressure.

An underconsolidated soil deposit is one which is still not fully consolidated under the existing overburden pressure.

Determination of preconsolidation pressure.

In Fig. 11.4 is shown the relation between void ratio and effective stress on a semilog plot, typical for a laterally confined undisturbed preconsolidated soil specimen. The initial portion of the curve resembles the recompression curve of a remoulded soil specimen. The lower portion is the virgin compression curve.

The preconsolidation pressure is the greatest effective stress to which the soil has been subjected to in the past and underwhich it has undergone full consolidation. A Casagrande (1936) has given the following method for determining the approximate value of preconsolidation pressure.

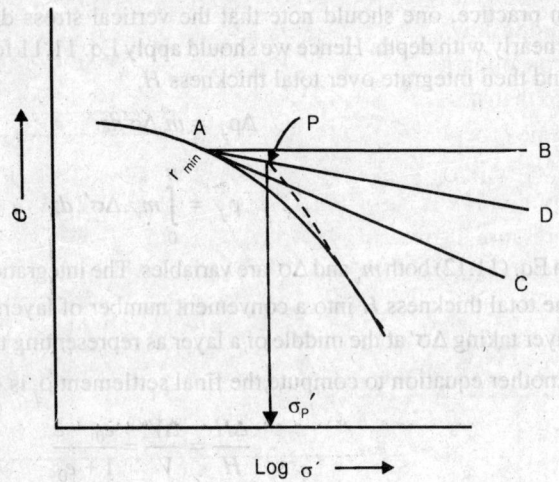

Fig. 11.4. Determination of preconsolidation pressure (After A. Casagrande)

Referring to Fig. 11.4 the point A of maximum curvature (minimum radius) is selected and horizontal line AB is drawn. Line AC is drawn tangential to the curve at A and angle BAC is bisected by AD. The straight portion of the virgin compression curve is extended back to intersect the bisector AD at P. The effective stress corresponding to point P is the preconsolidation pressure σ_p' .

11.5 Terzaghi's Theory of One Dimensional Consolidation

Terzaghi (1923) derived the basic differential equation of consolidation which represents the first step in the theoretical analysis of the consolidation process.

Following are the assumptions made in Terzaghi's one-dimensional consolidation theory.

1.   The soil mass is homogeneous and fully saturated.

2.   The soil particles and water are incompressible.

3.   Darcy's law for flow of water through soil mass is applicable during consolidation.

4.   Coefficient of permeability is constant during consolidation.

5. Load is applied in one direction only and deformation occurs only in the direction of load application.

6. The deformation is due entirely to decrease in volume.

7. The drainage of pore water occurs only in one direction.

8. A boundary drainage face offers no resistance to flow of water from soil.

9. During consolidation the change in thickness is continuous but final value of compression is related to initial thickness only.

10. The time lag in consolidation is due entirely to permeability of soil. Any secondary time effect is disregarded.

Let a saturated clay layer of thickness H lie between two layers of sand which serve as two drainage faces. When the clay layer is subjected to a pressure increment $\Delta\sigma$, the pressure increment is first borne by pore water so that at initial time t_0 the excess pore pressure $\bar{u} = \Delta\sigma$ at all points along the depth of clay layer and is plotted as line AB in Fig. 11.5 (b). Drainage of pore water into the sand layers starts and the excess pore pressure at the top and bottom boundaries of clay layer drops down to zero and remains so at all times, during the consolidation process. At the end of consolidation process, say, at $t = t_f$ the excess pore pressure will have been completely dissipated so that $\bar{u} = 0$ at all points and is represented by the line CD in Fig. 11.5 (b). At any intermediate time t, between t_0 and t_f, part of consolidating pressure $\Delta\sigma$ is transferred to soil particles so that $\Delta\sigma = \Delta\sigma' + \bar{u}$. The distribution of excess pore pressure $\bar{u}$ at any intermediate time t is represented by a curve such as CFD in Fig. 11.5 (b). A number of such curves representing excess pore pressure distribution along the depth of clay layer at different instants of time $t = t_1, t_2, ...$ can be drawn and they are known as isochrones. The slope of an isochrone at any point at a given time gives the rate of change of $\bar{u}$ with depth.

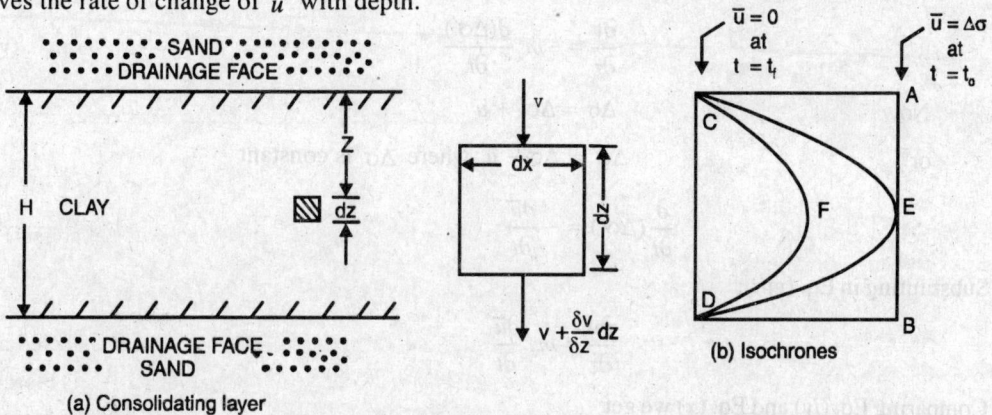

(a) Consolidating layer

(b) Isochrones

Fig. 11.5. One-dimensional consolidation

At any time t, the hydraulic head h corresponding to the excess pore pressure $\bar{u}$ is given by

$$h = \frac{\bar{u}}{\gamma_w} \qquad ...(i)$$

The hydraulic gradient i is given by

$$i = \frac{\partial h}{\partial z} = \frac{1}{\gamma_w} \cdot \frac{\partial \bar{u}}{\partial z} \qquad ...(ii)$$

Applying Darcy's law, the velocity of flow of pore water due to this hydraulic gradient is given by

$$v = ki = \frac{k}{\gamma_w} \cdot \frac{\partial \bar{u}}{\partial z} \qquad ...(iii)$$

The rate of change of velocity along the depth of the layer is given by

$$\frac{\partial v}{\partial z} = \frac{k}{\gamma_w} \cdot \frac{\partial^2 \bar{u}}{\partial z^2} \qquad \qquad ...(iv)$$

Let us consider a soil element of size dx, dz and of width dy perpendicular to the plane of figure. If v is the velocity of water at entry, the velocity at exit will be $\left(v + \frac{\partial v}{\partial z} \cdot dz \right)$ as indicated in Fig. 11.5. The quantity of water entering the soil element in unit time $= v.dx\, dy$. The quantity of water leaving the soil element in unit time $= \left(v + \frac{\partial v}{\partial z} \cdot dz \right) dx\, dy$. Hence the net quantity of water squeezed out of the soil element in unit time is given by

$$\Delta q = \left(v + \frac{\partial v}{\partial z} dz \right) dx\, dy - v\, dx\, dy$$

i.e.

$$\Delta q = \frac{\partial v}{\partial z} dx\, dy\, dz \qquad \qquad ...(v)$$

The decrease in the volume of soil element is equal to the volume of water squeezed out.

Also, we have

$$\Delta V = - m_v V_0 \Delta\sigma' \qquad \qquad ...(vi)$$

where V_0 = volume of soil element at time t_0 = $dx\, dy\, dz$

∴ Change in volume per unit time is given by

$$\frac{\partial}{\partial t}(\Delta V) = - m_v (dx\, dy\, dz) \frac{\partial(\Delta\sigma')}{\partial t} \qquad \qquad ...(vii)$$

Comparing Eq. (*v*) and (*vii*), we get

$$\frac{\partial v}{\partial z} = - m_v \frac{\partial(\Delta\sigma')}{\partial t} \qquad \qquad ...(viii)$$

Now

$$\Delta\sigma = \Delta\sigma' + \bar{u}$$

or

$$\Delta\sigma' = \Delta\sigma - \bar{u} \text{ where } \Delta\sigma \text{ is constant}$$

∴

$$\frac{\partial}{\partial t}(\Delta\sigma') = - \frac{\partial \bar{u}}{\partial t} \qquad \qquad ...(ix)$$

Substituting in Eq. (*viii*),

$$\frac{\partial v}{\partial z} = m_v \cdot \frac{\partial \bar{u}}{\partial t} \qquad \qquad ...(x)$$

Comparing Eq. (*iv*) and Eq. (*x*) we get

$$\frac{\partial \bar{u}}{\partial t} = \frac{k}{m_v \cdot \gamma_w} \cdot \frac{\partial^2 \bar{u}}{\partial z^2} \qquad \qquad ...(xi)$$

or

$$\frac{\partial \bar{u}}{\partial t} = C_v \frac{\partial^2 \bar{u}}{\partial z^2} \qquad \qquad ...(11.15)$$

where

$$C_v = \frac{k}{m_v \gamma_w} \qquad \qquad ...(11.16)$$

C_v denotes coefficient of consolidation.

Eq. 11.15 is the basic differential equation of consolidation which relates the rate of dissipation of excess pore pressure with the rate of expulsion of pore water from a unit volume of soil.

The coefficient of consolidation C_v as defined in Eq 11.16 indicates the combined effects of permeability and compressibility of soil on the rate of volume change. If k is expressed in m/sec, m_v in m²/kN and γ_w in kN/m³, unit of C_v will be m²/sec.

The mathematical steps involved in obtaining the solution, by means of Fourier series, of the differential equation of consolidation is presented in Appendix II. However, the following points can be understood even without going through the detailed solution.

The hydraulic boundary conditions to be satisfied by the solution of the differential equation of consolidation are :

(i)   at $t = 0$, at any distance z, $\bar{u} = \bar{u}_0 = \Delta\sigma$

(ii)   at $t = \infty$, at any distance z, $\bar{u} = 0$ and

(iii)   at any intermediate time t, at $z = 0$, $\bar{u} = 0$ and at $z = H$, $\bar{u} = 0$

If ρ_f denotes final settlement under pressure increment $\Delta\sigma$ and ρ the settlement at any intermediate time t, then the degree of consolidation attained at that time t is given by

$$U(\%) = \frac{\rho}{\rho_f} \times 100$$

...(11.17)

The degree of consolidation is a function of time factor T_v.

$$U(\%) = f(T_v)$$

...(11.18)

The time factor T_v is a dimensionless parameter defined by the following equation.

$$T_v = \frac{C_v t}{d^2}$$

...(11.19)

where $d =$ drainage path. The drainage path represents the maximum distance a water particle has to travel within the layer to reach a drainage face. When a clay layer is bound by two drainage faces, double drainage occurs. When the clay layer is bound by a drainage face at one end and an impervious boundary at other end, single drainage occurs.

For the case of double drainage, $d = \dfrac{H}{2}$

For the case of single drainage, $d = H$

where $H =$ thickness of layer.

The time factor, $\qquad T_v = \dfrac{C_v t}{d^2} = \dfrac{k}{m_v \gamma_w} \cdot \dfrac{t}{d^2}$

We notice that the time factor, and hence the degree of consolidation, depends upon (i) coefficient of permeability, k (ii) coefficient of volume compressibility, m_v (iii) thickness of layer and (iv) number of drainage faces. In addition it is found to depend upon the consolidating pressure and its manner of distribution across the depth of layer. The time factor contains the physical constants influencing the time-rate of consolidation. The values of time factor T_v corresponding to various values of degree of consolidation U for the two types of drainage conditions and different distributions of consolidating pressure are presented in Tables 11.1 and 11.2. However, the following approximate expressions may be used to compute T_v, in the absence of the tables.

When $\qquad\qquad U < 60\%, T_v = \dfrac{\pi}{4}\left(\dfrac{U}{100}\right)^2$ ...(11.20)

and when $\qquad U > 60\%, T_v = -0.9332 \log_{10}\left(1 - \dfrac{U}{100}\right) - 0.0851$ ...(11.21)

Table 11.1 Values of Time Factor (Taylor, 1948) (For single drainage or double drainage with uniform distribution of initial excess hydrostatic pressure)

$U(\%)$	T_v
10	0.008
20	0.031
30	0.071
40	0.126
50	0.197
60	0.287
70	0.403
80	0.567
90	0.848
100	∞

Table 11.2 Values of Time Factor (For single drainage with triangular distribution of initial excess hydrostatic pressure)

$U(\%)$	T_v	
	(a)	(b)
10	0.047	0.003
20	0.102	0.009
30	0.158	0.024
40	0.221	0.049
50	0.294	0.092
60	0.383	0.160
70	0.500	0.272
80	0.685	0.440
90	0.940	0.720
100	∞	∞

Note : In both Tables 11.1 and 11.2, the value of U is assumed to be the average for the entire stratum.

Case (a) $\bar{u}_0 = 0$ at pervious boundary

$\bar{u}_0 = \Delta\sigma'$ at impervious boundary

Case (b) $\bar{u}_0 = 0$ at impervious boundary

$\bar{u}_0 = \Delta\sigma'$ at pervious boundary

11.6. Consolidation Test

The apparatus used in the laboratory consolidation test is known as consolidometer (or oedometer). It consists essentially of a loading frame and a consolidation cell. The soil specimen is kept in the consolidation cell. To simulate double drainage condition two porous plates, one on top and the other at bottom of specimen are used. In the case of single drainage condition, only one porous plate is used, the other being replaced by a non-porous plate.

There are two types of consolidation cells as shown in Figs. 11.6 and 11.7.

In the fixed ring cell the bottom porous plate is fixed relative to the top plate and only the top plate is free to move downwards and compress the specimen. In the floating ring cell both top and bottom porous plates are relatively free to compress the specimen towards the middle.

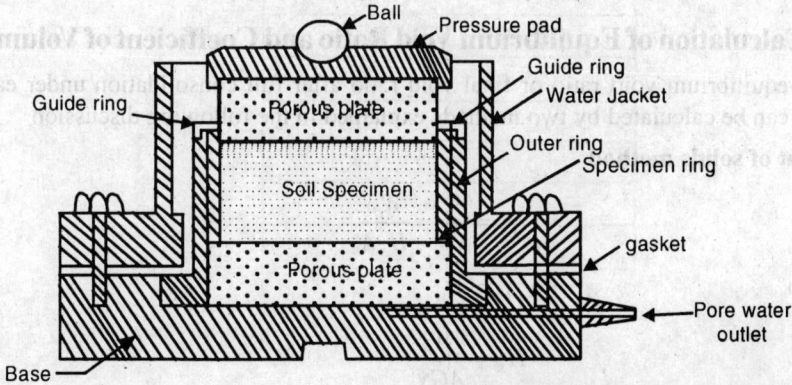

Fig. 11.6. Fixed ring consolidation cell

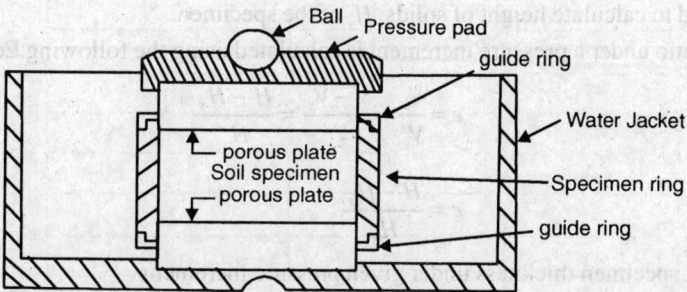

Fig. 11.7. Floating ring consolidation cell

The floating ring cell has the advantage of having smaller effects of friction between the specimen ring and the soil specimen, whereas direct measurement of permeability of the specimen at any stage of loading can be made only in the fixed ring cell.

The loading frame is equipped to apply vertical pressure on the soil specimen in convenient increments. During the test the specimen is allowed to consolidate fully under different vertical pressures such as 10, 20, 50, 100, 200, 400, 800, 1600 kN/m². Each pressure increment is maintained constant until the compression ceases, generally for 24 hours. The vertical compression of specimen is measured with the help of a dial gauge and dial gauge readings are taken after application of each pressure increment, at the end of elapsed time intervals of 0.25, 1.00, 2.25, 4.00, 6.25, 9.00, 12.25, 16.00, 20.25, 25, 36, 49, 60 minutes and 2, 4, 8 and 24 hours. Attention is paid to record the final compression under each pressure increment. After completion of consolidation under the desired maximum vertical pressure, the specimen is unloaded and allowed to swell. The final dial reading corresponding to the completion of swelling is recorded. The specimen is taken out and dried to determine its water content and the weight of soil solids. The consolidation test data are used to determine the following :

1. Void ratio and Coefficient of volume change.
2. Coefficient of consolidation, and
3. Coefficient of permeability.

11.6.1 Calculation of Equilibrium Void Ratio and Coefficient of Volume Change

The equilibrium void ratio or final void ratio after full consolidation under each pressure increment can be calculated by two methods explained in the following discussion

1.   Height of solids method

We have,
$$\gamma_s = \frac{W_s}{V_s} = \frac{W_d}{H_s A}$$

Also
$$\gamma_s = G\gamma_w$$

$\therefore$
$$H_s = \frac{W_d}{AG\gamma_w} \qquad\qquad ...(11.22)$$

where H_s = height of solids

$\quad\quad W_d$ = weight of dried specimen

$\quad\quad A$ = cross sectional area of specimen

Eq. 11.22 is used to calculate height of solids, H_s of the specimen.

The final void ratio under a pressure increment is calculated using the following Eq. 11.23

$$e = \frac{V_v}{V_s} = \frac{V - V_s}{V_s} = \frac{H - H_s}{H_s}$$

$$e = \frac{H - H_s}{H_s} \qquad\qquad ...(11.23)$$

where H = final specimen thickness under given pressure increment

$\quad\quad = H_1 + \Delta H$

$\quad\quad H_1$ = specimen thickness at the beginning of application of the pressure increment

$\quad\quad \Delta H$ = compression or decrease in thickness under the pressure increment

Application of this method to a problem is illustrated in Table 11.3.

Table 11.3 Illustration of Height of Solids Method

Initial thickness of specimen, $H_0 = 25$ mm

Area of cross section, $A = 50$ cm^2

Mass of oven-dried specimen, $M_d = 179.7$ gm

Specific gravity of soil solids, $G = 2.7$

Height of solids, $H_s = \dfrac{M_d}{AG\rho_w} = \dfrac{179.7}{(50)(2.7)(1)} = 13.31$ mm

Applied Pressure (kN/m²)	Final dial gauge reading (10^{-2} mm)	Change in thickness ΔH (mm)	Specimen height $H = H_1 + \Delta H$ (mm)	Void ratio $e = \dfrac{H - H_s}{H_s}$
0	100		25.00	0.878
		−0.71		
25	171		24.29	0.825
		−1.19		
50	290		23.10	0.735
		−0.89		
100	379		22.21	0.669
		−1.06		
200	485		21.15	0.589
		−1.17		
400	602		19.98	0.501
		−0.88		
800	690		19.10	0.435
		+1.06		
0	584		20.16	0.515

2. Change in void ratio method

Assuming the specimen is fully saturated, the final void ratio e_f at the end of the test is computed from the relation.

$$e_f = \omega_f . G$$

where, ω_f = final water content at the end of the test

The change in void ratio Δe under each pressure increment is obtained from the relation :

$$\frac{\Delta H}{H} = \frac{\Delta V}{V} = \frac{\Delta e}{1 + e}$$

$$\therefore \qquad \Delta e = \frac{1+e_f}{H_f} \Delta H \qquad\qquad (11.24)$$

where H_f = final thickness of specimen, at the end of test.

After calculating Δe using Eq. 11.24 and working backwards from the known value of e_f, the equilibrium void ratio corresponding to each pressure increment can be evaluated.

The application of the method to a problem is illustrated in Table 11.4.

Table 11.4 Illustration of Change in Void Ratio Method

Initial thickness of specimen, $H_0 = 25$ mm

Area of cross section, $A = 50$ cm²

Final thickness of specimen, $H_f = 18.63$ mm

Final water content, $\omega_f = 31.6\%$

Specific gravity of soil solids, $G = 2.7$

$$e_f = \omega_f G = 0.316 \times 2.7 = 0.853$$

$$\Delta e = \frac{1+e_f}{H_f} . \Delta H = \frac{1+0.853}{18.63} \Delta H = 0.099 \Delta H$$

Applied Pressure (kN/m²)	Final dial gauge reading (10⁻²mm)	Change in thickness ΔH(mm)	Specimen height H = H₁+ΔH (mm)	Change in void ratio Δe = 0.099ΔH	Void ratio e
0	50		25.00		1.482
		−0.85		−0.084	
25	135		24.15		1.398
		−0.94		−0.093	
50	229		23.21		1.305
		−1.74		−0.172	
100	403		21.47		1.133
		−1.46		−0.144	
200	549		20.01		0.989
		−1.53		−0.151	
400	702		18.48		0.838
		−1.21		−0.120	
800	823		17.27		0.718
		+1.36		+0.135	
0	687		18.63		0.853

Coefficient of volume change

The coefficient of volume change m_v can be calculated by the following two methods :

1. **Change in void ratio method :** In this method the coefficient of volume change is computed using the following equation.

$$m_v = -\frac{\Delta e}{1+e_0}.\frac{1}{\Delta\sigma'}$$

2. **Change in thickness method:** In this method the coefficient of volume change is calculated using the following equation.

$$m_v = -\frac{\Delta H}{H_0}.\frac{1}{\Delta\sigma'}$$

11.6.2 Determination of Coefficient of Consolidation

The coefficient of consolidation C_v is determined by methods based on the comparison between the characteristics of the theoretical relation between time factor T_v and degree of consolidation U to the relation between elapsed time t and degree of consolidation U obtained for soil specimen in laboratory test. Two commonly used methods are (i) square root of time fitting method and (ii) logarithm of time fitting method.

Square root of time fitting method (Taylor method)

Based on the expression for U as a function of T_v, the theoretical curve between $\sqrt{T_v}$ and U is obtained as shown in Fig. 11.8. The curve is straight upto U = 60% and the abscissa corresponding to U = 90% point on curve is 1.15 times the abscissa of point of intersection of straight line portion of curve produced with the horizontal line at U = 90%. Taylor (1948) made use of this characteristic of the theoretical curve to determine the R_{90} point on the laboratory consolidation curve obtained by plotting dial reading (R) as ordinate against square root of time ($\sqrt{t}$) as abscissa for any pressure increment. Here R_{90} corresponds to U = 90%.

The initial dial reading at $t = 0$ is denoted by R_0 and it corresponds to U = 0. In the Fig. 11.9, line A is drawn coinciding with the straight portion of laboratory curve. Let it intersect the R-axis at R_c. Then R_c represents the corrected zero reading. The consolidation between R_0 and R_c is called initial consolidation. From point R_c another line B is drawn such that every point on it has abscissa 1.15 times that of corresponding point on A-line. Let the line B intersect the laboratory curve at point P.

Then point P corresponds to $U = 90\%$ and its coordinates redesignated as $\sqrt{t_{90}}$ and R_{90}. The coefficient of consolidation is calculated from the following equation.

$$(T_v)_{90} = \frac{C_v \, t_{90}}{d^2}$$

which gives

$$C_v = \frac{(T_v)_{90} \cdot d^2}{t_{90}}$$

where $(T_v)_{90}$ = time factor corresponding to $U = 90\%$ which can be obtained from Table 11.1

d = average drainage path for the pressure increment under consideration

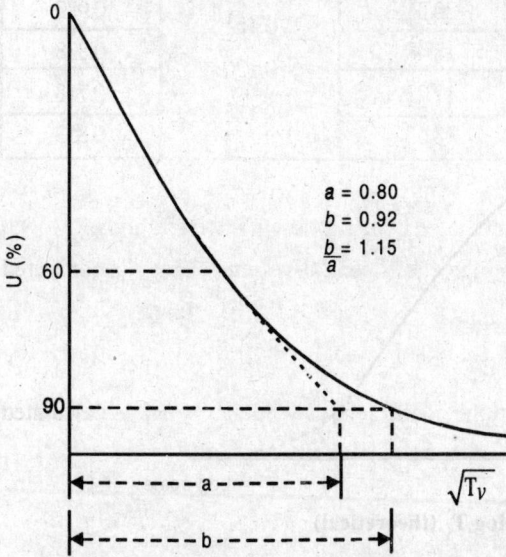

Fig. 11.8. U versus $\sqrt{T_v}$ (theoretical)

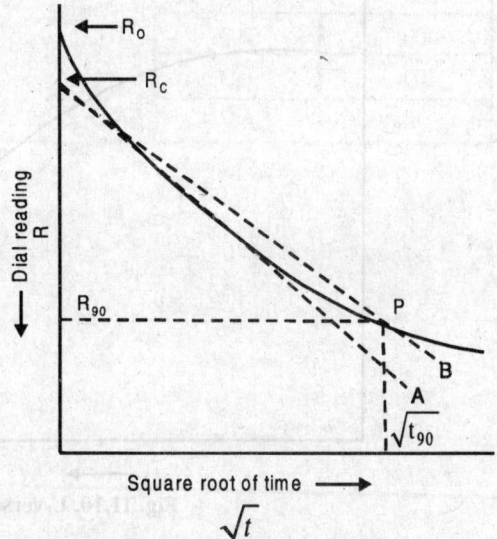

Fig. 11.9. R versus $\sqrt{t}$ (from laboratory data)

Note : $d = \left[\dfrac{H_1 + H_2}{2}\right]$ for single drainage

$d = \dfrac{1}{2}\left[\dfrac{H_1 + H_2}{2}\right]$ for double drainage

H_1 and H_2 represent thickness of specimen at the beginning and end of consolidation under the pressure increment.

Logarithm of time fitting method (Casagrande method)

Based on the theoretical relation between degree of consolidation U and time factor T_v, the curve shown in Fig. 11.10 is obtained by plotting U as ordinate against $\log T_v$ as abscissa. A notable characteristic of this curve is that the intersection of tangent drawn at the point of inflexion and the asymptote of the lower portion of curve intersect at a point which corresponds to U = 100%. A Casagrande (1930) suggested that this characteristic can be used to determine the point whose coordinates are U = 100% and R_{100} on the laboratory consolidation curve obtained by plotting dial reading R against logarithm of time ($\log t$) as shown in Fig. 11.11.

The corrected zero reading R_c is obtained on the assumption that the initial portion of the curve can be approximated by a parabola. Two points A and B are marked on the initial portion of the curve corresponding to say, $t_1 = 1$ min and $t_2 = \dfrac{t_1}{4} = \dfrac{1}{4}$ min. respectively. If z is the vertical distance between

A and B, a horizontal line is drawn at a distance z above B to get R_c which corresponds to U = 0. The two straight portions of the curve are produced to intersect at point P which corresponds to U = 100%. The coordinates of point P are R_{100} and log t_{100}. The consolidation from R_c to R_{100} is the primary consolidation. It may be noted that the straigt portion of curve from R_{100} to R_f represents a part of secondary consolidation. After locating R_c and R_{100} the point corresponding to U = 50% is marked on the curve using the value of R_{50} computed as shown below.

$$R_{50} = R_c + \frac{(R_{100} - R_c)}{2}$$

The value of t_{50} is read out from the graph and coefficient of consolidation C_v is computed.

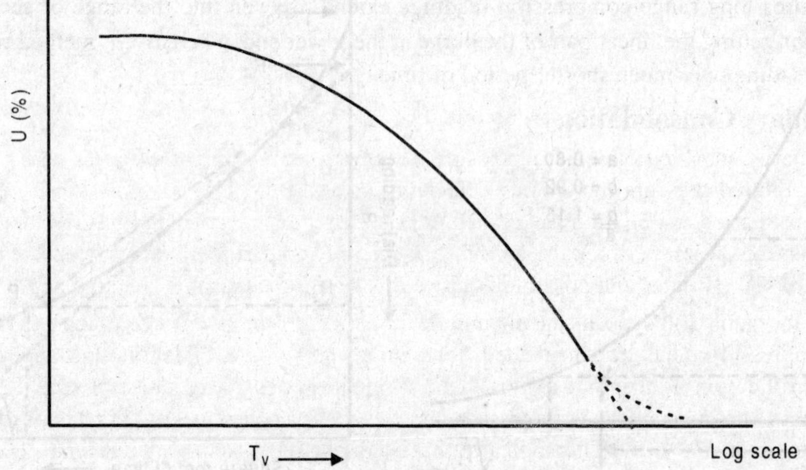

Fig. 11.10. U versus log T$_v$ (theoretical)

$$C_v = \frac{(T_v)_{50}d^2}{t_{50}}$$

where drainage path d is computed as noted in the preceding method.

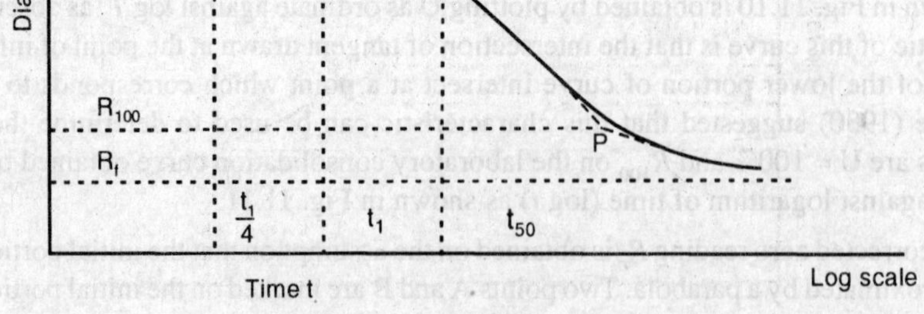

Fig. 11.11. R versus log t (from laboratory data)

Determination of coefficient of permeability

We have
$$C_v = \frac{k}{m_v \gamma_w}$$

$\therefore$
$$k = C_v . m_v . \gamma_w$$

After finding m_v and C_v, k can be computed using the above equation.

Also k can be determined by performing a falling head test on the consolidation test specimen, taken in a fixed ring consolidometer to which a stand pipe is attached. The test is performed after the specimen is fully consolidated under a particular pressure increment.

It should be noted that a straight line portion is not always obtained in the Taylor method ($\sqrt{t}$ method). In such cases, Casagrande method (log t method) will have to be used. However, the log t method requires long range compression readings extending even into the range of secondary compression for getting the linear part of the curve at the lower end, whereas $\sqrt{t}$ method requires compression reading over much shorter period of time.

11.7 Secondary Consolidation

The primary consolidation under a pressure increment ceases when the excess pore pressure caused by the applied pressure increment is fully dissipated. But some compression is observed even after the primary consolidation has ceased. It is referred to as secondary consolidation and is due to highly viscous water between the points of contact of soil particles being forced out, change in orientation of soil particles and possible fracture of some of the particles due to creep.

In many inorganic soil deposits the magnitude of secondary compression is much less than that of primary compression and is often neglected. Terzaghi's theory of consolidation is not applicable to secondary consolidation as it is not governed by dissipation of excess pore pressure. It can be observed that any experimental time-compression curve will be in agreement with Terzaghi's theoretical curve only upto about U = 60%. This indicates that secondary consolidation comes into play even before the primary consolidation ends and continues thereafter. The secondary consolidation is represented by a series of straight lines with different slopes on e versus log t plot. It is of much significance in the case of highly organic soils, micaceous soils and some loosely deposited clays.

Referring to Fig. 11.12, the straight line representing secondary compression on e-log t plot may have equation of the following form:
$$\Delta e = -C_\alpha \log_{10} \frac{t_2}{t_1}$$

in which
$$\Delta e = e_1 - e_2$$

C_α = coefficient of secondary compression

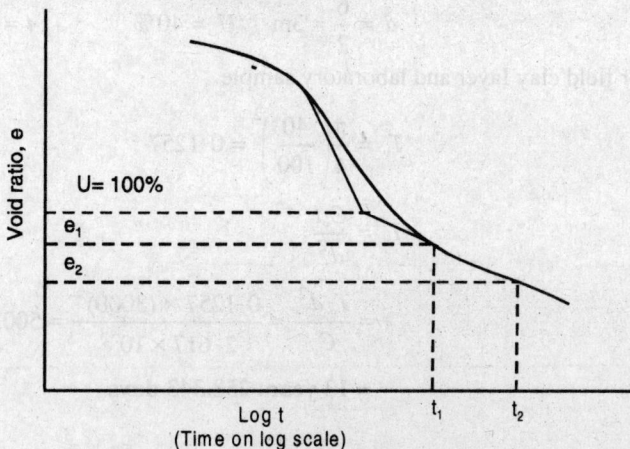

Fig. 11.12. Secondary consolidation

Example 11.1. In a consolidation test void ratio decreased from 0.70 to 0.65 when the load was changed from $50 \, kN/m^2$ to $100 \, kN/m^2$. Compute compression index and coefficient of volume change

Solution : $\sigma_1 = 50 \, kN/m^2$ $\sigma_2 = 100 \, kN/m^2$

$e_1 = 0.70$ $e_2 = 0.65$

$$C_c = \frac{\Delta e}{\Delta \log \sigma} = \frac{e_1 - e_2}{\log \dfrac{\sigma_2}{\sigma_1}} = \frac{0.70 - 0.65}{\log \dfrac{100}{50}} = 0.166$$

$$m_v = \frac{\Delta V}{V_0} \cdot \frac{1}{\Delta \sigma} = \frac{\Delta e}{1 + e_0} \cdot \frac{1}{\Delta \sigma}$$

$$= \frac{0.70 - 0.65}{1 + 0.70} \times \frac{1}{(100 - 50)}$$

$$= 5.88 \times 10^{-4} \, m^2/kN$$

Example 11.2. A soil sample 20 mm thick takes 20 minutes to reach 20 percent consolidation. Find the time taken for a clay layer 6m thick to reach 40 % consolidation. Assume double drainage in both cases.

Solution:

For laboratory sample :

drainage path, $d = \dfrac{20}{2} = 10$ mm

Degree of consolidation, $U = 20 \%$ at time, $t = 20$ min.

Time factor, $T_v = \dfrac{\pi}{4} \left(\dfrac{U}{100} \right)^2 = \dfrac{\pi}{4} \left(\dfrac{20}{100} \right)^2 = 0.0314$

Also, $T_v = C_v \dfrac{t}{d^2}$

$\therefore$ $C_v = T_v \cdot \dfrac{d^2}{t} = \dfrac{0.0314 \times 10^2}{20 \times 60} = 2.617 \times 10^{-3} \, mm^2/sec$

For field clay layer :

$$d = \frac{6}{2} = 3m \qquad U = 40\% \qquad t = ?$$

C_v is same for field clay layer and laboratory sample.

$$T_v = \frac{\pi}{4} \left(\frac{40}{100} \right)^2 = 0.1257$$

$$T_v = \frac{C_v t}{d^2}$$

$\therefore$ $t = \dfrac{T_v d^2}{C_v} = \dfrac{0.1257 \times (3000)^2}{2.617 \times 10^{-3}} = 5003.343$ days

$$= \textbf{13 years 258.343 days}$$

Example 11.3. In the laboratory a 2 cm thick soil sample takes 25 minutes to reach 30 % degree of consolidation. Find the time taken for a 5 m thick clay layer in field to reach 40 % consolidation. Assume double drainage in both cases.

Solution :

For
$$U = 30\,\%, T_v = \frac{\pi}{4}\left(\frac{U}{100}\right)^2 = \frac{\pi}{4}\left(\frac{30}{100}\right)^2 = 0\cdot0707$$

For
$$U = 40\,\%, T_v = \frac{\pi}{4}\left(\frac{U}{100}\right)^2 = \frac{\pi}{4}\left(\frac{40}{100}\right)^2 = 0\cdot1257$$

We have
$$T_v = C_v \frac{t}{d^2}$$

For laboratory sample,
$$d = \frac{20}{2} = 10 \text{ mm} \quad t_{30} = 25 \text{ min} \quad \text{for } U = 30\,\%.$$

For $U = 40\,\%, t_{40} = ?$

$$\frac{C_v}{d^2} = \frac{T_v}{t} = \frac{T_{v30}}{t_{30}} = \frac{T_{v40}}{t_{40}}$$

$$\therefore \quad t_{40} = \frac{T_{v40}}{T_{v30}}\cdot t_{30} = \frac{0\cdot1257}{0\cdot0707}(25) = 44\cdot45 \text{ min}$$

For field layer,
$$d_2 = \frac{5000}{2} = 2500 \text{ mm}$$

For 40 % consolidation, $t_2 = ?$

For laboratory sample, $d_1 = 10$ mm and $t_1 = 44.45$ min for 40 % consolidation

$$\frac{T_v}{C_v} = \frac{t_1}{d_1^2} = \frac{t_2}{d_2^2}$$

$$\therefore \quad t_2 = \frac{d_2^2}{d_1^2}\cdot t_1 = \left(\frac{2500}{10}\right)^2 44\cdot45$$

$$= 2778125 \text{ minutes}$$

$$= \mathbf{1929 \text{ days } 6.1 \text{ hr}}$$

Example 11.4. Following data are obtained from consolidation tests on two specimens A and B.

Pressure (kN/m²)	Equilibrium void ratio	
	A	B
100	0.535	0.630
150	0.480	0.615

The initial thickness of specimen A was 30 mm and that of B 20 mm. If the time taken for specimen A to reach 50 per cent degree of consolidation is 1/3 of that required by specimen B to reach the same degree of consolidation, find the ratio of coefficients of permeability of the two clay specimens.

Solution :

We have
$$m_v = \frac{\Delta e}{1 + e_0} \cdot \frac{1}{\Delta \sigma'}$$

$$(m_v)_A = \frac{0 \cdot 630 - 0 \cdot 480}{1 + 0 \cdot 535} \times \frac{1}{50} = 7 \cdot 1661 \times 10^{-4} \ kN/m^2$$

$$(m_v)_B = \frac{0 \cdot 630 - 0 \cdot 615}{1 + 0 \cdot 630} \times \frac{1}{50} = 1 \cdot 8405 \times 10^{-4} \ kN/m^2$$

Also,
$$T_v = C_v \frac{t}{d^2} = (C_v)_A \frac{t_A}{d_A^2} = (C_v)_B \frac{t_B}{d_B^2}$$

$$\frac{(C_v)_A}{(C_v)_B} = \frac{t_B}{t_A} \cdot \left(\frac{d_A}{d_B}\right)^2 = 3\left(\frac{30}{20}\right)^2 = 6.75$$

Further,
$$C_v = \frac{k}{m_v \gamma_w} \qquad \therefore \quad k = C_v \cdot m_v \cdot \gamma_w$$

Hence
$$\frac{k_A}{k_B} = \frac{(C_v)_A}{(C_v)_B} \cdot \frac{(m_v)_A}{(m_v)_B} = 6 \cdot 75 \times \frac{7 \cdot 1661}{1 \cdot 8405}$$

$$= 26.28$$

Example 11.5. In a consolidation test the void ratio of soil sample decreases from 1.20 to 1.10 when the pressure is increased from 200 to 400 kN/m². Calculate the coefficient of consolidation if the coefficient of permeability is 8.0×10^{-7} mm/sec.

Solution :

$$e_0 = 1.20 \qquad\qquad e_1 = 1.10$$
$$\sigma_0 = 200 \ kN/m^2 \qquad\qquad \sigma_1 = 400 \ kN/m^2$$

$$m_v = \frac{\Delta V}{V_0} \cdot \frac{1}{\Delta \sigma} = \frac{\Delta e}{1 + e_0} \cdot \frac{1}{\Delta \sigma} = \frac{(1 \cdot 10 - 1 \cdot 20)}{1 + 1 \cdot 20} \cdot \frac{1}{(400 - 200)}$$

$$= 2.27 \times 10^{-4} \ m^2/kN$$

$$k = 8.0 \times 10^{-7} \ mm/sec = 8.0 \times 10^{-10} \ m/sec$$

$$C_v = \frac{k}{m_v \gamma_\omega} = \frac{8.0 \times 10^{-10}}{2.27 \times 10^{-4} \times 9.81} = 0.359 \times 10^{-6} \ m^2/sec$$

Example 11.6. A 5m thick saturated soil stratum has a compression index of 0.25 and coefficient of permeability 3.2×10^{-3} mm/sec. If the void ratio is 1.9 at vertical stress of 0.15 N/mm², compute the void ratio when the vertical stress is increased to 0.2 N/mm². Also calculate settlement due to above stress increase and time required for 50 % consolidation.

Solution :

$$H_0 = 5m \qquad C_c = 0.25 \qquad k = 3.2 \times 10^{-3} \ mm/sec$$
$$\sigma_0 = 0.15 \ N/mm^2 \qquad \sigma_1 = 0.2 \ N/mm^2$$
$$e_0 = 1.9 \qquad e_1 = ?$$

$$C_c = \frac{\Delta e}{\Delta \log \sigma} = \frac{e_0 - e_1}{\log \sigma_1 - \log \sigma_0}$$

$$e_0 - e_1 = C_c \log \frac{\sigma_1}{\sigma_0} = 0 \cdot 25 \log \frac{0 \cdot 2}{0 \cdot 15} = 0 \cdot 0312$$

$$\therefore \qquad\qquad e_1 = 1.9 - 0.0312 = 1.869$$

Settlement, $\qquad\qquad \Delta H = H_0 \cdot \frac{C_c}{1 + e_0} \log \frac{\sigma_1}{\sigma_0}$

$$= \frac{5000 \times 0 \cdot 25}{1 + 1 \cdot 9} \log \frac{0 \cdot 2}{0 \cdot 15} = 53 \cdot 8 \text{ mm}$$

For $\qquad\qquad U = 50\%, \ T_v = \frac{\pi}{4} \left(\frac{U}{100} \right)^2 = \frac{\pi}{4} \left(\frac{50}{100} \right)^2 = 0 \cdot 196$

$$m_v = \frac{\Delta V}{V_0} \cdot \frac{1}{\Delta \sigma} = \frac{\Delta e}{1 + e_0} \cdot \frac{1}{\Delta \sigma}$$

$$= \frac{1 \cdot 9 - 1 \cdot 869}{1 + 1 \cdot 9} \cdot \frac{1}{(0 \cdot 2 - 0 \cdot 15)} = 0 \cdot 214 \text{ mm}^2/\text{N}$$

$$C_v = \frac{k}{m_v \, \gamma_w} = \frac{3 \cdot 2 \times 10^{-3}}{0 \cdot 214 \times 9 \cdot 81 \times 10^{-6}} = 1524 \text{ mm}^2/\text{sec}$$

Also, $\qquad\qquad T_v = \frac{C_v \, t}{d^2}$ and $d = H$ for single drainage

$$\therefore \qquad\qquad t = \frac{T_v \cdot d^2}{C_v} = \frac{0 \cdot 196 \, (5000)^2}{1524} = 3215 \cdot 2 \text{ sec}$$

$$= 53.58 \text{ min}$$

Example 11.7. A clay layer whose total settlement under a given load is expected to be 250 mm, settles by 50 mm in 15 days after the application of a load increment. How many days will be required for it to reach a settlement of 125 mm. How much settlement will occur in 300 days? The layer has double drainage.

Solution :

$\rho_f = 250 \text{ mm}$

$\rho_1 = 50 \text{ mm in time } t_1 = 15 \text{ days}$

$\rho_2 = 125 \text{ mm in time } t_2 = ?$

For $\qquad \rho_1 = 50 \text{ mm}, \ U = \frac{50}{250} \times 100 = 20\%$

For $\qquad U = 20\%, \ T_{v_1} = \frac{\pi}{4} \left(\frac{U}{100} \right)^2 = \frac{\pi}{4} \left(\frac{20}{100} \right)^2 = 0 \cdot 0314$

For $\qquad \rho_2 = 125 \text{ mm}, \ U = \frac{125}{250} \times 100 = 50\%$

For $\qquad U = 50\%, \ T_{v_2} = \frac{\pi}{4} \left(\frac{50}{100} \right)^2 = 0 \cdot 1963$

We have $\qquad\qquad T_v = C_v \frac{t}{d^2}.$

Since C_v and d are same for both cases we can write

$$\frac{Tv_1}{t_1} = \frac{Tv_2}{t_2}$$

$$\therefore \qquad t_2 = \frac{Tv_2}{Tv_1}.t_1 = \left(\frac{0.1963}{0.0314}\right)15 = 93.77 \text{ days}$$

For $t_3 = 300$ days, $\qquad \dfrac{Tv_3}{t_3} = \dfrac{Tv_1}{t_1}$

$$Tv_3 = \frac{Tv_1}{t_1}.t_3 = \left(\frac{0.0314}{15}\right)(300) = 0.628$$

$$Tv_3 = \frac{\pi}{4}\left(\frac{U}{100}\right)^2 \qquad \therefore \quad U = \sqrt{\frac{0.628 \times 4 \times 10000}{\pi}} = 89.4\%$$

Since $U > 60\%$ we recalculate U from

$$T_v = 1.781 - 0.9332 \log_{10}(100 - U)$$

$$\log_{10}(100 - U) = \frac{1.781 - 0.628}{0.9332} = 1.2355$$

$$100 - U = \text{antilog } 1.2355 = 17.2$$

$$\therefore \qquad U = 100 - 17.2 = 82.8\%$$

Settlement occurring in 300 days $= \dfrac{82.8}{100} \times 250 \ = \textbf{207 mm}$

Example 11.8. A 1 cm thick laboratory soil sample reaches 60% consolidation in 32.5 seconds under double drainage condition. Find how much time will be required for a 10 m thick layer in the field to reach the same degree of consolidation if it has drainage face on one side only.

Solution :

For $U = 60\%$, $\qquad\qquad T_v = \dfrac{\pi}{4}\left(\dfrac{U}{100}\right)^2 = \dfrac{\pi}{4}\left(\dfrac{60}{100}\right)^2$

$$= 0.2827$$

For laboratory sample,

$$d_1 = \frac{10}{2} = 5 \text{ mm for double drainage}$$

$$T_v = C_v.\frac{t}{d^2} \text{ and } t_1 = 32.5 \text{ sec for } U = 60\%.$$

$$\therefore \qquad C_v = T_v.\frac{d^2}{t} = \frac{0.2827(5)^2}{32.5} = 0.2175 \text{ mm}^2/\text{sec}$$

For field layer,

$$d = 10000 \text{ mm for single drainage}$$

For $\qquad\qquad U = 60\%,\ t_2 = ?$

$$t_2 = \frac{T_v}{C_v}.d^2 = \frac{0.2827}{0.2175}(10000)^2$$

$$= 1.2998 \times 10^8 \text{ sec}$$

$$= \textbf{1504.4 days}$$

Example 11.9. A 6m thick bed of clay is overlain by 9m thick layer of sand with water table at 4m below ground surface. For the clay layer specific gravity of soil particles is 2.7, average liquid limit 45% and natural water content 40%. For the sand layer the bulk unit weights above and below water table are 18 kN/m³ and 20.5 kN/m³ respectively. Calculate the settlement of a building constructed on sand layer if it causes an increase in effective vertical stress of 100 kN/m² at the middle of clay layer.

Solution :

For sand layer, $\gamma = 18$ kN/m³ above water table

and $\gamma' = 20.5 - 9.81 = 10.69$ kN/m³ below water table

For clay layer, $e_0 = w_{sat} \, G = 0.4 \times 2.7 = 1.08$

$$\gamma' = \frac{(G-1)\gamma_w}{1+e} = \frac{(2 \cdot 7 - 1) \, 9 \cdot 81}{1 + 1 \cdot 08}$$

$$= 8.02 \text{ kN/m}^3$$

At middle of clay layer,

$$\sigma_0' = 4 \times 18 + 5 \times 10.69 + 3 \times 8.02 = 149.51 \text{ kN/m}^2$$

For clay layer,

$$C_c = 0.009 \, (w_L - 10) = 0.009 \, (45 - 10) = 0.315$$

Settlement,

$$\rho = \frac{H}{1 + e_0} \, C_c \log_{10} \frac{\sigma_0' + \Delta\sigma'}{\sigma_0'}$$

$$= \frac{6}{1 + 1 \cdot 08} \times 0 \cdot 315 \log_{10} \frac{149 \cdot 51 + 100}{149 \cdot 51}$$

$$= 0.2021 \text{ m} = \textbf{202.1 mm}$$

Example 11.10.

A clay layer lies sandwiched between two sand strata as shown in following figure. Find the settlement due to 1000 kN load which may be treated as concentrated load.

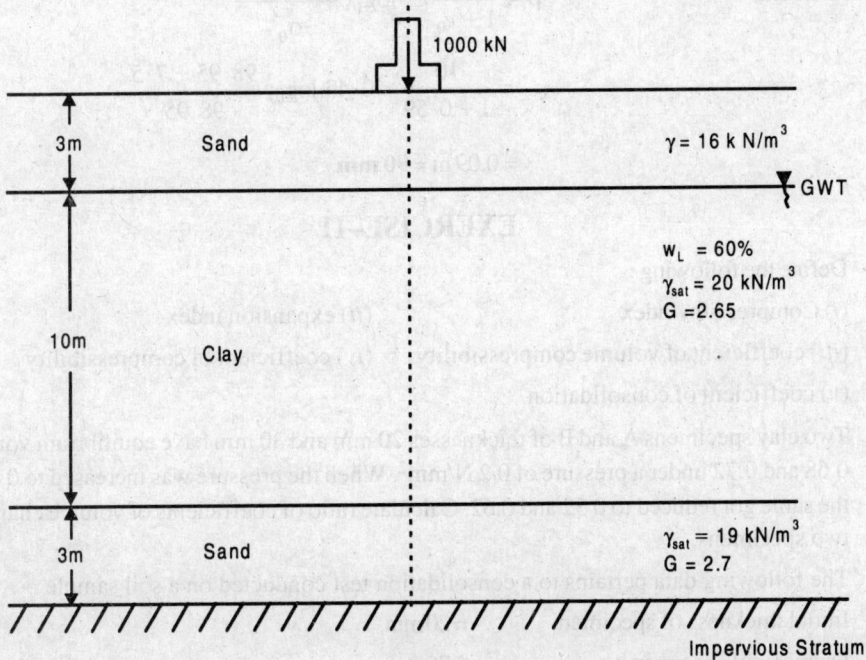

Solution :

For clay layer,

Initial thickness, $H = 10$ m

$$\gamma_{sat} = \frac{(G + e)\gamma_w}{1 + e}$$

$$20 = \frac{(2 \cdot 65 + e)}{1 + e} 9.81$$

$$20 + 20e = 26 + 9.81e$$

Initial void ratio, $e_0 = \frac{26 - 20}{20 - 9 \cdot 81} = 0 \cdot 59$

$$C_c = 0.009 (w_L - 10) = 0.009 (60 - 10) = 0.45$$

Initial overburden pressure at the middle of clay layer,

$$\sigma_0' = 3 \times 16 + 5 \times (20 - 9.81) = 98.95 \text{ kN/m}^2$$

Using Boussinesq equation,

Vertical stress at middle of clay layer, due to 1000 kN point load

$$\Delta\sigma' = \frac{3Q}{2\pi} \frac{z^3}{(r^2 + z^2)^{5/2}} = \frac{3Q}{2\pi z^2} \quad \text{(since } r = 0 \text{ on axis of load)}$$

$$= \frac{3(1000)}{2\pi(8)^2} = 7 \cdot 5 \text{ kN/m}^2$$

Settlement, $\rho = \dfrac{H}{1 + e_0} C_c \log_{10} \dfrac{\sigma_0' + \Delta\sigma'}{\sigma_0'}$

$$= \frac{10}{1 + 0.59} \times 0 \cdot 45 \log_{10} \frac{98 \cdot 95 + 7 \cdot 5}{98 \cdot 95}$$

$$= 0.09 \text{ m} = \textbf{90 mm}$$

EXERCISE–11

11.1   Define the following :

   (i) Compression index   (ii) expansion index

   (iii) coefficient of volume compressibility   (iv) coefficient of compressibility

   (v) coefficient of consolidation.

11.2   Two clay specimens A and B of thicknesses 20 mm and 30 mm have equilibrium void ratio of 0.68 and 0.72 under a pressure of 0.2 N/mm². When the pressure was increased to 0.4 N/mm² the same got reduced to 0.52 and 0.62. Calculate ratio of coefficients of volume change of the two specimens.

11.3   The following data pertains to a consolidation test conducted on a soil sample :

   Initial thickness of specimen $= 20$ mm

   Diameter of specimen $= 75$ mm

Specific gravity of soil particles   = 2.76

Weight of dry specimen   = 0.78 N

Pressure (N/mm^2)	Final dial reading $(x\ 0.001\ mm)$
0	100
0.05	358
0.10	497
0.20	630
0.40	765
0.80	896
0	755

Calculate equilibrium void ratio for each pressure increment, plot on semi-log sheet and find compression index. Also calculate degree of saturation of speciment at end of test.

11.4   Under a consolidation pressure of 0.2 N/mm² the thickness and water content of a saturated specimen were found to be 20.5 mm and 16 per cent. On increasing the pressure to 0.4 N/mm² the specimen got compressed by 1.05 mm. Find compression index of soil taking G = 2.76.

11.5   A clay layer subjected to an effective pressure increment of 0.1 N/mm² got compressed by 52 mm. If the initial thickness of layer was 6.8 m, find average coefficient of volume change for the soil.

11.6   A clay stratum 8m thick has a void ratio of 0.85 at an initial pressure of 1.5 N/mm² and void ratio of 0.72 at a pressure of 3.0 N/mm². The liquid limit of soil is 55%. Determine the compression of this stratum due to the pressure increase.

11.7   A saturated clay stratum 8.0 m thick lies sandwiched between two thick sand layers. In addition to the thick sand strata there is a thin continuous sand layer 2m below the top of the clay layer. Consolidation tests on undisturbed samples of clay show it to have a coefficient of consolidation of 5 × 10⁻² mm²/sec. Application of a uniform load at the ground surface will cause a uniform increase in vertical stress throughout the clay layer.

(i) Determine the time required to reach 50% degree of consolidation.

(ii) How much time will be required for the same degree of consolidation without the thin sand layer.

11.8   List the assumptions made in deriving Terzaghi's one dimensional consolidation theory.

11.9   Explain any one method of determining preconsolidation pressure.

11.10 Distinguish between :

(a) Under consolidated, normally consolidated and over consolidated soil deposits

(b) Primary compression and secondary compression

(c) Degree of consolidation and over consolidation ratio.

(d) Coefficient of compressibility and coefficient of consolidation (*e*) Excess hydrostatic pressure and hydrodynamic lag.

11.11   State whether the following statements are true or false.

     (*i*) Coefficient of consolidation is used for calculating time rate of settlement.

    (*ii*) Time factor is a function of degree of consolidation.

   (*iii*) Coefficient of compressibility and coefficient of volume compressibility are one and the same.

   (*iv*) Basic differential equation of consolidation relates rate of dissipation of pore pressure with rate of expulsion of pore water.

    (*v*) Time factor is independent of drainage conditions.

   (*vi*) Solution of Terzaghi's one-dimensional consolidation equation indicates how pore pressure dissipates with time at different depths.

CHAPTER 12

SHEAR STRENGTH OF SOILS

12.1 Introduction

At a point inside a soil mass subjected to non-isotropic stress system one can find shear stress along with normal stress induced on all planes passing through that point, except on principal planes. On the potential failure plane shear stress induced will tend to exceed the shear strength of soil causing excessive shear deformation leading to shear failure. The shear strength is the maximum shearing resistance that is mobilized on the potential failure plane and is equal to the ultimate shear stress in the limiting equilibrium condition. The development of shear strength is attributed to one or more of the following :

  (*i*)   The frictional resistance between the particles at their points of contact (combination of sliding friction and rolling friction).

  (*ii*) Cohesion or force of attraction between particles.

  (*iii*) The structural resistance to displacement because of interlocking of particles and cementation or adhesion between particles.

Shear strength of soil is one of the three engineering properties of soil, the other two being permeability and compressibility. Knowledge of shear strength is required in all engineering problems involving stability analysis such as in the design of foundations of structures, retaining walls and stability of earth slopes.

12.2 Mohr Circle For Two-Dimensional Stress System

A two dimensional stress system is one in which all stresses act in a plane. A general two-dimensional stress system at a point inside soil mass can be represented by a block diagram as shown in Fig. 12.1(a). The Mohr circle for the two-dimensional stress system shown in the Fig. 12.1(a) is drawn as shown in Fig. 12.1(b).

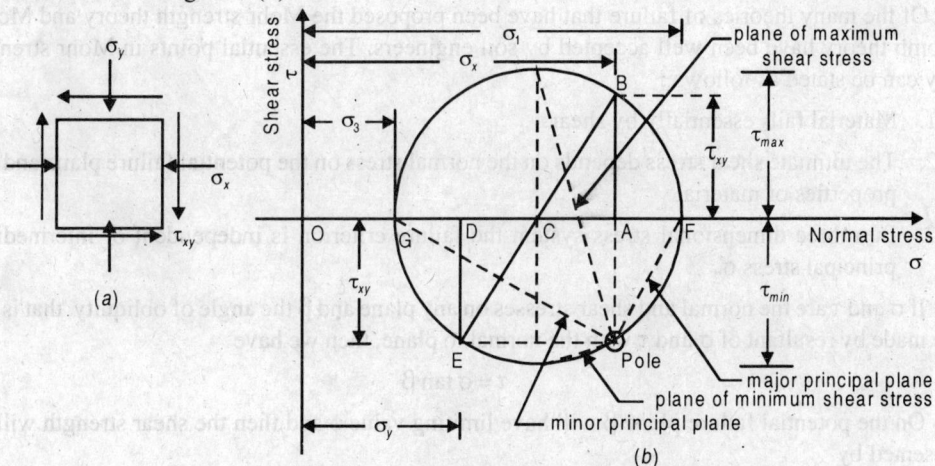

Fig. 12.1 Mohr circle for two-dimensional stress system

Notations

σ_x = normal stress on plane perpendicular to X-axis

σ_y = normal stess on plane perpendicular to Y-axis

201

τ_{xy} = shear stress in XY plane

σ_1 = major principal stress

σ_3 = minor principal stress

τ_{max} = maximum shear stress

τ_{min} = minimum shear stress

Sign conventions

A normal stress is considered positive if it is compressive.

A shear stress is considered positive if its moment about any point inside the element is clockwise.

The co-ordinates of point B on the Mohr circle represents the normal and shear stresses acting on the vertical plane in Fig 12.1(a). If through the point B we draw a line parallel to the vertical plane, it cuts the circle at a point called the pole of Mohr circle. Instead of considering the vertical plane we can also consider the horizontal plane in Fig 12.1(a). The normal and shear stresses acting on this plane are represented by the coordinates of point E on Mohr circle. If through the point E we draw a line parallel to the horizontal plane, it cuts the circle at the pole. The pole of a Mohr circle is a unique point which is such that if a line is drawn parallel to a given plane of the element, then this line will cut the Mohr circle at a point whose co-ordinates will represent the normal and shear stresses acting on that plane. In other words, if the pole is joined to a point on the Mohr circle, then that line will be parallel to the plane on which normal and shear stresses represented by the coordinates of that point act. All other uses of Mohr circle are indicated in Fig 12.1(b)

In a three dimensional stress system we will have, normal stress σ_z in addition to σ_x, σ_y; shear stresses τ_{yz} and τ_{zx} in addition to τ_{xy}; and intermediate principal stress σ_2 in addition to σ_1, σ_3.

Analysis of many problems in geotechnical engineering is simplified assuming two-dimensional state of stresses and been found adequate.

12.3   Theories of Failure for Soils

Of the many theories of failure that have been proposed the Mohr strength theory and Mohr-Coulomb theory have been well accepted by soil engineers. The essential points in Mohr strength theory can be stated as follows :

1.   Material fails essentially by shear.

2.   The ultimate shear stress depends on the normal stress on the potential failure plane and the properties of material.

3.   In a three dimensional stress system the failure criterion is independent of intermediate principal stress σ_2

If σ and τ are the normal and shear stresses on any plane and β the angle of obliquity, that is the angle made by resultant of σ and τ with the normal to plane, then we have

$$\tau = \sigma \tan \beta$$

On the potential failure plane β will have limiting value ϕ and then the shear strength will be represented by

$$\tau = \sigma \tan \phi \qquad \qquad ...(12.1)$$

We note that Mohr did not consider cohesion of soil and hence this theory is applicable only for cohesionless soils. ϕ is called angle of internal friction. The line represented by Eq. 12.1 is called Mohr strength envelope. In practice, for a large range of values of σ, the Mohr envelope is actually slightly curved as ϕ decreases slightly with increase in σ.

Mohr-Coulomb theory

This theory was first proposed by Coulomb (1776) and later generalized by Mohr. If we plot shear stress at failure as ordinate against normal stress as abscissa we obtain a curve called the strength envelope. It can be represented by the equation

$$\tau_f = F(\sigma)$$

Coulomb assumed the relation between τ_f and σ to be linear and gave the following equation popularly known as Coulomb's equation

$$\tau_f = c + \sigma \tan \phi \qquad \qquad ...(12.2)$$

where c is the intercept of the strength envelope on the τ axis and tan ϕ the slope of the strength envelope. c is known as cohesion and ϕ angle of internal friction or more comprehensively angle of shearing resistance. c and ϕ together are called shear strength parameters and are variable for any soil depending on conditions of testing such as drainage conditions and rate of strain.

Mohr generalized the strength envelope, also known as failure envelope, as a curve which becomes flatter with increasing normal stress, as shown in Fig 12.2 (b).

In conclusion it can be stated that the strength envelope will be a straight line if ϕ is assumed to be constant. In the case of Mohr generalized envelope, a straight line can be fitted within a range of σ values. The strength envelope will be tangential to any Mohr circle at failure as shown in Fig 12.2.

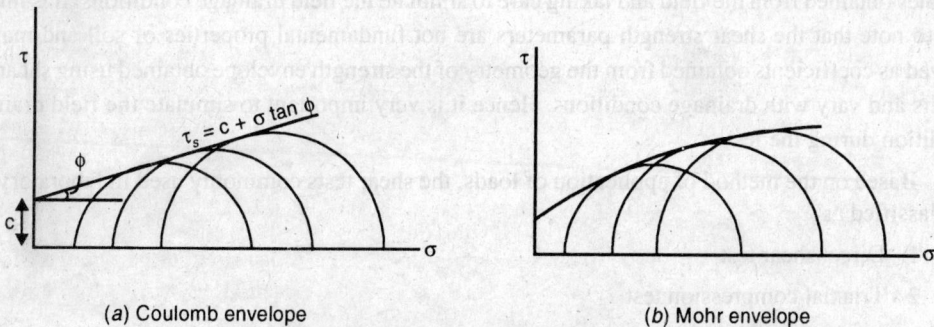

(a) Coulomb envelope   (b) Mohr envelope

Fig. 12.2 Strength Envelopes (Coulomb and Mohr)

Based on values of shear strength parameters, soils can be described as (*i*) cohesive soil (*ii*) cohesionless soil and (*iii*) purely cohesive soil. The strength envelopes for the three cases are shown in Fig 12.3.

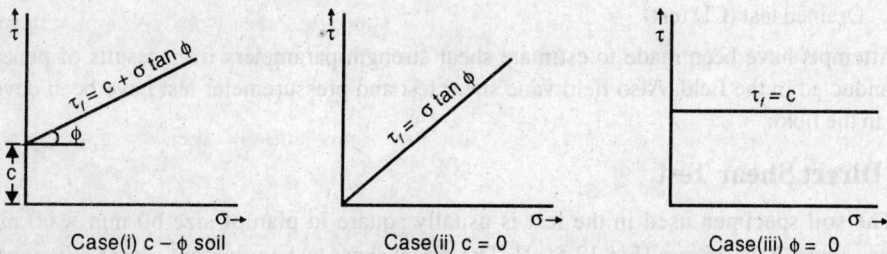

Case(i) c – ϕ soil   Case(ii) c = 0   Case(iii) ϕ = 0

Fig. 12.3 Strength envelopes for the three types of soils

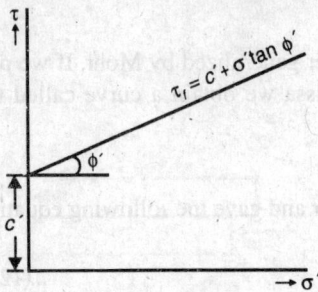

Fig. 12.4 Effective strength parameters

12.4 Terzaghi's Effective Stress Principle

According to Terzaghi, the shear strength of soil in the case of saturated soil is a function of effective normal stress on the potential failure plane.

The Mohr strength envelope can then be represented by the following equation.

$$\tau_f = c' + \sigma' \tan \phi' \qquad \qquad ...(12.3)$$

where $\sigma' = (\sigma - u)$

c' and ϕ' are referred to as effective cohesion and effective angle of shearing resistance respectively.

Note : While c' and ϕ' are referred to as effective shear strength parameters, c and ϕ are known as apparent or total shear strength parameters.

12.5 Determination of Shear Strength Parameters

Shear strength parameters are determined in the laboratory by conducting shear tests and using the results of these tests to obtain the failure envelope. The tests are conducted on undisturbed soil samples obtained from the field and taking care to simulate the field drainage conditions. It is important to note that the shear strength parameters are not fundamental properties of soil and may be viewed as coefficients obtained from the geometry of the strength envelope obtained using shear test results and vary with drainage conditions. Hence it is very important to simulate the field drainage condition during the test.

Based on the method of application of loads, the shear tests commonly used in laboratory can be classified as

1. Direct shear test
2. Triaxial compression test
3. Unconfined compression test
4. Vane shear test

Based on drainage conditions the shear tests are classified as

1. Undrained test (UU test)
2. Consolidated undrained test (CU test)
3. Drained test (CD test)

Attempts have been made to estimate shear strength parameters from results of penetration tests conducted in the field. Also field vane shear test and pressuremeter test have been developed for use in the field.

12.6 Direct Shear Test

The soil specimen used in the test is usually square in plan of size 60 mm × 60 mm and thickness about 20 to 25 mm [Fig 12.5(a)]. The direct shear test equipment essentially consists of (i) shear box, (ii) loading yoke for applying normal force, (iii) geared jack for applying shear force and facilities for measuring shear force, shear displacement and vertical deformation for volume change [Fig 12.6]

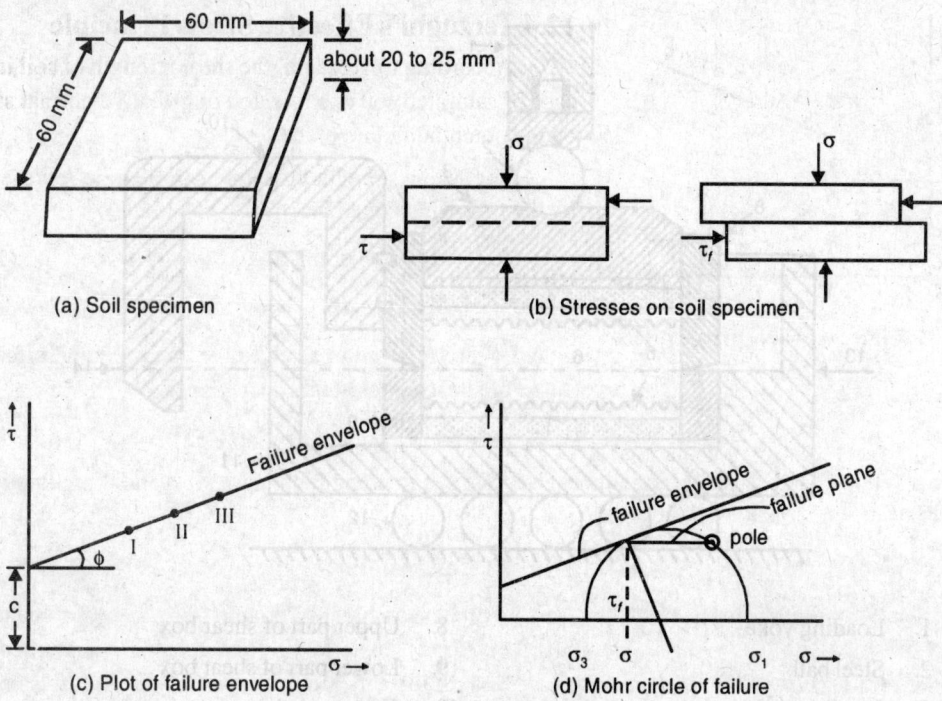

Fig. 12.5 Direct sheat test

The shear box consists of two halves; the lower half is in contact with the shear box container which freely slides on rollers and to which the shear force is applied by means of geared jack. The soil specimen is placed in the shear box such that it gets sheared on a horizontal plane exactly at its mid-height. The specimen is sandwiched between a pair of metal grid plates and a pair of porous plates (or non-porous plates) as shown in Fig 12.6. The grid plates provided with serrations are placed with serrations at right angles to direction of shearing to provide grip on the specimen. For conducting drained test perforated grid plates and porous stones are used.

The principle of direst shear test is illustrated in Fig 12.5 (b) and (c). A normal stress σ is applied on the specimen and is kept constant throughout the test. The shear stress τ is caused by application of shear force through geared jack and is transmitted to the top half of the shear box, which bears against shear force measuring device (such as proving ring dial gauge), through the soil specimen. The shear stress is gradually increased until the specimen fails and there will be no transmission of shear force from lower half to top half of shear box. If test continues beyond 20% strain it is usual to stop the test and define failure point as corresponding to any desired level of strain upto 20%. The test is conducted on preferably minimum of three specimens subjected to three different values of σ. By plotting τ_f against σ the failure envelope is obtained. c and ϕ are obtained by measurement from the plot. Fig 12.5 (d) illustrates the Mohr circle at failure drawn for a specimen sheared under a normal stress σ.

The shear box test can be either strain controlled or stress controlled. In the strain controlled shear box test the shear strain is made to increase at a constant rate and the shear stress is measured. In the stress controlled shear box test the arrangement is for increasing the shear stress at constant rate and measuring the shear strain.

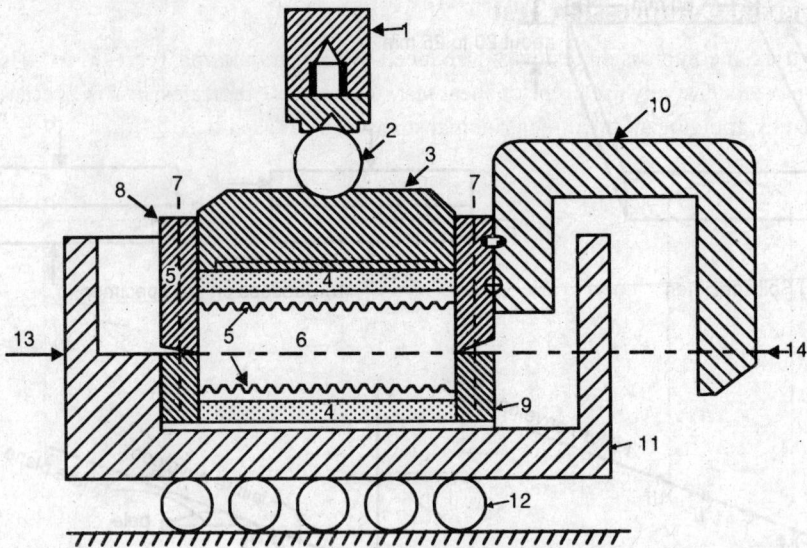

Fig. 12.6 Shear box with accessories

1.	Loading yoke	8.	Upper part of shear box
2.	Steel ball	9.	Lower part of shear box
3.	Loading pad	10.	U-arm
4.	Porous stones	11.	Container for shear box
5.	Metal grids	12.	Rollers
6.	Soil specimen	13.	Shear force (applied by jack)
7.	Pins to fix two halves of shear box	14.	Shear resistance (measured by proving ring dial gauge)

Advantages of direct shear test

1. The direct shear test is a simple test compared to the triaxial compression test.

2. Since the thickness of the sample is small quick drainage and hence rapid dissipation of pore pressure is possible.

Disadvantages of direct shear test

1. The shear stress is not uniformly distributed being more at the edge than at the center. Because of this the entire shear strength is not mobilized simultaneously at all points on the failure plane and this leads to progressive failure of the specimen.

2. The failure plane is predetermined. Therefore the specimen is not allowed to fail along its weakest plane.

3. Shear displacement causes reduction in area under shear. Corrected area should be used in computing normal and shear stresses.

4. The side walls of the shear box can cause lateral restraint on the edges of the specimen.

5. There is little control on drainage of pore water as compared with triaxial compression test.

6. Measurement of pore pressure is not possible.

12.7   Triaxial Compression Test

The triaxial compression test was introduced by Casagrande and Terzaghi in 1936 and to this day is the most extensively used type of shear test. As the name indicates, in this test the specimen is compressed by applying all the three principal stresses, σ_1, σ_2 and σ_3.

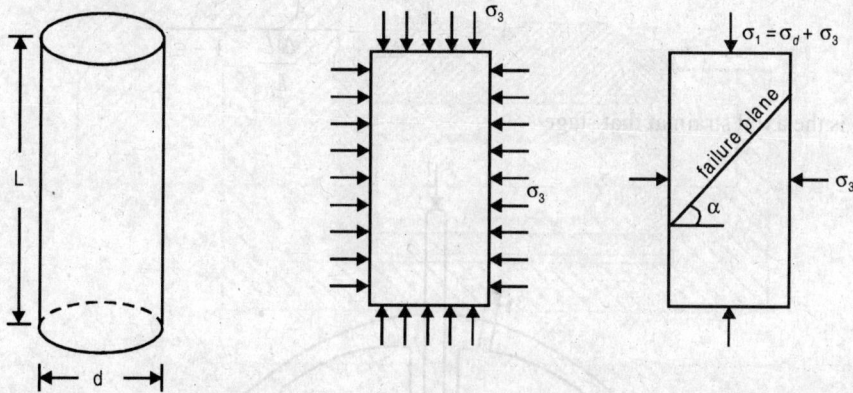

Fig. 12.7

The soil specimen used in the test is cylindrical in shape with length 2 to 2.5 times the diameter. The triaxial compression test equipment essentially consists of (*i*) triaxial cell, (*ii*) loading frame with accessories for applying gradually increasing axial load on specimen at constant rate of strain, (*iii*) provision for measuring axial force and axial displacement, (*iv*) constant pressure system to apply and maintain constant cell pressure, (*v*) pore pressure measuring apparatus and (*vi*) volume change gauge.

The triaxial cell (Fig 12.8) consists of a high pressure cylindrical cell, made of a transparent material like perspex, fitted between base and top cap and is provided at the base with inlet for cell fluid, outlets for drainage of pore water from specimen and measurement of pore pressure. At the top an air release valve to expel air from the cell and a steel plunger for applying axial force on specimen are provided.

The soil specimen is kept inside the triaxial cell with porous plates (or non porous plates for undrained test) at top and bottom. The loading cap is placed on top porous plate. The specimen is enclosed in a rubber membrane to prevent its contact with the cell fluid. After filling the cell with fluid (usually water) required cell pressure (σ_3) is applied by means of constant pressure system. The additional axial force called the deviator force is applied through the plunger and the deviator force corresponding to different axial deformations at regular intervals are noted. The test is continued until the specimen fails. If the test continues even after 20% strain, it may be stopped and failure point defined at desired strain level upto 20%. The deviator stress σ_d at any stage of the test is given by

$$\sigma_d = \frac{F}{A_c}$$

where,

F = deviator force i.e., additional axial force applied through plunger,

A_c = corrected area of cross section of specimen at that stage.

If A_i = initial area of cross section of specimen

L_i = initial length of specimen

A_c = corrected area of specimen when the axial compression is ΔL and change in volume is ΔV, we have, initial volume $V_i = A_i L_i$ and volume at any stage of compression, $(V_i + \Delta V) = A_c (L - \Delta L)$

$$\therefore \quad A_c = \frac{V_i + \Delta V}{L_i - \Delta L}$$

In the case of undrained test on saturated soil sample, $\Delta V = 0$ and

$$A_c = \frac{A_i L_i}{L_i - \Delta L} = \frac{A_i}{1 - \dfrac{\Delta L}{L_i}} = \frac{A_i}{1 - \in}$$

where $\in$ is the axial strain at that stage

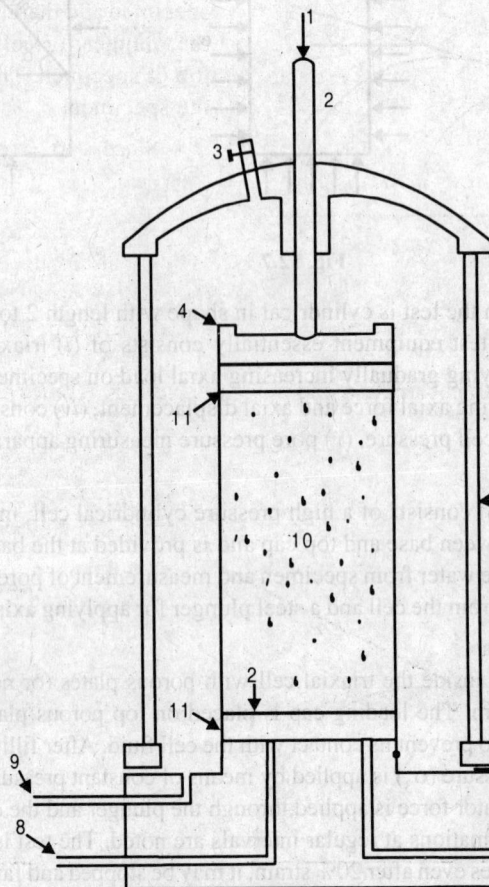

1. Axial load (measured by proving ring
    dial gauge)

2. Loading ram   7.   Pore water outlet
3. Air release valve   8.   Additional pore water outlet
4. Top cap   9.   Cell fluid inlet
5. Perspex Cylinder   10. Soil specimen (enclosed in tubber membrane with
6. Sealing ring   'o' rings at the ends)
                                         11. Porous disc

Fig. 12.8. Triaxial Cell

After finding deviator stress σ_d at failure, we have major principal stress at failure $\sigma_1 = (\sigma_d + \sigma_3)$. With this set of (σ_1, σ_3) values, Mohr circle at failure is drawn. The test is conducted on preferably a minimum of three specimens subjected to different values of cell pressure σ_3. The Mohr circle at failure is drawn for each specimen and the common tangent touching all the circles will be failure envelope (Fig 12.9). c and ϕ are read out from the plot. For the benefit of student it is repeated that deviator stress is the additional axial stress applied on the specimen through the plunger. This is because the cell pressure σ_3 not only acts on the sides of the specimen but also acts on top of the specimen and there will be equal reaction at the base. This

Fig. 12.9 Mohr circles at failure

is due to the fact that area of plunger is smaller than area of cross section of specimen and there will be space around the plunger for cell pressure to act on top of specimen. Thus σ_3 acts all round the specimen

Since $\sigma_1 = (\sigma_d + \sigma_3)$

we have $\sigma_d = (\sigma_1 - \sigma_3)$

Further, in the triaxial compression test on cylindrical soil specimen we have $\sigma_2 = \sigma_3$. Special triaxial cell has been developed, to test soil specimen cubical in shape, with separate chambers for applying σ_2 and σ_3 separately.

Relation between shear strength parameters and principal stresses at failure.

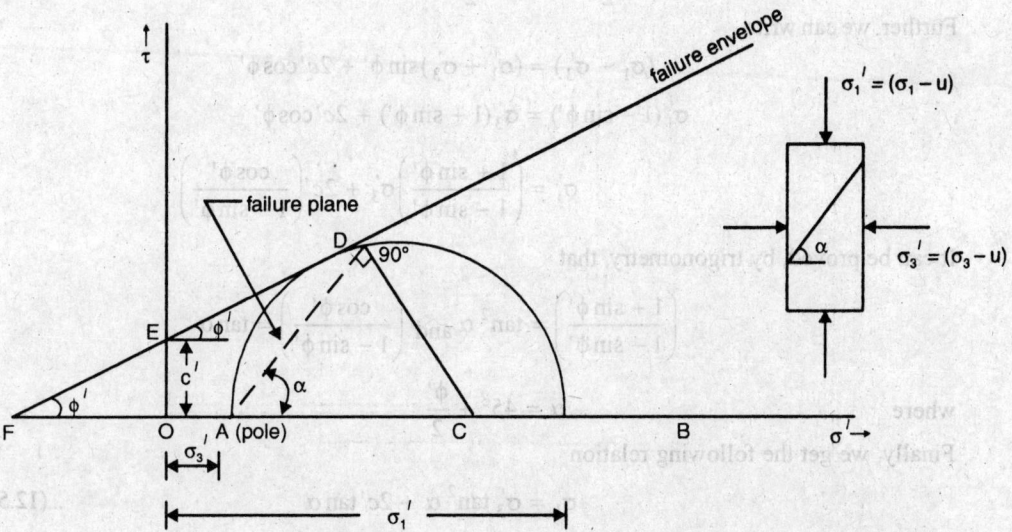

Fig. 12.10 Mohr circle at failure (for deriving relation between σ_1', σ_3', c' and ϕ')

In Fig 12.10, Mohr circle at failure is drawn for specimen subjected to triaxial compression test. Point A is the pole of Mohr circle, AD the failure plane and $\angle DAC = \alpha$, the angle made by failure plane with the major principal plane. Let the failure envelope be produced backwards to intersect σ-axis at F.

In the right angled triangle DFC,

$$\sin \phi' = \frac{DC}{FC}$$

$$DC = \text{radius of Mohr circle} = \frac{\left(\sigma_1' - \sigma_3'\right)}{2}$$

$$FC = FO + OC$$

$$OC = OA + AC = \sigma_3' + \frac{\left(\sigma_1' - \sigma_3'\right)}{2} = \frac{\left(\sigma_1' + \sigma_3'\right)}{2}$$

In the right-angled triangle EOF

$$\frac{FO}{EO} = \cot \phi'$$

$$FO = EO \cot \phi' = c' \cot \phi'$$

$$FC = OC + FO = \frac{\left(\sigma_1' + \sigma_3'\right)}{2} + c' \cot \phi'$$

Hence

$$\sin \phi' = \frac{DC}{FC} = \frac{\left(\sigma_1' - \sigma_3'\right)/2}{\left(\sigma_1' + \sigma_3'\right)/2 + c' \cot \phi'}$$

Simplifying,

$$\frac{\left(\sigma_1' - \sigma_3'\right)}{2} = \frac{\left(\sigma_1' + \sigma_3'\right)}{2} \sin \phi' + c' \cos \phi' \qquad \text{...(12.4)}$$

Further, we can write

$$\left(\sigma_1' - \sigma_3'\right) = \left(\sigma_1' + \sigma_3'\right) \sin \phi' + 2c' \cos \phi'$$

$$\sigma_1' (1 - \sin \phi') = \sigma_3'(1 + \sin \phi') + 2c' \cos \phi'$$

$$\sigma_1' = \left(\frac{1 + \sin \phi'}{1 - \sin \phi'}\right) \sigma_3' + 2c' \left(\frac{\cos \phi'}{1 - \sin \phi'}\right)$$

It can be proved, by trigonometry, that

$$\left(\frac{1 + \sin \phi'}{1 - \sin \phi'}\right) = \tan^2 \alpha \text{ and } \left(\frac{\cos \phi'}{1 - \sin \phi'}\right) = \tan \alpha$$

where

$$\alpha = 45° + \frac{\phi'}{2}$$

Finally, we get the following relation

$$\sigma_1' = \sigma_3' \tan^2 \alpha + 2c' \tan \alpha \qquad \text{...(12.5)}$$

Note 1: Eq. 12.5 is also written in the form

$$\sigma_1' = \sigma_3' N_\phi + 2c' \sqrt{N_\phi} \qquad \text{...(12.6)}$$

where

$$N_\phi = \tan^2 \alpha$$

Note 2 : We can show that

$$\alpha' = 45° + \frac{\phi'}{2}$$

Referring to Fig. 12.10, in right-angled triangle CDF, $\angle DCF = (90 - \phi')$

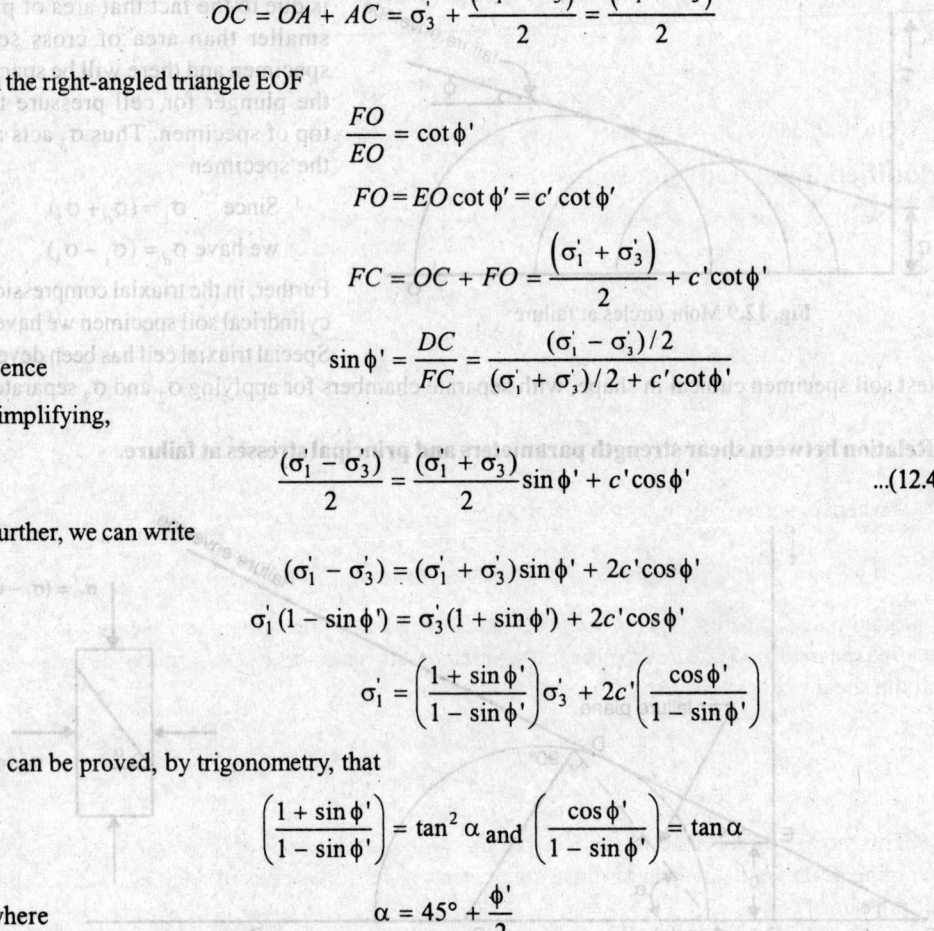

In Δ^{le} CAD,

$$\angle CAD = \angle CDA = \alpha$$

$$\angle CAD + \angle CDA + \angle DCA = 180°$$

$$\alpha + \alpha + (90 - \phi') = 180°$$

$$2\alpha = 90° + \phi'$$

$$\alpha = 45° + \frac{\phi'}{2}$$

Eq. 12.5 or 12.6 is found very useful for analytical solution of problems. In undrained test, pore pressure u is measured by means of pore pressure measuring apparatus. σ_1' and σ_3' are given by

$$\sigma_1' = \sigma_1 - u$$

$$\sigma_3' = \sigma_3 - u$$

In drained test $u = 0$ so that $\quad \sigma_1' = \sigma_1$ and $\sigma_3' = \sigma_3$

Modified failure envelope

We have the following relation between shear strength parameters and principal stresses at failure.

$$\frac{(\sigma_1' - \sigma_3')}{2} = \frac{(\sigma_1' + \sigma_3')}{2} \sin\phi' + c'\cos\phi'$$

We rewrite the above equation as

$$\frac{(\sigma_1' - \sigma_3')}{2} = \left(\frac{\sigma_1' + \sigma_3'}{2}\right) \tan\psi + d \qquad ...(12.7)$$

where $d = c' \cos\phi'$ and $\tan\psi = \sin\phi'$

If we plot $\dfrac{(\sigma_1' - \sigma_3')}{2}$ against $\dfrac{(\sigma_1' + \sigma_3')}{2}$ for different sets of observations (σ_1', σ_3') we get a set of plotted points. The best fitting straight line through the points will be represented by Eq 12.7 and is called the modified failure envelope. The intercept d on the vertical axis and angle ψ are measured and the shear strength parameters computed from the relations

$$\phi' = \sin^{-1}(\tan\psi)$$

$$c' = \frac{d}{\cos\phi'}$$

This modified procedure was introduced by Lambe and Whitman (1969) and provides a means for averaging scattered data when tests are conducted on a large number of samples with wide range of cell pressures.

Types of failure of soil specimens in the triaxial compression test.

Depending on the soil type and its physical properties a soil specimen can exhibit one of the three failure patterns indicated in Fig. 12.11

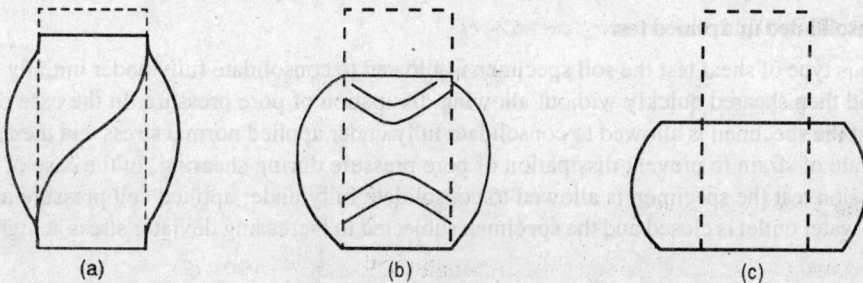

(a)   (b)   (c)

Fig. 12.11 Failure patterns

(*i*) Fig 12.11(a) is an example of a brittle failure with a well defined failure plane and little lateral bulging.

(*ii*) In Fig 12.11(b) is shown a semi-plastic failure with shear cones and some lateral bulging.

(*iii*) Fig 12.11(c) is a typical plastic failure with excessive lateral bulging and absence of failure plane. If this type of failure is noticed, it is usual to stop the test when the strain exceeds 20% and the failure point is defined corresponding to a strain level upto 20% depending on practical consideration.

Advantages of triaxial compression test

The advantages of triaxial compression test particularly when compared with direct shear test are outlined below.

1. The specimen is free to fail along the weakest plane, unlike in the direct shear test in which the specimen is forced to fail along a predetermined plane.

2. The stress distribution on the failure plane is uniform. The shear strength is mobilized uniformly at all points on the failure plane, unlike in the direct shear test in which progressive failure takes place.

3. There is complete control of drainage conditions. This enables better simulation of field drainage conditions during the test as compared with the direct shear test

4. Precise measurements of pore pressure and volume change are possible during the test.

5. The stresses induced on any plane within the specimen at any stage of the test can be determined.

6. As failure usually occurs near the middle of sample, the effect of end restraint is not a serious disadvantage.

12.8 Types Of Shear Tests Based On Drainage Conditions

The shear strength parameters in the case of saturated soils depend very much upon the drainage conditions and therefore in the laboratory shear test the drainage condition expected in the field for a particular problem should be simulated. Based on drainage condition the shear tests are classified as

(*i*)   Unconsolidated Undrained test or simply undrained test (UU test)

(*ii*)   Consolidated Undrained test (CU test)

(*iii*)   Consolidated Drained test or simply drained test (CD test)

(i) Undrained test

Drainage is not permitted throughout the test. In the case of direct shear test drainage is not permitted during the application of both normal stress and shear stress. In the case of triaxial compression test drainage is not permitted during the application of both cell pressure and deviator stress. Since the test is conducted fast allowing no time for either consolidation of sample initially or dissipation of pore pressure in later stage, the test is also called quick test.

(ii) Consolidated undrained test

In this type of shear test the soil specimen is allowed to consolidate fully under initially applied stress and then sheared quickly without allowing dissipation of pore pressure. In the case of direct shear test the specimen is allowed to consolidate fully under applied normal stress and then sheared at high rate of strain to prevent dissipation of pore pressure during shearing. In the case of triaxial compression test the specimen is allowed to consolidate fully under applied cell pressure and then the pore water outlet is closed and the specimen subjected to increasing deviator stress at high rate of strain

(iii) Drained test

In this type of shear test drainage is allowed throughout the test. The specimen is allowed to consolidate fully under the applied initial stress and then sheared at low rate of strain giving sufficient time for the pore water to drain out at all stages. The test may continue for several hours to several days.

12.9 Unconfined Compression Test

Unconfined compression test can be regarded as a special case of triaxial compression test in which no lateral pressure or confining pressure is applied so that $\sigma_2 = \sigma_3 = 0$. The soil specimen is cylindrical in shape with length about 2 to 2.5 times its diameter. The laboratory equipment for conducting unconfined compression test has facilities for compressing the specimen at uniform rate of strain and measuring the axial deformation and corresponding axial compressive force. The maximum compressive stress resisted by specimen before failure is called unconfined compressive strength. It is denoted by q_u and computed as shown below.

$$q_u = \frac{F}{A_c}$$

where F = axial compressive force at failure

A_c = corrected area of cross section of specimen at failure

$$= \frac{A_0}{1 - \epsilon}$$

A_o = initial area of cross section of specimen.

$$\epsilon = \frac{\Delta L}{L_o} = \text{axial stain at failure point}$$

The unconfined compression test is a quick test in which no drainage is allowed. The test is conducted on saturated clay and volume change is assumed to be zero. The undrained shear strength parameters obtained are denoted by c_u and ϕ_u. The test results are acceptable for soils having no friction or little friction. The failure envelopes for the two cases are shown in Fig. 12.12 and Fig 12.13.

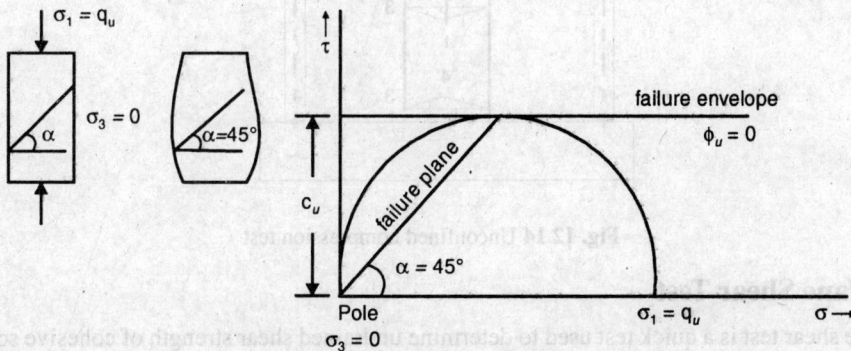

Fig. 12.12 Unconfined compression test ($\phi_u = 0$ soil)

The angle α which the failure plane makes with the horizontal is measured after carefully sketching the failed specimen. In Fig 12.14 is shown the equipment for conducting unconfined compression test in the laboratory.

Unconfined compression test can also be conducted in field. The soil specimen is placed between two conical seatings attached to two metal plates. The soil specimen is loaded through a calibrated spring by manually operated screw jack at the top of the machine. Then a graph of load versus deformation can be plotted.

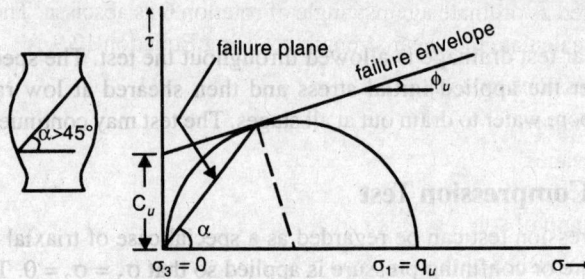

Fig. 12.13 Unconfined compression test (c-ϕ soil)

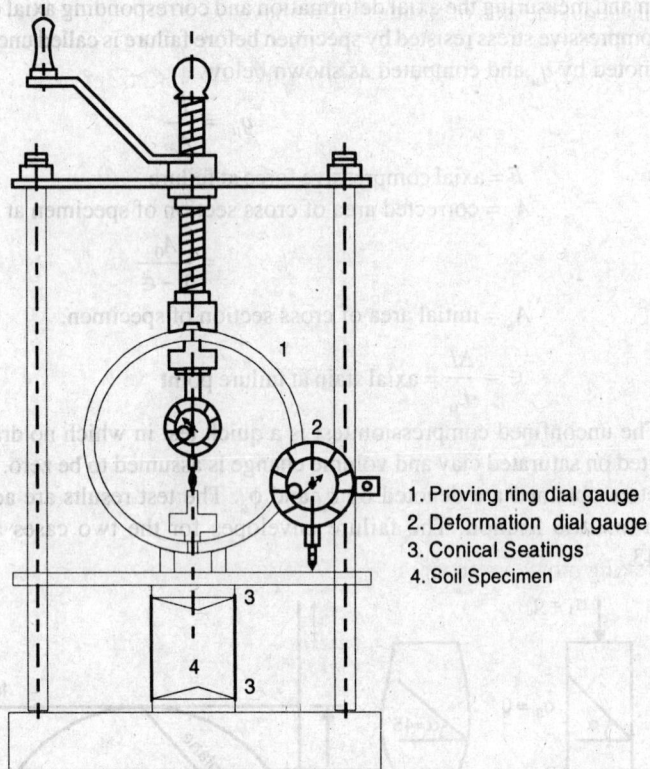

1. Proving ring dial gauge
2. Deformation dial gauge
3. Conical Seatings
4. Soil Specimen

Fig. 12.14 Unconfined compression test

12.10 Vane Shear Test

Vane shear test is a quick test used to determine undrained shear strength of cohesive soils. The equipment essentially consists of four high tensile steel plates called vanes which are welded orthogonally to the bottom end of a steel rod called the torque rod with an arrangement to measure the torque and rotation. A typical arrangement consists of a calibrated torsion spring attached to the top of torque rod which is rotated by a combination of worm gear and worm wheel. The vane shear test can be conducted both in the laboratory and in field. A typical laboratory set of vanes has 20 mm height, diameter of 12 mm across vanes with blade thickness of 0.5 to 1 mm. The field set of vanes usually is 100 to 200 mm in height, 50 to 100 mm in diameter across vanes with blade thickness of 2.5 mm. To conduct the test the vanes are gently pushed into the soil and the torque rod is rotated at a uniform rate of usually 1° per minute. The torque T corresponding to angle of rotation θ at uniform interval are noted. Torque T is plotted as ordinate against angle of rotation θ as abscissa. The torque T_f at failure is found and is used to calculate the shear strength τ_f, using Equation 12.9.

Derivation of expressions for torque.

Let T_f = torque at failure

H = height of vanes

d = diameter across vanes

τ_f = shear strength of soil

Case (*i*): The vane is pushed with its top end below the surface of soil so that both top and bottom ends partake in shearing.

We note that shearing takes place along cylindrical surface of diameter d and height H. Taking moments about the axis of torque rod, we have

$$\tau_f = (\pi d H \tau_f)\frac{d}{2} + 2\int_0^{d/2} (2\pi r dr \tau_f) . r$$

$$= \frac{\pi d^2 H \tau_f}{2} + 4\pi\tau_f \int_0^{d/2} r^2 dr$$

$$= \frac{\pi d^2 H \tau_f}{2} + 4\pi\tau_f \left[\frac{r^3}{3}\right]_0^{d/2}$$

$$= \frac{\pi d^2 H \tau_f}{2} + \frac{\pi d^3}{6}\tau_f$$

$$= \pi d^2 \tau_f \left[\frac{H}{2} + \frac{d}{6}\right] \qquad \qquad ...(12.8)$$

Case (*ii*) The vane is pushed inside the soil with its top end flush with surface of soil so that only bottom end partakes in shearing.

Taking moments about the axis of torque rod we have

$$T_f = (\pi d H \tau_f)\frac{d}{2} + \int_0^{d/2} (2\pi r dr \tau_f). r$$

$$T_f = \tau_f \pi d^2 \left[\frac{H}{2} + \frac{d}{12}\right] \qquad \qquad ...(12.9)$$

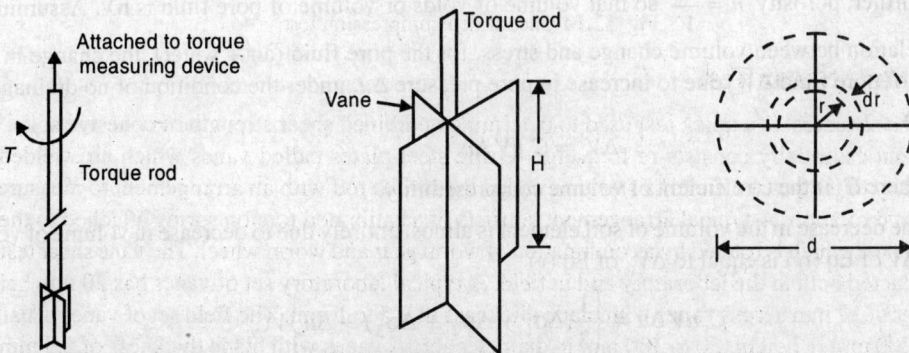

Fig. 12.15 Vane shear test

12.11 Skempton's Pore Pressure Parameters

During undrained shear a change in applied stress causes change in pore pressure and the relation between the two is expressed in terms of empirical coefficients called pore pressure parameters. The change in pore pressure is a fraction of the change in applied stress. The dimensionless quantity representing that fraction is called a pore pressure parameter.

Let $\Delta\sigma_1$, $\Delta\sigma_2$ and $\Delta\sigma_3$ be the increase in the three principle stresses acting on a soil element. Let ΔV and Δu represent change in volume and increase in pore pressure respectively.

The increase in the effective principal stresses will then be

$$\Delta\sigma_1' = \Delta\sigma_1 - \Delta u$$
$$\Delta\sigma_2' = \Delta\sigma_2 - \Delta u \qquad \qquad ...(i)$$
$$\Delta\sigma_3' = \Delta\sigma_3 - \Delta u$$

The strains in the three directions are given by

$$\epsilon_1 = \frac{1}{E}[\Delta\sigma_1' - \mu(\Delta\sigma_2' + \Delta\sigma_3')]$$

$$\epsilon_2 = \frac{1}{E}[\Delta\sigma_2' - \mu(\Delta\sigma_1' + \Delta\sigma_3')]$$

$$\epsilon_3 = \frac{1}{E}[\Delta\sigma_3' - \mu(\Delta\sigma_1' + \Delta\sigma_2')] \qquad \qquad ...(ii)$$

By addition, we have volumetric strain

$$\epsilon_v = \epsilon_1 + \epsilon_2 + \epsilon_3 = \frac{\Delta V}{V} = \frac{(1 - 2\mu)}{E}(\Delta\sigma_1' + \Delta\sigma_2' + \Delta\sigma_3')$$

$$= \frac{3(1 - 2\mu)}{E}\frac{1}{3}(\Delta\sigma_1' + \Delta\sigma_2' + \Delta\sigma_3') \qquad \qquad ...(iii)$$

We put $\dfrac{3(1 - 2\mu)}{E} = C_c$ which represents the compressibility of the soil element and substituting from Eq. (i) in Eq. (iii) obtain

$$\frac{\Delta V}{V} = C_c\frac{1}{3}(\Delta\sigma_1 + \Delta\sigma_2 + \Delta\sigma_3 - 3\Delta u)$$

i.e.,
$$\frac{\Delta V}{V} = C_c\left\{\frac{1}{3}(\Delta\sigma_1 + \Delta\sigma_2 + \Delta\sigma_3) - \Delta u\right\} \qquad \qquad ...(iv)$$

Further, porosity $n = \dfrac{V_v}{V}$ so that volume of voids or volume of pore fluid is nV. Assuming a linear relation between volume change and stress, for the pore fluid (air + water), the change in the volume of pore fluid ΔV_w due to increase in pore pressure Δu, under the condition of no drainage is given by

$$\Delta V_w = C_v nV\Delta u \qquad \qquad ...(v)$$

where C_v is the co efficient of volume compressibility.

The decrease in the volume of soil element is almost entirely due to decrease in volume of voids so that ΔV of Eq (iv) is equal to ΔV_w of Eq (v)

$$C_v nV\Delta u = \left\{\frac{1}{3}(\Delta\sigma_1 + \Delta\sigma_2 + \Delta\sigma_3) - \Delta u\right\}VC_c$$

$$(C_v n + C_c)\Delta u = C_c\frac{1}{3}(\Delta\sigma_1 + \Delta\sigma_2 + \Delta\sigma_3)$$

$$\Delta u = \frac{C_c}{C_c + C_v n}\left[\frac{1}{3}(\Delta\sigma_1 + \Delta\sigma_2 + \Delta\sigma_3)\right]$$

$$\Delta u = \frac{1}{1 + n\dfrac{C_v}{C_c}}\left[\frac{1}{3}(\Delta\sigma_1 + \Delta\sigma_2 + \Delta\sigma_3)\right] \qquad ...(vi)$$

In the conventional triaxial test. we have $\Delta\sigma_2 = \Delta\sigma_3$

$$\Delta u = \frac{1}{1 + n\dfrac{C_v}{C_c}}\left[\frac{1}{3}(\Delta\sigma_1 + 2\Delta\sigma_3)\right]$$

$$\Delta u = \frac{1}{1 + n\dfrac{C_v}{C_c}}\left[\frac{1}{3}(\Delta\sigma_1 - \Delta\sigma_3) + \Delta\sigma_3\right] \qquad ....(vii)$$

Clearly, Eq (vii) is obtained assuming the soil element to be elastic and isotropic but this assumption is gross over simplification in this case. Hence Skempton suggested that Eq (vii) may be written in the following form

$$\Delta u = B\{\Delta\sigma_3 + A(\Delta\sigma_1 - \Delta\sigma_3)\} \qquad ...(12.10)$$

B and A are known as Skempton's pore pressure parameters and are to be determined experimentally

To determine the pore pressure parameters, stress changes are made in two stages in an undrained triaxial test. In the first stage cell pressure in increased. In the second stage the deviator stress is increased keeping cell pressure constant. Let Δu_1 and Δu_2 represent the changes in pore pressure in the two stages. Then the total change in pore pressure in given by

$$\Delta u = \Delta u_1 + \Delta u_2 \qquad ...(12.11)$$

Comparing Eq 12.10 and 12.11, we have

$$\Delta u_1 = B\Delta\sigma_3$$

or

$$\frac{\Delta u_1}{\Delta\sigma_3} = B \qquad ...(12.12)$$

and

$$\Delta u_2 = BA\,(\Delta\sigma_1 - \Delta\sigma_3)$$

$$= \bar{A}\,(\Delta\sigma_1 - \Delta\sigma_3)$$

or

$$\bar{A} = \frac{\Delta u_2}{(\Delta\sigma_1 - \Delta\sigma_3)} \qquad ...(12.14)$$

Factors affecting Skempton's pore pressure parameters are highlighted in the following discussion.

We observe that

$$B = \frac{1}{1 + n\dfrac{C_v}{C_c}}$$

In the case of fully saturated soil the pore fluid being water, C_v is too small compared to C_c so that the ratio $\dfrac{C_v}{C_c}$ is very small and can be approximately taken as zero. Hence for a fully saturated soil $B = 1$. In the case of dry soil, the pore fluid being only air, C_v is too large compared to C_c so that the ratio

$\dfrac{C_v}{C_c}$ approaches infinity. Hence for a dry soil $B = 0$. For partially saturated soil B lies between 0 and 1. For soils compacted at optimum moisture content as determined from standard Proctor test, B is found to vary typically from 0.1 to 0.5.

Fig 12.16 illustrates variation of B with degree of saturation.

A typical plot of parameter A against axial strain is shown in Fig 12.17 and it indicates that A depends on strain level.

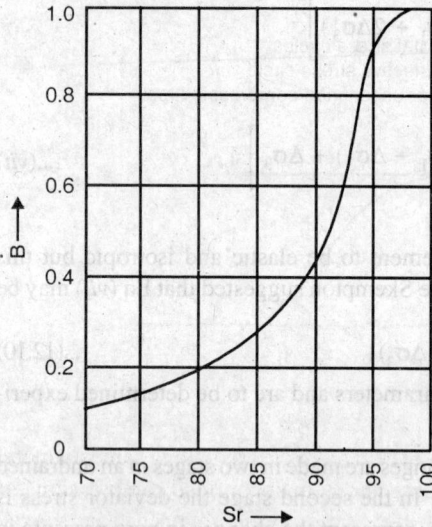

Fig. 12.17 A versus strain

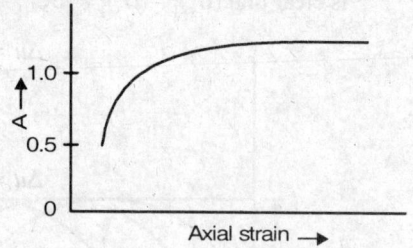

Fig. 12.16 Variation of B with S_r

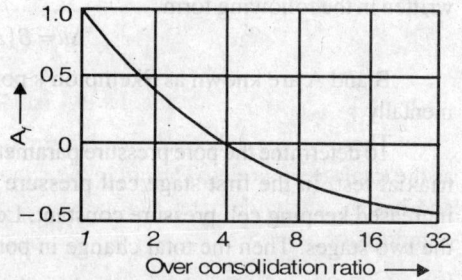

Fig. 12.18 Effect of preconsolidation on A

In practice it is usual to quote A at failure (maximum deviator stress) or at maximum effective principal stress ratio. The value of A at quoted stress level also depends upon whether the total stresses are increasing or decreasing. The stress history of the soil has a significant influence on parameter A. Preconsolidation of test specimen is found to reduce considerably the parameter A as is clear from Fig 12.18. Type of shear, degree of sample disturbance and nature of fluid can be cited as other factors which affect A. In Table 12.1 typical value of A (at failure) are presented for different types of soils.

Table 12.1 Typical values of A (at failure)

Soil type	A (at failure)
Very loose fine saturated sand	2 to 3
Saturated clay :	
Extra sensitive to quick	1.2 to 2.5
Normally consolidated	0.7 to 1.3
Lightly preconsolidated	0.3 to 0.7
Heavily preconsolidated	–0.5 to 0
Compacted sandy clays	0.25 to 0.75
Compacted clayey gravel	–0.25 to 0.25

12.12 Brief discussion on shear strength of different soil types

I Shear strength of fully saturated cohesive soils

(a)   Undrained test : When undrained tests are conducted on identical specimens of a fully saturated clay with different cell pressures, the Mohr circles obtained will all be of same diameter as shown in Fig 12.19. The failure envelope is horizontal giving $\phi_u = 0$ and $\tau_f = c_u$. It is clear that $(\sigma_1)_I - (\sigma_3)_I = (\sigma_1)_{II} - (\sigma_3)_{II}$.

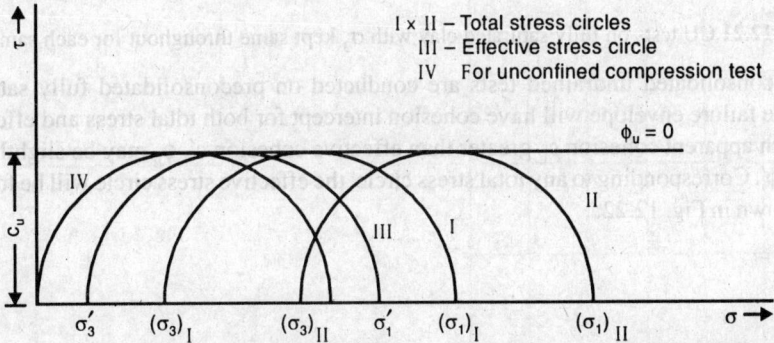

Fig. 12.19 Undrained test on fully saturated clay

There will be only one effective stress circle as represented by circle III, having same diameter as the total stress circles. The circle IV is the Mohr circle obtained from unconfined compression test and has the same diameter as total stress circles and effective stress circle obtained from triaxial compression test.

(b)   Consolidated-undrained test : If consolidated undrained tests are conducted on remoulded fully saturated clay specimens initially consolidated with same cell pressure and subsequently sheared under undrained condition with different cell pressures the Mohr Circles obtained will be of same diameter as shown in Fig. 12.20. The failure envelope is horizontal with $\phi_u = 0$ and $\tau_f = c_u$. Further if the tests are conducted with specimens initially consolidated with same but increased cell pressure and then sheared with different cell pressures, the failure envelope obtained is still horizontal with $\phi_u = 0$ but c_u will be greater than in the previous case.

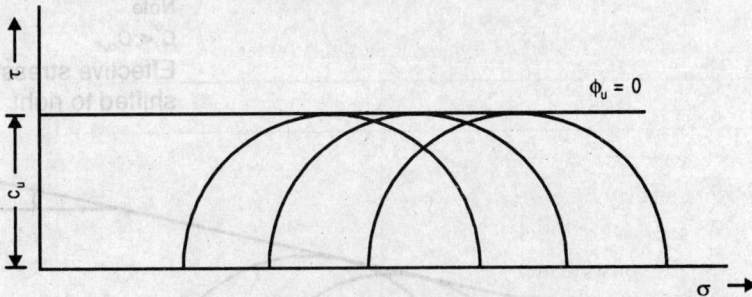

Fig. 12.20 CU test on fully saturated clay

Further it is usual in laboratory practice to initially consolidate a specimen with a certain cell pressure and then shear the specimen under undrained condition at the same cell pressure. This test repeated for different specimens with different values of cell pressure will give rise to total stress envelopes and effective stress envelopes both passing through the origin of stresses, as shown in Fig. 12.21.

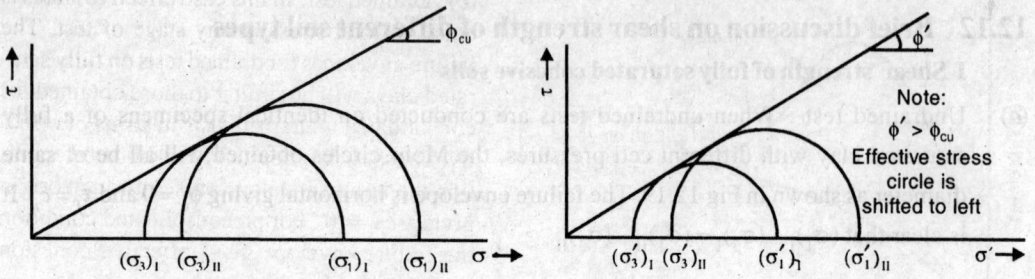

Fig. 12.21 CU tests on fully saturated clay with σ_3 kept same throughout for each sample

When consolidated undrained tests are conducted on preconsolidated fully saturated clay specimens the failure envelope will have cohesion intercept for both total stress and effective stress plottings, with apparent cohesion c_u greater than effective cohesion c'. ϕ_u may be slightly greater or smaller than ϕ'. Corresponding to any total stress circle, the effective stress circle will be found shifted to right as shown in Fig. 12.22.

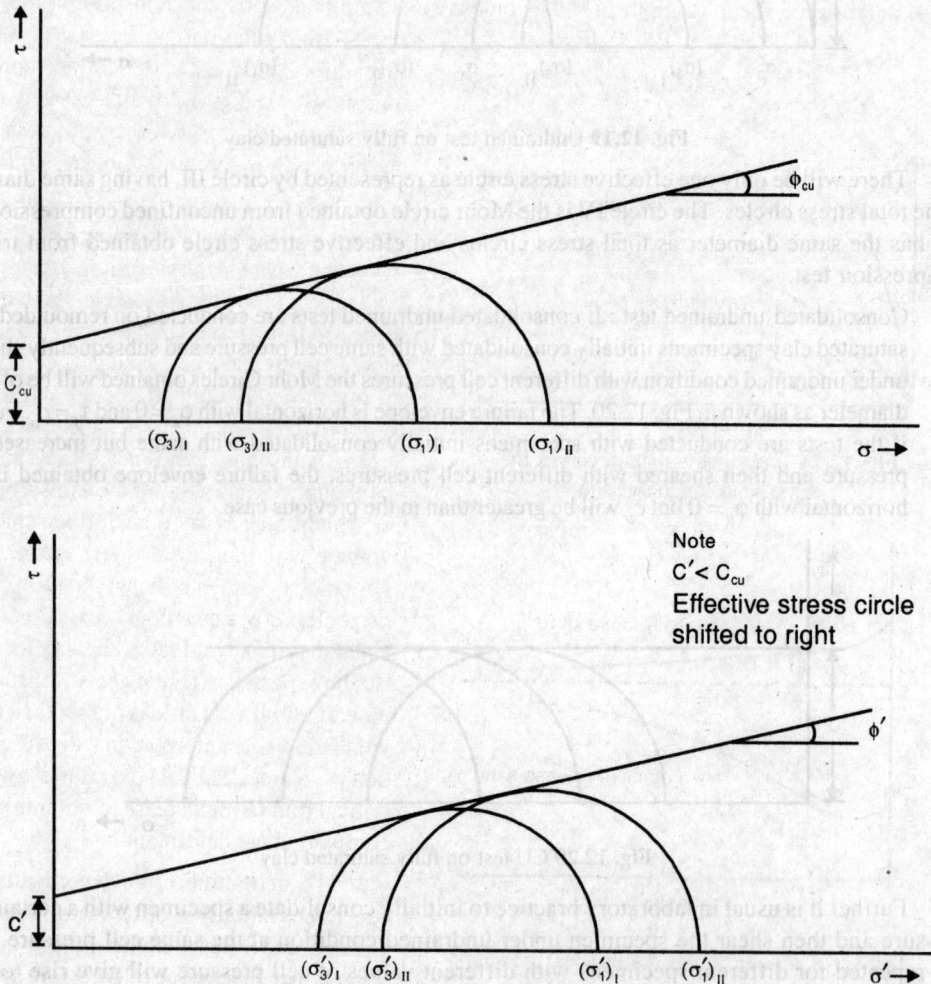

Fig. 12.22 CU tests on preconsolidated fully saturated clay

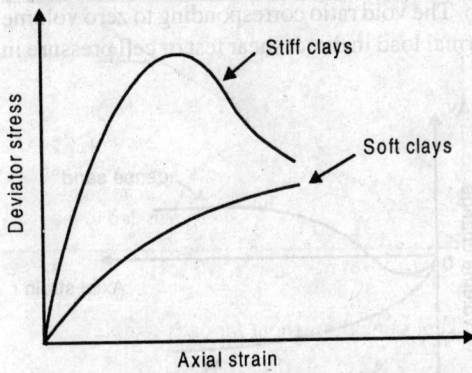

Fig. 12.23 Typical stress-strain curves for clays

(c) Drained test: In this case effective stress is equal to total stress at any stage of test. The failure envelopes for drained tests on fully saturated clays will be similar to those obtained for consolidated undrained tests in terms of effective stresses. For normally consolidated condition the failure envelope passes through origin giving $c_d = 0$. For preconsolidated condition the failure envelope gives effective cohesion intercept, c'. The angle ϕ' will be slightly different for the two cases.

In Fig. 12.23 typical stress-strain curves for clays are shown.

Shear strength of cohesionless soils

A prior discussion of stress-strain and volume change characteristics is helpful for proper understanding of shear strength of cohesionless soils (granular soils and non plastic silts). The stress-strain relation can be easily obtained from direct shear test under drained condition on saturated specimen or alternatively on dry specimen. In Fig 12.24, typical stress strain curves are shown for cohesionless soils.

Volume change characteristic during shear is best understood by examining change in void ratio with increasing strain. In Fig. 12.25, void ratio is plotted against shear strain for both loose sand and dense sand. It is clear that dense sand expands and loose sand gets compressed during shear. At high strains the curves tend to approach each other. At sufficiently high strains both dense sand and loose sand may be thought of as attaining the same void ratio at which there will be no further change in volume with increase in shear strain. Such a void ratio is termed critical void ratio. If a cohesionless soil is having its initial void ratio equal to critical void ratio, shear deformation can be expected to take place at constant volume. If initial void ratio is higher than critical void ratio, shear deformation will cause reduction in volume. If initial void ratio is lower than critical roid ratio, increase in volume will accompany shear deformation.

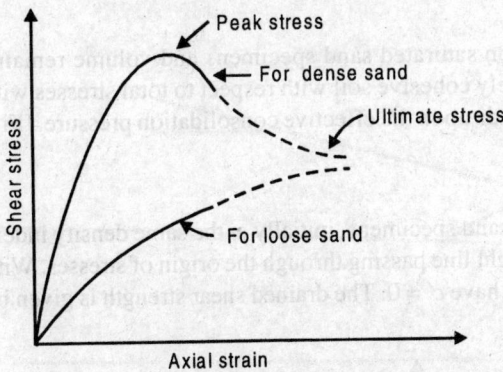

Fig. 12.24 Typical stress-strain curves for dense sand and loose sand

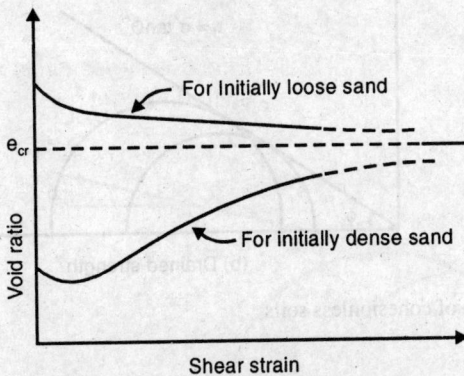

Fig. 12.25 Volume change during shear in sands

To determine critical void ratio, soil specimens initially at different void ratios are sheared under same normal stress in direct shear test or under same cell pressure in triaxial shear test. The initial void ratio values are plotted against measured volume change for each specimen. The void ratio corresponding to zero volume change gives the critical roid ratio for the particular normal load in direct shear test or cell pressure in triaxial shear test. (Fig. 12.26)

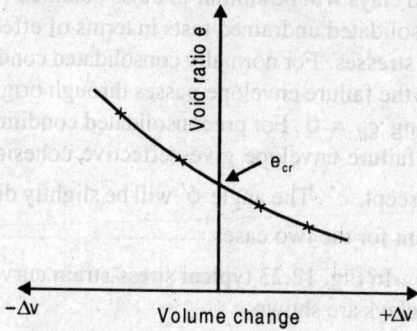

Fig 12.26 Determination of critical void ratio

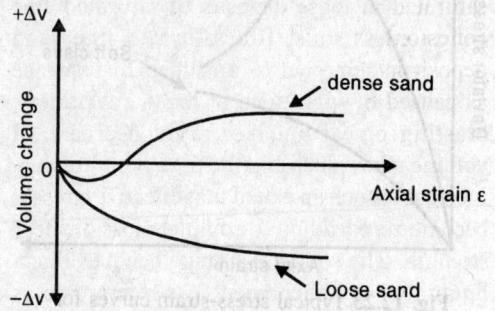

Fig. 12.27 Volume change in cohesionless soil

In Fig 12.27 volume change is plotted against axial strain for both dense sand and loose sand. In shearing of dense sand there will initially be slight decrease in volume and then sand expands or dilates. A loose sand gets compressed when sheared.

Undrained strength of saturated sand

If undrained triaxial tests are conducted on saturated sand specimens and volume remains essentially constant, the sand behaves like a purely cohesive soil with respect to total stresses with $\phi_u = 0$. The undrained shear strength $\tau_f = c_u$ depends upon the effective consolidation pressure. [Fig. 12.28 (a)]

Drained strength of saturated sand

If drained tests are conducted on saturated sand specimens, initially at the same density index, the failure envelope will be approximately a straight line passing through the origin of stresses. With effective stresses being equal to total stresses, we have $c' = 0$. The drained shear strength is given by $\tau_f = \sigma' \tan \phi'$ [Fig. 12.28 (b)]

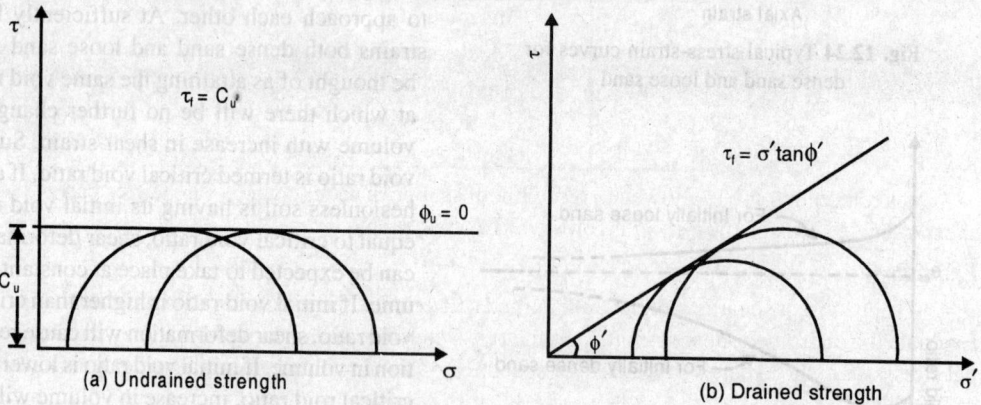

(a) Undrained strength

(b) Drained strength

Fig. 12.28 Shear strength of cohesionless soils

Angle of repose

When a cohesionless soil is poured from a small height it forms a heap. The angle between the sloping side of heap and horizontal is referred to as angle of repose. The angle of repose will be nearly equal to ϕ' the angle of shearing resistance obtained from shear tests on the same cohesionless soil in loose state.

Liquefaction

Liquefaction is a phenomenon which can saturated in loose deposits of saturated fine cohesionless soils. If a saturated fine sand deposit is subjected to a sudden disturbance as caused by vibrations of heavy machinery, blasting or earthquake, rapid decrease in volume takes place and the pore pressure may increase to such an extent that effective stresses become zero leading to complete loss of shear strength. The soil at this stage behaves like a liquid and the phenomenon is referred to as liquefaction.

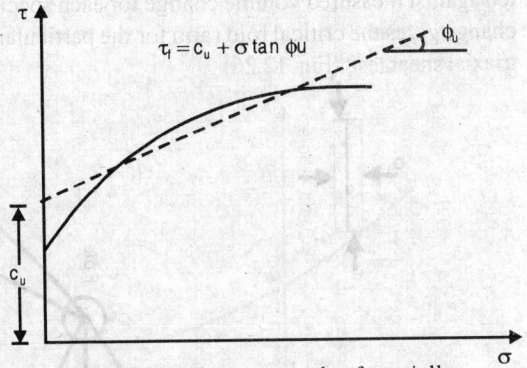

Fig. 12.29 Shear strength of partially saturated cohesive soil

Shear strength of partially saturated clays and composite soils

The failure envelopes obtained by conducting shear tests on partly saturated clays and composite soils will be curved as shown in Fig. 12.29. For any specific pressure range the curved envelope can be approximated by a straight line to obtain c and φ in terms of total stresses.

12.13 Sensitivity and Thixotropy

When undisturbed saturated clays are disturbed or remoulded without change in water content, they lose part of their shear strength. This phenomenon is referred to as sensitivity. The degree of sensitivity is given by the ratio of undisturbed shear strength to the remoulded shear strength under undrained condition.

$$\text{Sensitivity} = \frac{\tau_f (\text{undisturbed})}{\tau_f (\text{remoulded})} = \frac{c_u (\text{undisturbed})}{c_u (\text{remoulded})}$$

The sensitivity of clays is found to vary from about 1 to over 100. A typical classification of clays based on sensitivity is given in the following table.

Sensitivity	Classification of clay
< 2	Insensitive
2 – 4	Medium sensitive
4 – 8	Sensitive
8 – 16	Extra sensitive
> 16	Quick

If a remoulded soil is allowed to rest without change in water content, it regains a part of the lost shear strength. This phenomenon is referred to as thixotropy.

Example 12.1.

From a direct shear test on an undisturbed soil sample, the following data have been obtained. Evaluate the undrained shear strength parameters. Determine shear strength, major and minor principal stresses and their planes in the case of specimen of same soil sample subjected to a normal stress of $100 \, \text{kN/m}^2$.

Normal stress (kN/m²)	70	96	114
Shear stress at failure (kN/m²)	138	156	170

Answer : From plot on page 224,

$c = 84 \, \text{kN/m}^2$, $\phi = 36°$

When $\sigma = 100 \, \text{kN/m}^2$, $\tau_f = 160 \, \text{kN/m}^2$

$\sigma_1 = 410 \, \text{kN/m}^2$, $\sigma_3 = 20 \, \text{kN/m}^2$

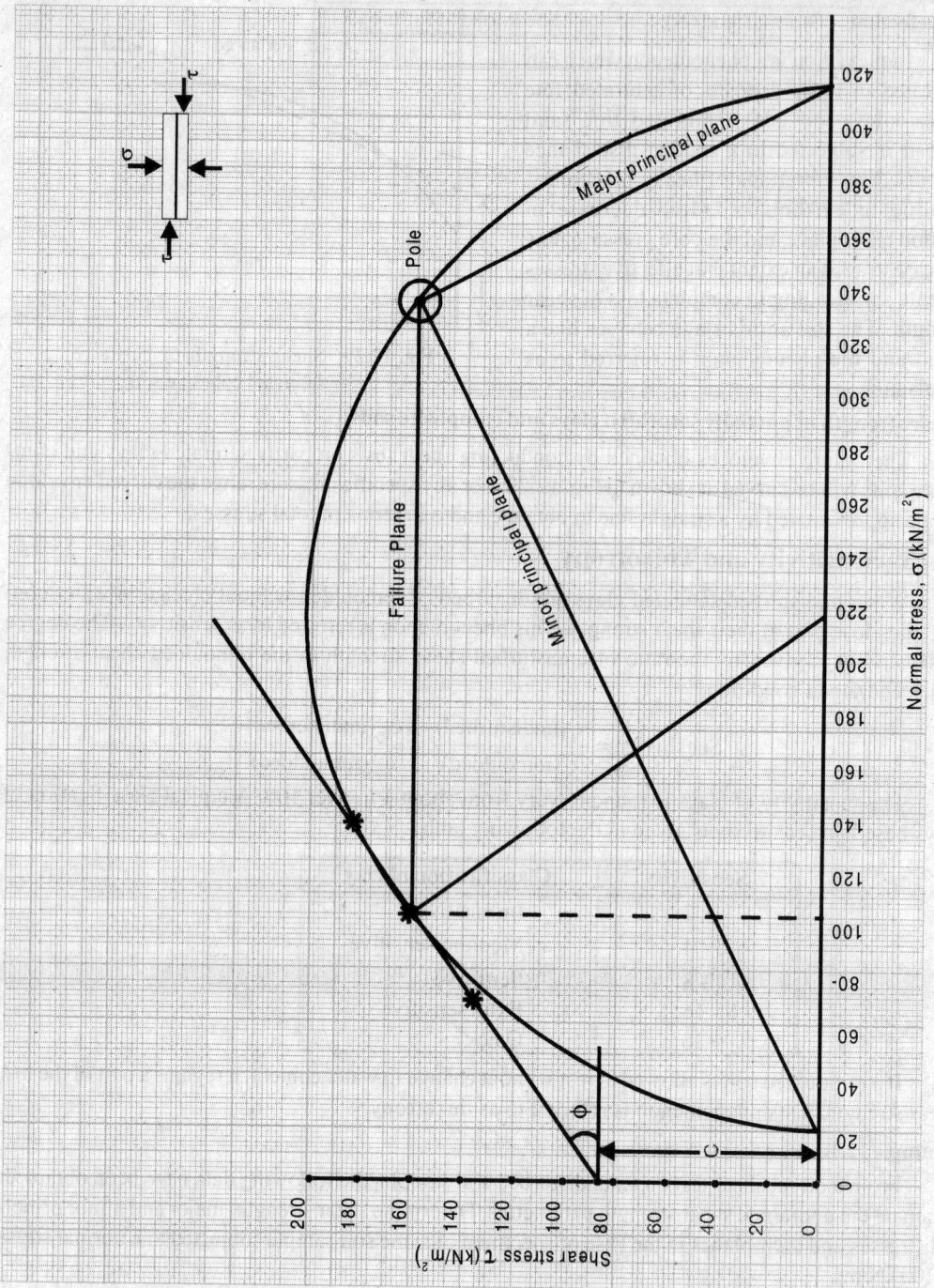

Example 12.2.

The following table gives data obtained from a direct shear test conducted on samples of compacted sand. The shear box dimensions are 60mm x 60mm

Normal load (N)	Shear load at failure (N)	
	Peak	Ultimate
110	95	65
225	195	135
340	294	200

Determine peak and ultimate angles of shearing resistance of sand.

Solution :

Area of specimen $= 60 \times 60 = 3600$ mm^2

Normal stress (kN/m^2)	Shear stress at failure (kN/m^2)	
	Peak	Ultimate
30.5	26.4	18.0
62.5	54.2	37.5
94.4	81.2	55.5

Answer :

From the plot of shear stress versus normal stress,

$\phi_{peak} = 40°$

$\phi_{ult} = 30°$

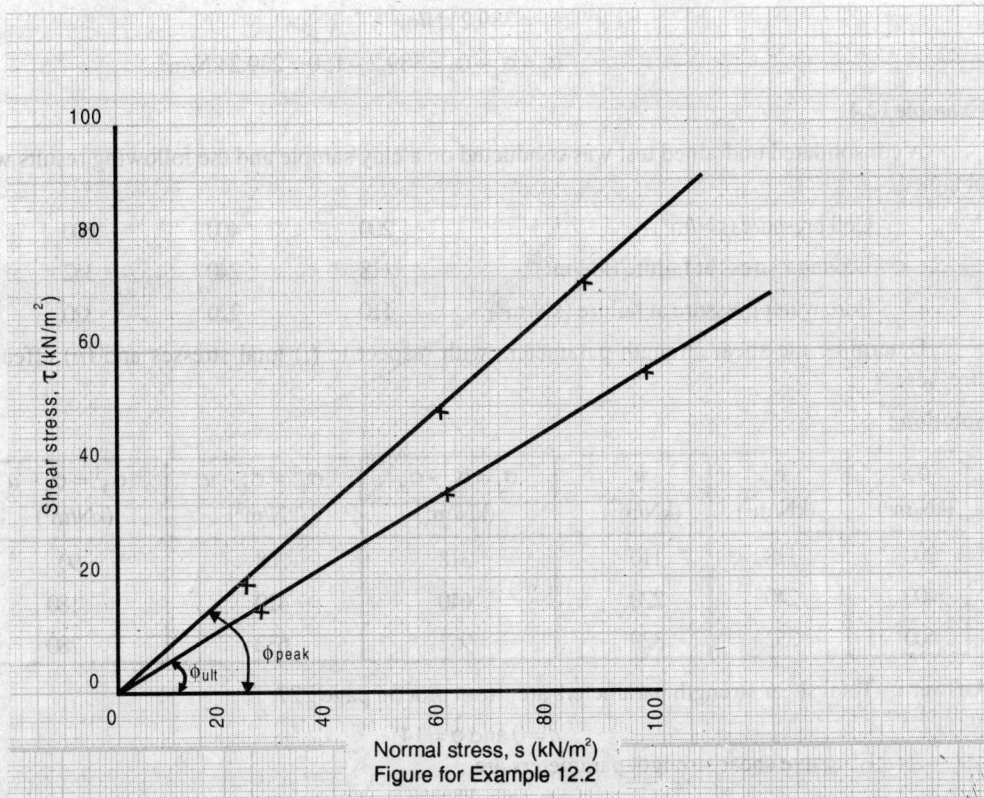

Figure for Example 12.2

Example 12.3

A direct shear test was carried out on a cohesive soil sample and the following results were obtained.

Normal stress (kN/m^2)	150	250
Shear stress at failure (kN/m^2)	110	120

What would be the deviator stress at failure if a triaxial test is carried out on the same soil with cell pressure of 150 kN/m^2.

Solution :

$$\tau_f = c + \sigma \tan \phi$$

By substitution, we get

$$110 = c + 150 \tan \phi$$

$$120 = c + 250 \tan \phi$$

Solving the above set of simultaneous equations, we obtain

$$c = 93.7 \text{ kN/m}^2$$

$$\phi = 5.7°$$

Further,

$$\sigma_1 = \sigma_3 \tan^2 \alpha + 2c \tan \alpha$$

where

$$\alpha = 45° + \frac{\phi}{2} = 45° + \frac{5.7°}{2} = 47.8°$$

When

$$\sigma_3 = 150 \text{ kN/m}^2$$

$$\sigma_1 = 150 \tan^2 47.8° + 2 (93.7) \tan 47.8°$$

$$= 389.2 \text{ kN/m}^2$$

∴

$$\sigma_d = \sigma_1 - \sigma_3 = 389.2 - 150 = 239.2 \text{ kN/m}^2$$

Example 12.4

A consolidated undrained test was conducted on a clay sample and the following results were obtained.

Cell pressure (kN/m^2)	200	400	600
Deviator stress at failure (kN/m^2)	118	240	352
Pore water pressure at failure (kN/m^2)	110	220	320

Determine the shear strength parameters with respect to (*i*) total stresses and (*ii*) effective stresses.

Solution :

σ_3 (kN/m^2)	σ_d (kN/m^2)	u (kN/m^2)	$\sigma_1 = \sigma_3 + \sigma_d$ (kN/m^2)	$\sigma_1{}^1 = \sigma_1 - u$ (kN/m^2)	$\sigma_3{}^1 = \sigma_3 - u$ (kN/m^2)
200	118	110	318	208	90
400	240	220	640	420	180
600	352	320	952	632	280

Answer : Total shear strength parameters are, from plot on page 227,

$$c = 0 \text{ and } \phi = 14°$$

Effective shear strength parameters are

$$c' = 0 \text{ and } \phi' = 24°$$

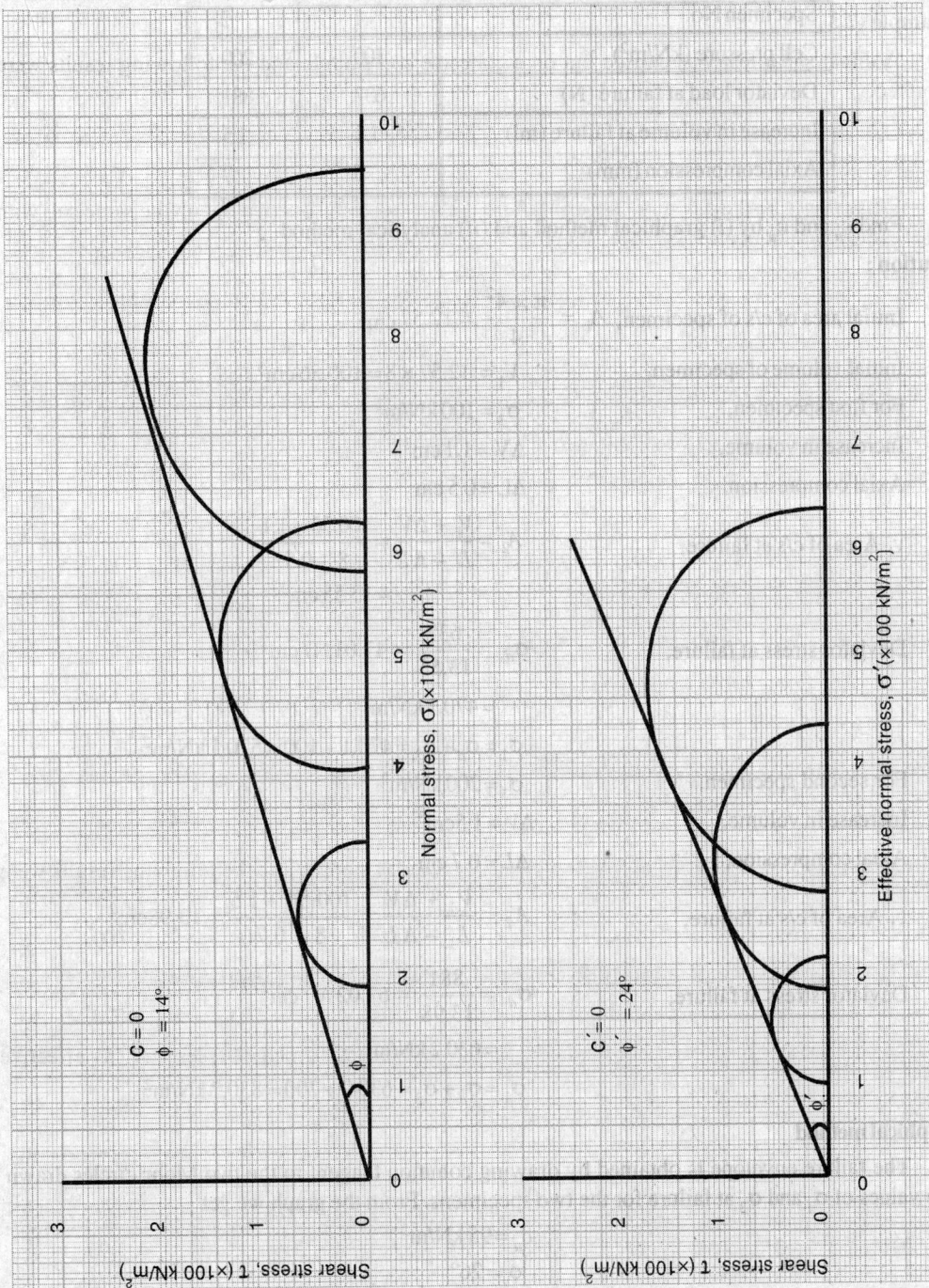

Example 12.5

The following table gives data obtained from triaxial compression test conducted under undrained conditions on two specimens of same soil sample. The diameter and height are 40 mm and 80 mm respectively for both samples.

Specimen No.	1	2
Cell pressure (kN/m^2)	100	200
Deviator load at failure (N)	637	881
Increase in volume at failure (ml)	1.1	1.5
Axial compression (mm)	5	7

Find c_u and ϕ_u by (i) graphical method and (ii) analytical method.

Solution :

Initial area of c/s of specimen, $A_1 = \dfrac{\pi \times 4^2}{4} = 12.57 \, cm^2$

Initial volume of specimen, $\quad V_1 = 12.57 \times 8 = 100.56 \, cm^3$

For first specimen, $\quad \sigma_3 = 100 \, kN/m^2$

Increase in volume, $\quad \Delta V = 1.1 \, cm^3$

Axial compression, $\quad \Delta L = 0.5 \, cm$

$\therefore$ Area of c/s at failure, $\quad A_2 = \dfrac{V_1 + \Delta V}{L_1 - \Delta L} = \dfrac{100.56 + 1.1}{8 - 0.5}$

$= 13.55 \, cm^2$

Deviator stress at failure, $\quad \sigma_d = \dfrac{637}{13.55} = 47.01 \, N/cm^2$

$= 470.1 \, kN/m^2$

$\sigma_1 = \sigma_d + \sigma_3 = 470.1 + 100 = 570.1 \, kN/m^2$

For second specimen, $\quad \sigma_3 = 200 \, kN/m^2$

Increase in volume, $\quad \Delta V = 1.5 \, cm^3$

Axial compression, $\quad \Delta L = 0.7 \, cm$

$\therefore$ Area of c/s at failure, $\quad A_2 = \dfrac{V_1 + \Delta V}{L_1 - \Delta L} = \dfrac{100.56 + 1.5}{8 - 0.7} = 13.98 \, cm^2$

Deviator stress at failure, $\quad \sigma_d = \dfrac{881}{13.98} = 63.02 \, N/cm^2$

$= 630.2 \, kN/m^2$

$\sigma_1 = \sigma_d + \sigma_3 = 630.2 + 200 = 830.2 \, kN/m^2$

Graphical method

The failure envelope is obtained by drawing common tangent to the two Mohr circles drawn with values of σ_1 and σ_3 at failure for the two specimens. From the graph we get

$c_u = 95 \, kN/m^2$

$\phi_u = 26°$

Analytical method

We have $\sigma_1 = \sigma_3 \tan^2 \alpha + 2c \tan \alpha$

Substituting values of σ_1 and σ_3, obtained from tests on the two specimens, in the above equation we get the following two equations.

$570.1 = 100 \tan^2 \alpha + 2c \tan \alpha \qquad (i)$

$830.2 = 200 \tan^2 \alpha + 2c \tan \alpha \qquad (ii)$

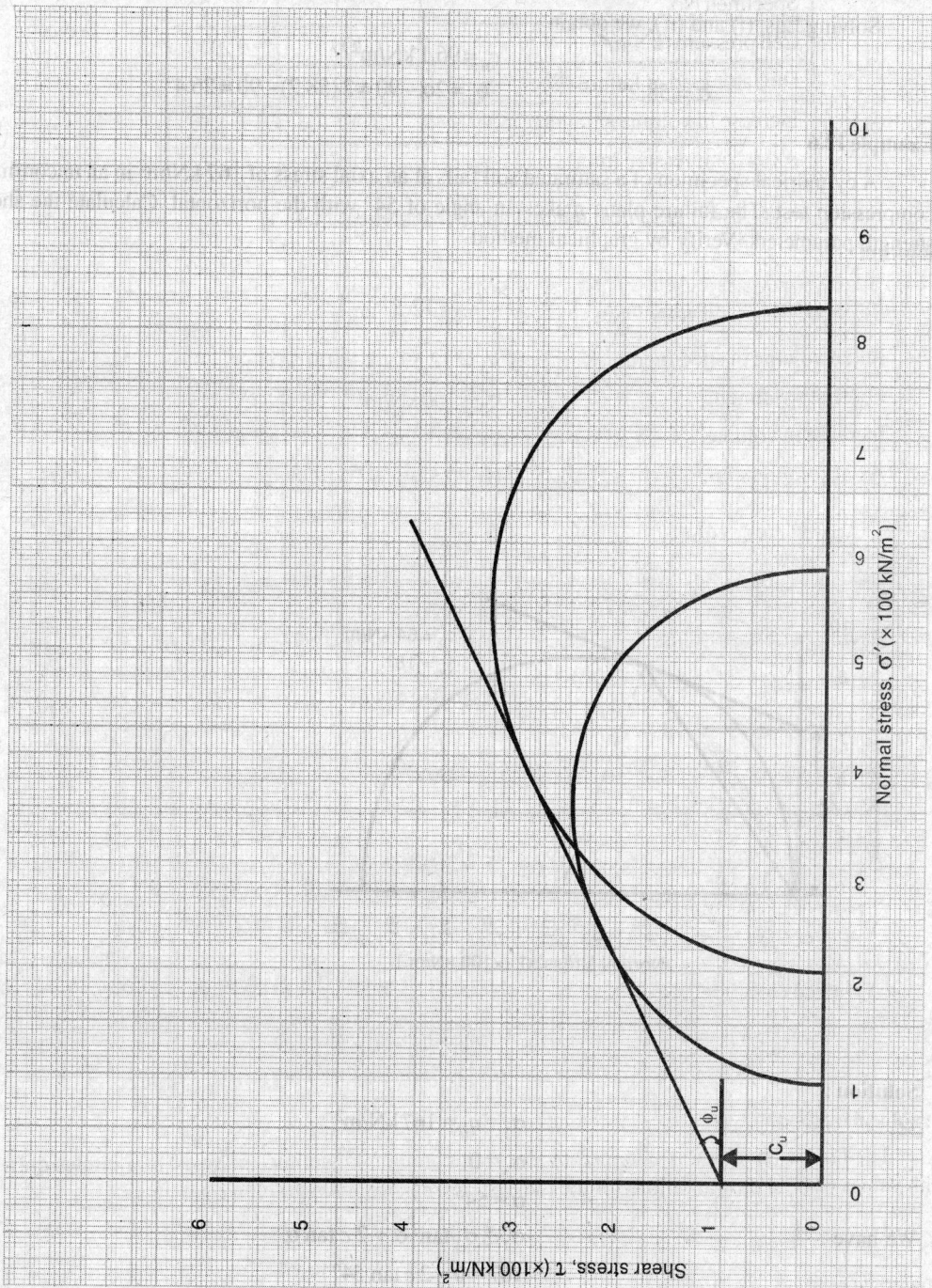

Figure for Example 12.5

Solving Eqs. (*i*) and (*ii*), we obtain

$$c_u = 96.1 \text{ kN/m}^2$$

$$\alpha = 58.2° \qquad \phi_u = 2\alpha - 90 = 2(58.2) - 90 = 26.4°$$

Example 12.6

A cylindrical specimen of a saturated soil fails at an axial stress of 167 kN/m^2 in an unconfined compression test. The failure plane makes an angle of 54° with the horizontal. Calculate the shear strength parameters. Verify by graphical method.

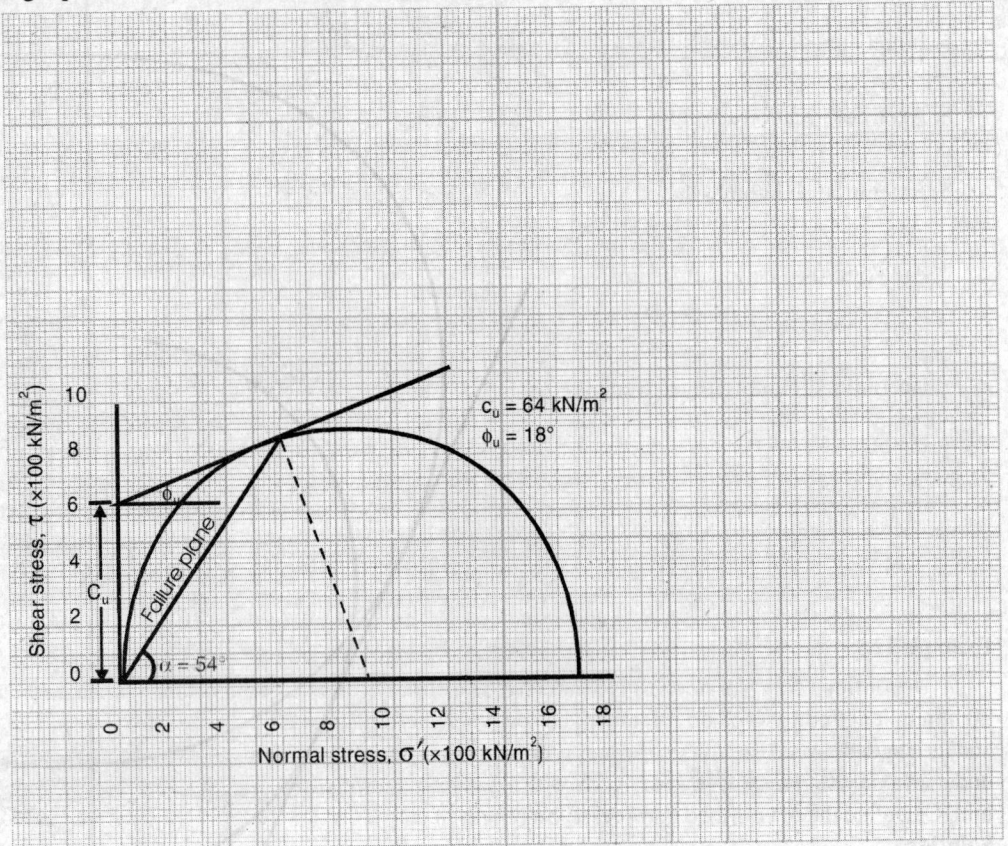

Solution :

$$\sigma_1 = q_u = 167 \text{ kN/m}^2$$

$$\sigma_3 = 0$$

$$\alpha = 54°$$

We have

$$\sigma_1 = \sigma_3 \tan^2 \alpha + 2c \tan \alpha$$

$$167 = 0 + 2c \tan 54°$$

∴

$$167 = 0 + 2c \tan 54°$$

$$\phi = 2\alpha - 90 = 2(54) - 90 = 18°, \quad c = \frac{167}{2 \tan 54°} = 64 \text{ kN/m}^2$$

Answer :

$$c_u = 64 \text{ kN/m}^2$$

$$\phi_u = 18°$$

Example 12.7

An unconfined compression test was conducted on an undisturbed sample of clay. The sample had a diameter of 38 mm and length 76 mm. The load at failure was 30 N and the axial deformation of the sample 11 mm. Determine the undrained shear strength parameters, if the failure plane made an angle of 50° with horizontal.

Solution :

Initial length of sample, $L_0 = 76$ mm

Initial diameter of sample, $d_0 = 38$ mm

Initial area of cross section, $A_0 = \dfrac{\pi(38)^2}{4} = 1134.1 \text{ mm}^2$

Axial deformation at failure, $\Delta L = 11 \text{ mm}$

Axial strain at failure, $\in = \dfrac{\Delta L}{L_0} = \dfrac{11}{76} = 0.145$

Area of cross section at failure, $A = \dfrac{A_0}{1 - \in} = \dfrac{1134.1}{1 - 0.145} = 1326.4 \text{ mm}^2$

Axial load at failure, $P = 30 \text{ N}$

Unconfined compressive strength, $q_u = \dfrac{P}{A} = \dfrac{30}{1326.4} = 0.023 \text{ N/mm}^2$

$\alpha = 50°$ $\therefore \phi_u = 2\alpha - 90° = 2(50) - 90 = 10°$

$$\sigma_1 = \sigma_3 \tan^2\alpha + 2c \tan\alpha$$

$$q_u = 0 + 2c \tan\alpha$$

$$c_u = \frac{q_u}{2\tan\alpha} = \frac{0.023}{2\tan 50°} = 0.0096 \text{ N/mm}^2$$

Example 12.8

Unconfined compression test is conducted on a saturated clay specimen 40 mm in diameter and 90 mm in length measured on its sides. The specimen has coned ends and its length between the apices of cones is 80 mm. The specimen fails under an axial compressive load of 460 N with axial deformation of 10 mm. Calculate the unconfined compressive strength of clay.

Solution :

Volume of specimen $= \dfrac{\pi(40)^2}{4}(90) - 2\left\{\dfrac{1}{3}\pi(20)^2(5)\right\} = 108908.5 \text{ mm}^3$

If L_1 is the length of cylinder having same volume but of uniform diameter throughout, we have

$$\frac{\pi(40)^2}{4} L_1 = 108908.5$$

$\therefore$ $L_1 = \dfrac{(108908.5)(4)}{\pi(40)^2} = 86.7 \text{ mm}$

Axial compression at failure, $\Delta L = 10 \text{ mm}$

Area of cross section at failure $A = \dfrac{A_0}{1 - \in} = \dfrac{\pi(40)^2/4}{1 - \dfrac{10}{86.7}} = 1420.4 \text{ mm}^2$

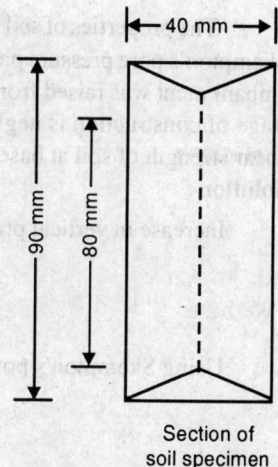

Unconfined compressive strength,

$$q_u = \frac{460}{1420.4} = 0.324 \, \text{N/mm}^2 = 324 \, \text{kN/m}^2$$

Also for fully saturated clay, $\phi_u = 0$

$$\therefore \qquad\qquad\qquad c_u = \frac{q_u}{2} = 162 \, \text{kN/m}^2$$

Example 12.9

In a vane shear test conducted in a soft clay deposit failure occured at a torque of 42 Nm. After wards the vane was allowed to rotate rapidly and the test was repeated in the remoulded soil. The torque at failure in the remoulded soil was 17 Nm. Calculate the sensitivity of soil. In both cases the vane was pushed completely inside soil. The height of vane and diameter across blades are 100 mm and 80 mm respectively.

Solution :

For vane, $H = 100$ mm and $d = 80$ mm. Since both ends of vane partake in shearing, we have

$$T = \pi d^2 \tau_f \left\{ \frac{H}{2} + \frac{d}{6} \right\}$$

(i)   For soil in natural state, $T = 42000 \, \text{N-mm}$

$$42000 = \pi (80)^2 \, \tau_f \left(\frac{100}{2} + \frac{80}{6} \right)$$

$$c = \tau_f = 0.033 \, \text{N/mm}^2$$

(ii)   For soil in remoulded state, $T = 17000 \, \text{N-mm}$

$$17000 = \pi (80)^2 \, \tau_f \left(\frac{100}{2} + \frac{80}{6} \right)$$

$$c = \tau_f = 0.013 \, \text{N/mm}^2$$

$$\therefore \qquad Sensitivity = \frac{(c) \, undisturbed}{(c) \, remoulded} = \frac{0.033}{0.013} = 2.54$$

Example 12.10

The properties of soil in a 3 m high embankment are $c' = 50 \, \text{kN/m}^2$, $\phi' = 20°$ and $\gamma = 16 \, \text{kN/m}^3$. Skempton's pore pressure parameters are found from triaxial test as $A = 0.5$ and $B = 0.9$. The height of embankment was raised from 3 m to 6 m. Assuming that the dissipation of pore pressure during this stage of construction is negligible and that lateral pressure is half of vertical pressure, estimate the shear strength of soil at base of embankment just after increasing the height of embankment.

Solution :

Increase in vertical pressure at base of embankment due to increase in height of embankment,

$$\Delta \sigma_1 = 16 \, (6 - 3) = 48 \, \text{kN/m}^2$$

$$\therefore \qquad \Delta \sigma_3 = \frac{1}{2} (\Delta \sigma_1) = \frac{1}{2} (48) = 24 \, \text{kN/m}^2$$

Using Skempton's pore pressure parameters, change in pore pressure is given by

$$\Delta u = B \, [\Delta \sigma_3 + A \, (\Delta \sigma_1 - \Delta \sigma_3)]$$

$$= 0.9 \, [24 + 0.5 \, (48 - 24)] = 32.4 \, \text{kN/m}^2$$

Initial vertical pressure at base of embankment,

$$\sigma_1 = 16 \, (3) = 48 \, \text{kN/m}^2$$

Effective stress at base of 6 m high embankment,

$$\sigma'_1 = (\sigma_1 + \Delta \sigma_1) - \Delta u$$

$$= (48 + 48) - 32.4 = 63.6 \, kN/m^2$$

Shear strength at base of 6m high embankment,

$$\tau_f = c^1 + \sigma \tan \phi^1$$
$$= 50 + 63.6 \tan 20° = 73.15 \, kN/m^2$$

Example 12.11

Following table gives deviator stress and pore pressure corresponding to different strains in an undrained triaxial test conducted on a compacted soil. The cell pressure was 300 kN/m^2 and pore pressure was zero before the application of cell pressure.

Strain (%)	Deviator Stress (kN/m²)	Pore Pressure (kN/m²)
0	0	250
2.5	560	280
5.0	945	150
7.5	1100	105
10.0	1150	70
12.5	1165	55
15.0	1140	50

Find

(i) Pore pressure parameter B,

(ii) Variation of A with strain

(iii) Variation of A with strain assuming the results are from consolidated undrained test. ($u = 0$ at $\sigma_d = 0$)

Solution :

(i) $\qquad B = \dfrac{\Delta u_1}{\Delta \sigma_3} = \dfrac{250}{300} = 0.83$

Table 1. Computation of A (UU test)

Strain %	u (kN/m²)	Δu_2 (kN/m²)	$(\Delta \sigma_1 - \Delta \sigma_3)$ (kN/m²)	$\bar{A} = \dfrac{\Delta u_2}{(\Delta \sigma_1 - \Delta \sigma_3)}$	$A = \dfrac{\bar{A}}{B}$
0	250	—	—		
2.5	280	30	560	0.053	0.64
5.0	150	– 100	945	– 0.106	– 0.128
7.5	105	– 145	1100	– 0.132	– 0.159
10.0	70	– 180	1150	– 0.156	– 0.188
12.5	55	– 195	1165	– 0.167	– 0.201
15.0	50	– 200	1140	– 0.175	– 0.211

Table 2. Computation of A_f (CU test)

Strain (%)	σ_3 (kN/m^2)	$(\Delta\sigma_d)_f$ (kN/m^2)	u_f (kN/m^2)	$A_f = \dfrac{u_f}{(\Delta\sigma_d)_f}$
0	300	0	0	–
2.5	"	560	280	0.5
5.0	"	945	150	0.159
7.5	"	1100	105	0.095
10.0	"	1150	70	0.061
12.5	"	1165	55	0.047
15.0	"	1140	50	0.044

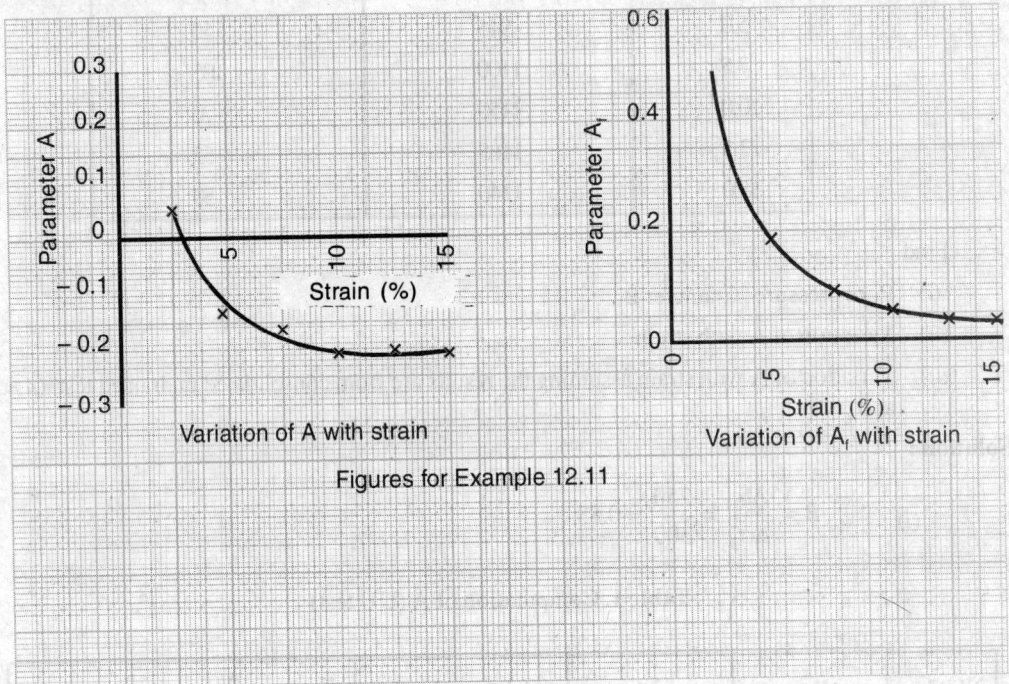

Variation of A with strain

Variation of A$_f$ with strain

Figures for Example 12.11

EXERCISE–12

12.1. Explain Mohr - Coulomb failure theory. Derive relation between principal stresses at failure and shear strength parameters.

12.2. Classify shear tests based on drainage conditions. How are these drainage conditions realised in the field.

12.3. Explain advantage of triaxial shear test over direct shear test.

12.4. On which types of soils unconfined compression test is conducted? Explain with the help of Mohr circles how shear strength parameters are determined in this type of test.

12.5. With the help of Mohr circles for each case, explain how shear strength parameters are determined by conducting shear tests on saturated samples under different drainage conditions.

SHEAR STRENGTH OF SOILS

12.6. Explain stress-strain and volume change characteristics of cohesionless soils during shear.

12.7. Define critical void ratio and explain how it can be determined for a cohesionless soil.

12.8. The results of a direct shear test on a 60 mm × 60 mm soil specimen are given below. Determine the shear strength parameters.

Normal load	(N)	300	400	500	600
Shear force at failure	(N)	195	263	324	399

12.9. Unconfined compression strength of a soil is found to be 150 kN/m². A sample of the same soil failed at a deviator stress of 200 kN/m², under cell pressure of 100 kN/m². Determine the shear strength parameters of soil.

12.10. In a triaxial test a soil specimen was consolidated under a cell pressure of 700 kN/m² and the increased pore pressure reading was 450 kN/m². The axial load was then increased to give a devialor stress of 570 kN/m² and pore pressure reading of 650 kN/m². Calculate the pore pressure parameters B and A.

CHAPTER 13

STABILITY OF SLOPES

13.1 Introduction

In geotechnical engineering the term slope commonly refers to a soil mass with its surface inclined to the horizontal, and there exist both natural slopes and man-made slopes. Examples of man-made slopes are side slopes of embankments for railways and highways, earth dams and levees. A slope failure involves downward and outward movement of a portion of slope called the sliding soil mass. A slope failure occurring in the case of a natural slope is referred to as landslide. Slope failures lead not only to economic loss but sometimes to catastrophes with loss of human life. Stability analysis of slopes assumes great importance in checking the stability of slopes and in suggesting remedial measures wherever necessary. In a slope failure the sliding soil mass slips along a surface called the slip surface. At every point on the slip surface shearing stresses are induced due to gravitational and seepage forces acting within the material above it and this is counteracted by shearing resistance mobilised in the opposite direction. The failure occurs when the shearing stresses induced exceed the shearing resistance mobilised. It is important to note that study of past landslides have indicated that slides may occur in almost every conceivable manner, slowly or suddenly, and with or without any apparent provocation. This point will become clear in the discussion on classification of land slides.

13.2 Infinite Slope and Finite Slope

For the purpose of stability analysis, slopes are broadly classified as infinite slopes and finite slopes.

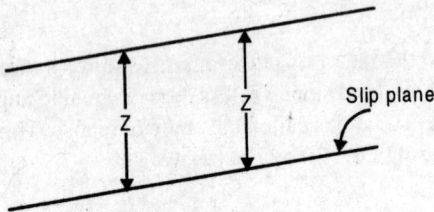

Fig. 13.1 Infinite Slope

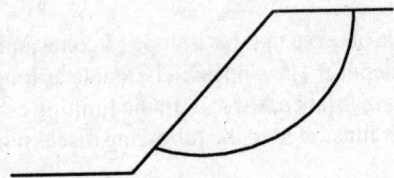

Fig. 13.2 Finite Slope

An infinite slope is very large in extent, theoretically infinite, and the properties of soil will be the same at identical depths so that the slip surface will be a plane parallel to the surface of slope, as shown in Fig. 13.1

A finite slope is limited in extent and the properties of soil will not be the same at identical depths so that the slip surface will be curved as shown in Fig. 13.2

13.3 Stability Analysis of Infinite Slope

Stability analysis of an infinite slope can be made by considering the forces acting on a prism of soil which forms a part of the sliding soil mass. The factor of safety against sliding is determined for different trial slip planes in order to locate the critical slip plane, for which the factor of safety will be a specified minimum.

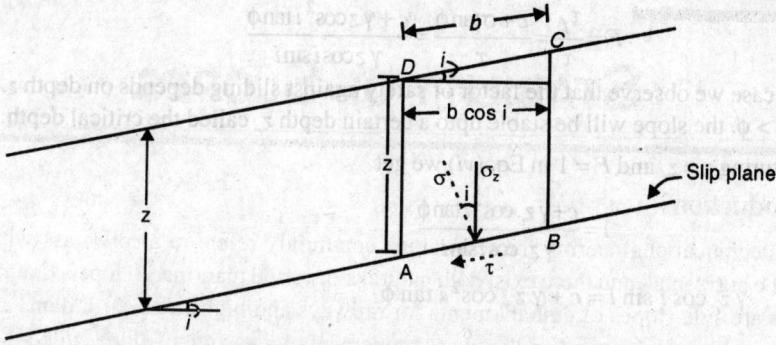

Fig. 13.3 Stability analysis of infinite slope

Let a slip plane be at depth z below the surface of slope, with i as the angle of slope. We consider a prism of soil $ABCD$ with inclined width b, depth z and of unit thickness perpendicular to the plane of paper, as shown in Fig. 13.3.

Vertical stress on plane AB,

$$\sigma_z = \frac{\text{Weight of prism } ABCD}{\text{Area of plane } AB} = \frac{(zb\cos i)\gamma}{b} = \gamma z\cos i \qquad ...(i)$$

$\therefore$ Normal stress on slip plane, $\sigma = \sigma_z \cos i = \gamma z \cos^2 i$ \qquad ...(ii)

Shear stress on slip plane, $\tau = \sigma_z \sin i = \gamma z \cos i \sin i$ \qquad ...(iii)

Case (i) : Slope of cohesionless soil ($c = 0$)

Factor of safety against sliding

$$F = \frac{\tau_f}{\tau} = \frac{c + \sigma \tan\phi}{\tau} = \frac{\gamma z \cos^2 i \tan\phi}{\gamma z \cos i \sin i} = \frac{\tan\phi}{\tan i} \qquad ...(iv)$$

We observe that for a slope of cohesionless soil the factor of safety against sliding is independent of depth z. The slope will be stable as long as the angle of slope i is less than or equal to angle of shearing resistance ϕ of soil. In the limiting condition $i = \phi$ and is called the angle of repose. This fact is further illustrated in the following discussion. Dividing Eq. (iii) by Eq. (ii), we get

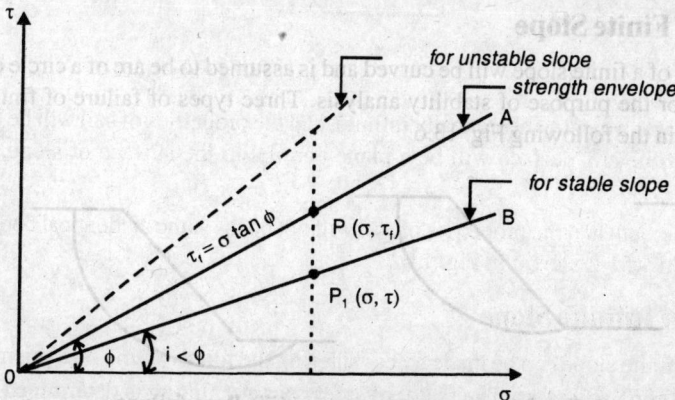

Fig. 13.4 Failure condition for infinite slope of cohesionless soil

$$\frac{\tau}{\sigma} = \frac{\gamma z \cos i \sin i}{\gamma z \cos^2 i} = \tan i$$

We see that even though τ and σ depend on z for a given slope angle i, their ratio is constant so that the plot of τ against σ is a straight line for a slope angle i.

From Fig. 13.4 it is clear that for any value of z and hence σ, τ will be less than τ_f for $i < \phi$.

Case (ii) Cohesive soil $(c - \phi$ soil)

Factor of safety against sliding,

$$F = \frac{\tau_f}{\tau} = \frac{c + \sigma \tan\phi}{\tau} = \frac{c + \gamma z \cos^2 i \tan\phi}{\gamma z \cos i \sin i} \qquad \qquad ...(vi)$$

In this case we observe that the factor of safety against sliding depends on depth z. For a given slope angle $i > \phi$, the slope will be stable upto a certain depth z_c called the critical depth.

Substituting $z = z_c$ and $F = 1$ in Eq. (vi) we get

$$1 = \frac{c + \gamma z_c \cos^2 i \tan\phi}{\gamma z_c \cos i \sin i}$$

$$\gamma z_c \cos i \sin i = c + \gamma z_c \cos^2 i \tan \phi$$

$$z_c = \frac{c}{\gamma (\cos i \sin i - \cos^2 i \tan\phi)} \qquad \qquad ...(vii)$$

Further

$$\frac{c}{\gamma z_c} = (\cos i \sin i - \cos^2 i \tan\phi) \qquad \qquad ...(viii)$$

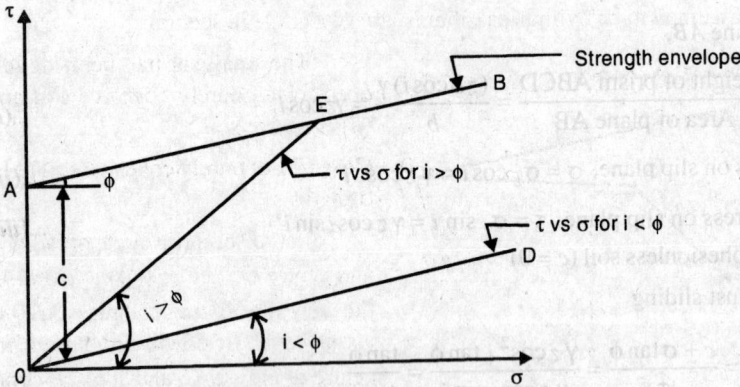

Fig. 13.5 Failure condition for infinite slope of cohesive soil.

The quantity $\dfrac{c}{\gamma z_c}$ is dimensionless, referred to as stability number and denoted by S_n. We note that it is a function of i and ϕ only. For $i < \phi$ the slope will always be stable irrespective of depth z. The above facts are further illustrated in the following discussion.

In Fig. 13.5 OD is a plot of τ vs σ for $i < \phi$ and AB the strength envelope. It is clear that for any value of z and hence σ, $\tau < \tau_f$ for $i < \phi$. OE is the plot of τ vs σ for $i > \phi$. It is clear that only upto critical depth represented by point E, $\tau < \tau_f$ for $i > \phi$.

13.4 Stability Analysis of Finite Slope

The slip surface in the case of a finite slope will be curved and is assumed to be arc of a circle or a logarithmic spiral in section for the purpose of stability analysis. Three types of failure of finite slopes are possible as illustrated in the following Fig. 13.6

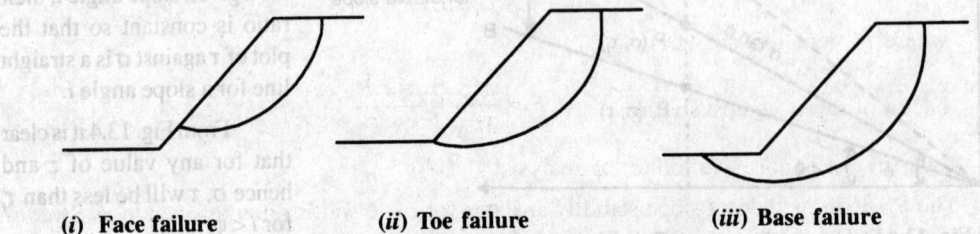

(i) Face failure   **(ii) Toe failure**   **(iii) Base failure**

Fig. 13.6 Types of failure of finite slopes

The term slope failure is applied collectively to the first two types.

In the case of face failure the slip surface cuts the surface of slope above the toe. This type of failure condition occurs when a stratum relatively strong compared to the top layers occurs above the toe level.

In the case of toe failure the slip surface passes through the toe. This type of failure condition occurs when a stratum relatively strong compared to the top layers occurs at toe level.

In the case of base failure the slip surface passes below the toe. This type of failure condition occurs when a stratum relatively hard compared to the top layers occurs below the toe level.

A number of methods are available for stability analysis of finite slopes. The commonly used methods are :

1. Swedish circle method
2. Friction circle method
3. By use of Taylor stability number
4. Bishop's method.

13.5 Swedish Circle Method

This method was developed at Swedish Geotechnical Commission headed by Fellenius. In this method the slip surface is assumed to be cylindrical, that is, arc of a circle in section.

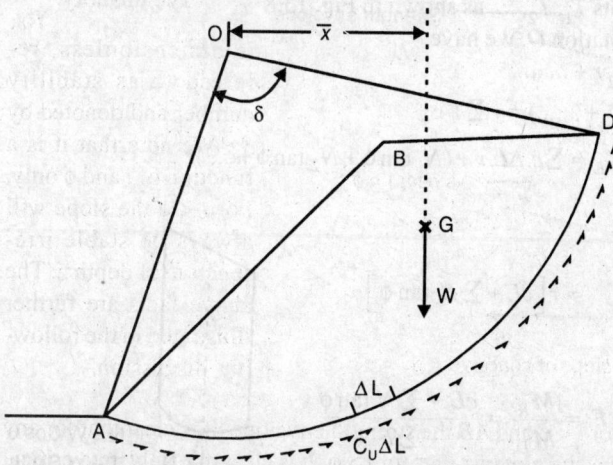

The analysis has been developed for (*i*) purely cohesive soil and (*ii*) cohesive soil.

Case (*i*) For purely cohesive soil ($\phi_u = 0$ analysis)

Let AD be a slip circle of radius r and centre O. Let $\angle AOD = \delta$ and W the weight of sliding soil mass $ABDA$, acting vertically downward through its centre of gravity G, and at distance $\bar{x}$ from O. The sliding of soil mass $ABDA$ along surface AD amounts to rotation about O. Taking moments about centre of rotation O we have

Fig. 13.7 Swedish circle method – ($\phi_u = 0$ analysis)

Driving moment, $\qquad\qquad M_D = W\bar{x}$

Resisting moment, $\qquad\qquad M_R = \sum c_u . \Delta L . r$

$$= c_u . r . \sum \Delta L$$

$$= c_u . r . \hat{L}$$

where $\hat{L} =$ length of arc AD.

Factor of safety against sliding $\qquad F = \dfrac{M_R}{M_D} = \dfrac{c_u . r . \hat{L}}{W\bar{x}}$

Case (*ii*) For cohesive soil ($c - \phi$ analysis)

The Swedish method of slope stability analysis for $c - \phi$ soil is often referred to as Swedish method of slices. In this analysis the following assumptions are made.

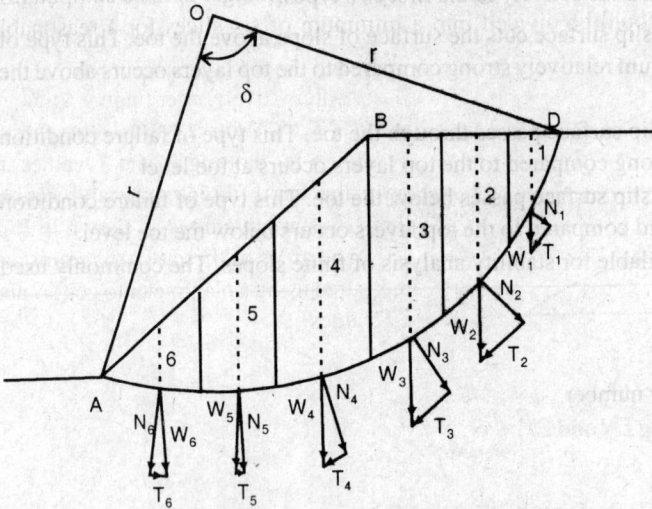

Fig. 13.8 Swedish method of slices

(i) The slip surface is cylindrical, that is, arc of a circle in section.

(ii) The sliding soil mass is assumed to consist of a number of vertical slices

(iii) The forces of interaction between adjacent slices are neglected.

Let AD be a slip circle of radius r, centre O, and central angle $\lfloor AOD = \delta$. Let the sliding soil mass $ABDA$ be divided into a number of vertical slices $1, 2, \ldots$. The weights $W_1, W_2, \ldots$ of slices $1, 2, \ldots$ acting through the centres of gravity of respective slices are resolved into normal components $N_1, N_2, \ldots$ and tangential components $T_1, T_2, \ldots$ as shown in Fig. 13.8.

Taking moments about centre of rotation O, we have

Driving moment, $\qquad M_D = T_1 r + T_2 r + \ldots$

$$= r(T_1 + T_2 + \ldots) = r \sum T$$

Restoring moment, $\qquad M_R = \sum c . \Delta L . r + (N_1 \tan \phi + N_2 \tan \phi + \ldots) r$

$$= cr . \sum \Delta L + (N_1 + N_2 + \ldots) r \tan \phi$$

$$= r \left[c\hat{L} + \sum N . \tan \phi \right]$$

where $\hat{L} = $ Length of arc AD.

Factor of safety against sliding, $\qquad F = \dfrac{M_R}{M_D} = \dfrac{c\hat{L} + \sum N \tan \phi}{\sum T}$

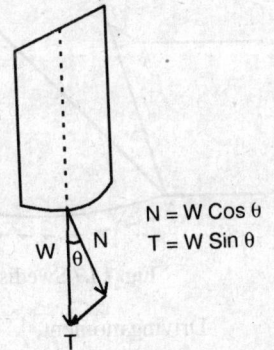

$N = W \cos \theta$
$T = W \sin \theta$

The Swedish method of slices is applicable not only to homogeneous soils but also to stratified soils, fully or partially submerged soils with considerations of seepage forces and pore pressures that may exist, and also non-uniform slopes.

$\sum N$ and $\sum T$ can be computed as shown in the following table.

Slice No.	Width of slice	Mid-ordinate	Weight of slice (W)	θ(deg)	$N = W \cos \theta$	$T = W \sin \theta$
1			W_1		N_1	T_1
2			W_2		N_2	T_2
.	.	.	.	.	.	.
.	.	.	.	.	.	.
.					$\sum N$	$\sum T$

The tangential component T is considered negative if it opposes sliding. For hand computation the section of sliding soil mass should be divided into a minimum of six slices for a reasonable accuracy.

Alternatively, after finding N_1, N_2.... and T_1, T_2..., N-curves and T-curves can be drawn by plotting N and T values as ordinates for different strips and joining them by smooth curves as shown in Fig. 13.9. The areas of these two diagrams are measured using a planimeter to get ΣN and ΣT values.

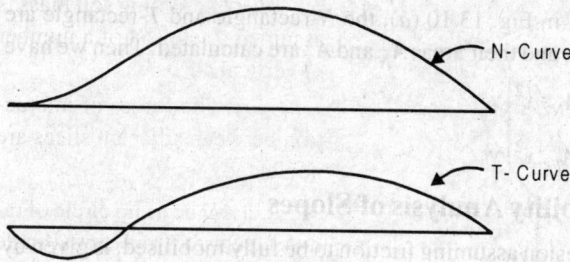

Fig. 13.9 N and T curves

Rectangular plot method for finding ΣN and ΣT.

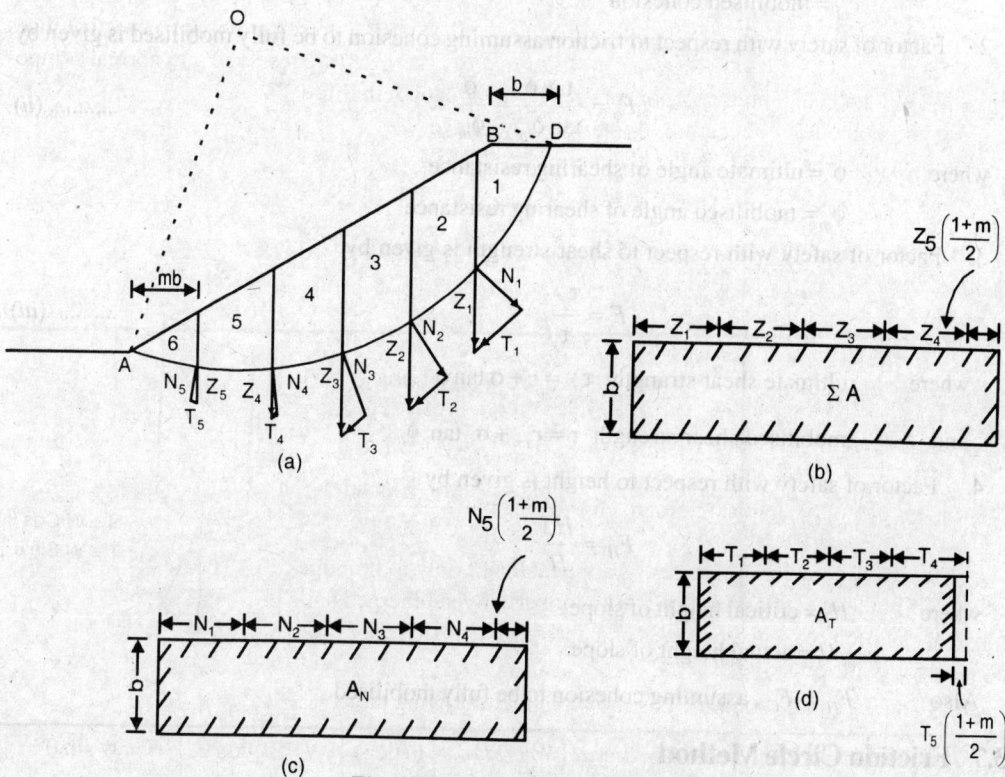

Fig. 13.10 Rectangular plot method

The rectangular plot method for finding ΣN and ΣT has been suggested by Alam singh (1962). Referring to Fig. 13.10, let $Z_1, Z_2, ...Z_n$ denote boundary ordinates between slices. If the width of each slice is b and that of the last slice mb, the total weight ΣW of the sliding soil mass $ABDA$ is given by

$\Sigma W = (\text{Area of section } ABDA) \times 1 \times \gamma$, considering unit length perpendicular to plane of figure.

i.e.,
$$\Sigma W = \left\{ Z_1 + Z_2 + ... + \frac{(1+m)}{2} Z_n \right\} b \times \gamma$$

In Fig. 13.1 (*b*) is shown how A-rectangle can be plotted. The area of this rectangle is ΣA in cm^2.

Then $\Sigma W = ($Area of A-rectangle$) \times x^2 \times \gamma$ if the scale of plotting is $1\ cm = x$ meters and γ is the unit weight of soil.

The same procedure can be adopted for ΣN and ΣT. If $N_1, N_2...$ and $T_1, T_2 ..$ are the normal and tangential components of $Z_1, Z_2....$ as shown in Fig. 13.10 (a), the N-rectangle and T-rectangle are constructed as shown in Fig.13.10 (c) and (d), and their areas A_N and A_T are calculated. Then we have

$$\Sigma N = A_N . x^2 . \gamma$$

$$\Sigma T = A_T . x^2 . \gamma$$

13.6 Factors of Safety Used in Stability Analysis of Slopes

1. Factor of safety with respect to cohesion assuming friction to be fully mobilised, is given by

$$F_c = \frac{c}{c_m} \qquad\qquad ....(i)$$

where c = ultimate cohesion

c_m = mobilised cohesion

2. Factor of safety with respect to friction assuming cohesion to be fully mobilised is given by

$$F_\phi = \frac{\tan \phi}{\tan \phi_m} \approx \frac{\phi}{\phi_m} \qquad\qquad ....(ii)$$

where ϕ = ultimate angle of shearing resistance

ϕ_m = mobilised angle of shearing resistance

3. Factor of safety with respect to shear strength is given by

$$F = \frac{\tau_f}{\tau} \qquad\qquad ....(iii)$$

where   ultimate shear strangth, $\tau_f = c + \sigma \tan \phi$

and   mobilised shear strength $\tau = c_m + \sigma \tan \phi_m$

4. Factor of safety with respect to height is given by

$$F_H = \frac{H_c}{H}$$

where H_c = critical height of slope

H = actual height of slope.

Also $F_H = F_C$, assuming cohesion to be fully mobilised.

13.7 Friction Circle Method

In the friction circle method the slip surface is assumed to be cylindrical $i.e.$, arc of a circle in section. The sliding soil mass is assumed to be acted upon by three forces keeping it in equilibrium, as shown in Fig. 13.11 (a)

(i)   The weight, W, of the sliding soil mass $ABDA$, acting vertically through its centre of gravity,

(ii)   The resultant cohesive force, $c_m \bar{L}$, acting parallel to chord AD and at distance a from centre

of rotation O, where $a = r.\dfrac{\hat{L}}{\bar{L}}$, $\hat{L}$ = length of arc AD and $\bar{L}$ = length of chord AD,

(iii) The resultant reaction R passing through the point of intersection of the above two forces and tangential to the friction circle.

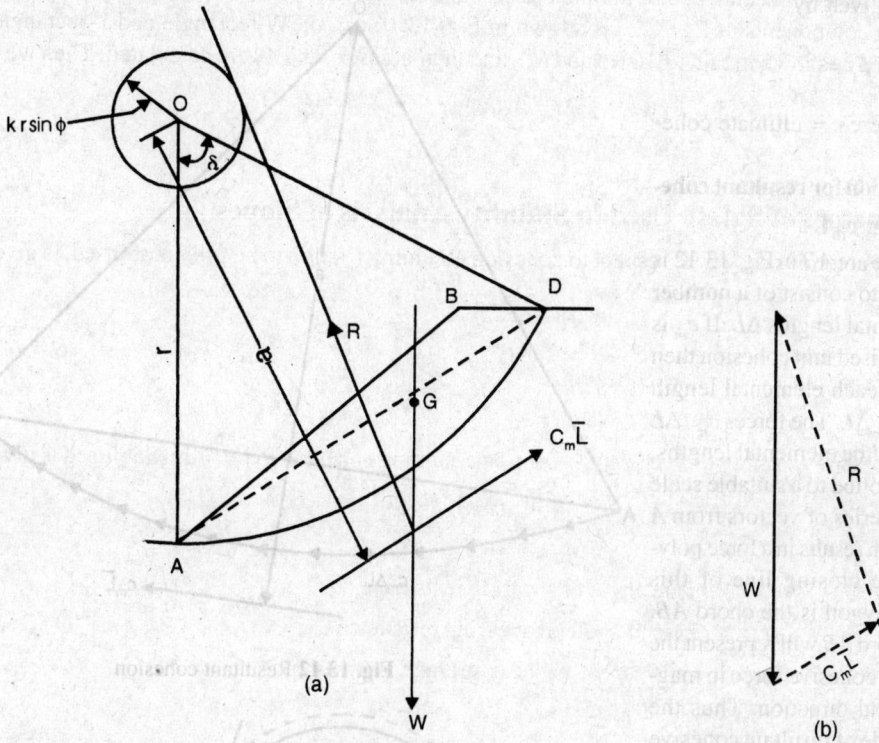

(a)

(b)

Fig. 13.11 Friction circle method

Procedure :

1. With centre O and radius r, the slip circle AD is constructed. The friction circle is drawn with centre O and radius $Kr \sin \phi$. K is taken as 1 unless otherwise given.

2. A vertical line is drawn through centroid of section $ABDA$, to get the line of action of weight W.

3. Chord AD is drawn. A line is drawn parallel to chord AD and at distance $a = r.\dfrac{\hat{L}}{\bar{L}}$ from O, to get the line of action of resultant cohesive force $c_m \bar{L}$. The length of arc AD, $\hat{L}$ is computed using the equation $\hat{L} = \dfrac{\pi r \delta}{180}$. The length of chord AD, $\bar{L}$ is obtained by measurement.

4. Through the point of intersection of the lines of action of forces W and $c_m \bar{L}$, a line is drawn tangential to the friction circle, to get the line of action of resultant reaction R.

5. The weight W of the sliding soil mass $ABDA$ is computed and plotted to scale as shown in Fig. 13.11 (b). Through the ends of the vector representing W, lines are drawn parallel to the lines of action of forces $c_m \bar{L}$ and R to complete the triangle of forces.

6.　The value of $c_m \bar{L}$ is obtained from the force triangle and divided by value of $\bar{L}$ to obtain the value of mobilised cohession c_m. The factor of safety with respect to cohesion, F_C is given by

$$F_c = \frac{c}{c_m}$$

where c = ultimate cohesion.

Explanation for resultant cohesive force, $c_m \bar{L}$.

　　The arc AB in Fig. 13.12 is assumed to consist of a number of elemental lengths ΔL. If c_m is the mobilised unit cohesion then force on each elemental length will be $c_m \Delta L$. The forces $c_m \cdot \Delta L$ acting on the elemental lengths, can be plotted to a suitable scale to get a series of vectors from A to B which results in a force polygon. The closing line of this force polygon is the chord AB. Then chord AB will represent the resultant cohesive force in magnitude and direction. Thus the magnitude of resultant cohesive force is $c_m \bar{L}$ where $\bar{L}$ is the length of chord AB. Let the resultant cohesive force acting parallel to chord AB, be at distance a from O. Taking moments about O and applying Varignon's theorem we get

$$(c_m \bar{L})(a) = \sum c_m . \Delta L . r$$

$$= c_m . r . \sum \Delta L$$

$$= c_m . r . \hat{L}$$

$$\therefore \quad a = \frac{\hat{L}}{L} r$$

Explanation for resultant reaction R

　　The arc AB is assumed to consist of a number of elemental lengths ΔL. The resultant reaction ΔR acting on an element ΔL will be inclined at angle ϕ to the normal at the mid-point of

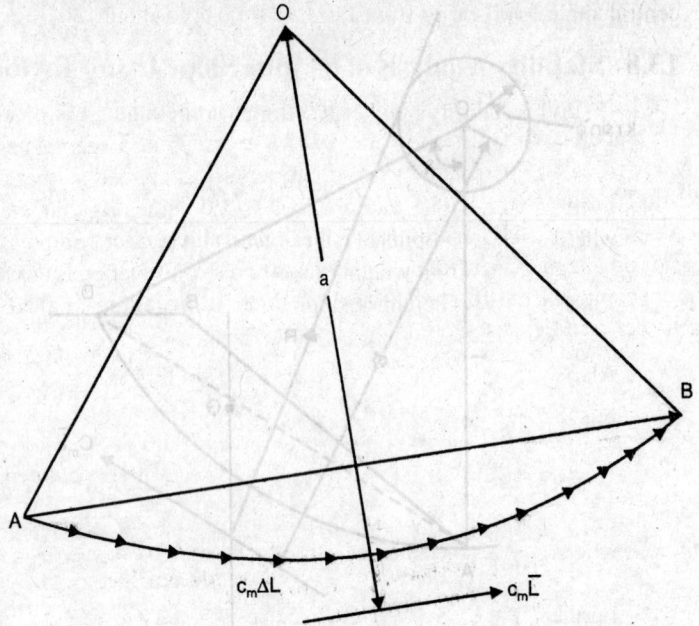

Fig. 13.12 Resultant cohesion

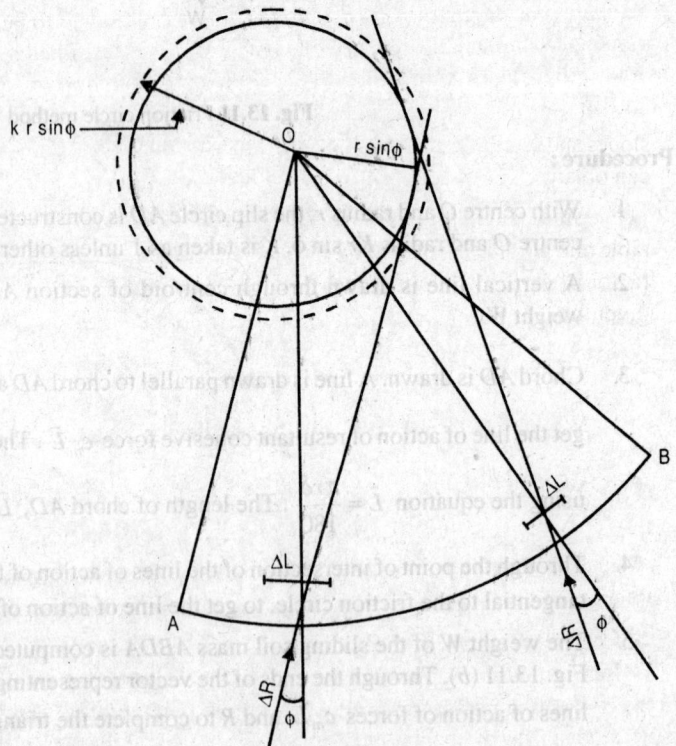

Fig. 13.13 Resultant reaction R.

the element, so that it will be tangential to the circle of radius $r \sin \phi$ drawn with centre O. It can be shown that the resultant reaction R for the entire arc AB will be tangential to the circle of radius $Kr \sin \phi$, drawn with centre O. This circle is called friction circle or ϕ - circle. The value of K depends on central angle δ and varies from 1 to 1.12 for δ upto about 120°.

13.8   Stability Analysis of a Finite Slope Using Taylor Stability Number

Taylor stability number is a dimensionless quantity denoted by S_n and defined as

$$S_n = \frac{c_m}{\gamma H} \qquad \qquad ...(i)$$

where c_m = mobilised cohesion on slip surface
$\qquad \quad \gamma$ = unit weight of soil
$\qquad \quad H$ = height of slope

Also $F_c = \dfrac{c}{c_m}$ so that $c_m = \dfrac{c}{F_c}$

Substituting in Eq. (i), we get

$$S_n = \frac{c}{F_c \gamma H} \qquad \qquad ...(ii)$$

where c = unit ultimate cohesion.

Further $F_c = F_H = \dfrac{H_c}{H}$ so that $F_c H = H_c$

Substituting in Eq. (ii)

$$S_n = \frac{c}{\gamma H_c} \qquad \qquad ...(iii)$$

where H_c = critical height of slope.

Using the friction circle method along with an analytical procedure, Taylor determined S_n for finite slopes and presented the results in the form of a table (Table 13.1) and a chart (Fig. 13.19) from which one can obtain value of S_n for different values of slope angle i and angle of shearing resistance ϕ.

Since Taylor stability number S_n is based on factor of safety with respect to cohesion, F_c, the table and chart give S_n only for the case where ϕ is assumed to be fully mobilised. But in cases where factor of safety is applicable to both cohesion and friction, we have mobilised shearing resistance given by

$$\tau_m = \frac{\tau_f}{F} = \frac{c}{F} + \frac{\sigma \tan \phi}{F}$$

While obtaining S_n from chart, mobilised angle of shearing resistance ϕ_m should be used

We have $\tan \phi_m = \dfrac{\tan \phi}{F}$

$$\therefore \qquad \phi_m = \tan^{-1}\left(\frac{\tan \phi}{F}\right)$$

As an approximation ϕ_m may be taken equal to $\dfrac{\phi}{F}$.

For a cohesionless soil ($c = 0$) the Taylor stability number is zero and Taylor's chart is not applicable. The factor of safety is given by $F = \dfrac{\tan \phi}{\tan i}$ and is independent of height of slope.

For long term stability c' and ϕ' obtained from drained test should be used. Use of Taylor's stability number gives an approximate idea of long term stability, if seepage effect can be neglected and no change in water content can be assumed.

In the case of fully submerged slopes, γ' should be used in the expression for S_n. When the slope is saturated, as for example by capillary water, γ_{sat} should be used in the expression for S_n. In the case of sudden drawdown γ_{sat} should be used in the expression for S_n and value of S_n should be

obtained from Taylor's chart, corresponding to weighted frictional angle ϕ_w given by $\phi_w = \dfrac{\gamma'}{\gamma_{sat}} \phi$

Taylor also determined stability number S_n for different values of slope angle i and depth factor D_f. The depth factor D_f is defined as the ratio of depth to hard strata below top of slope to the height of slope. This is illustrated in Fig. 13.14.

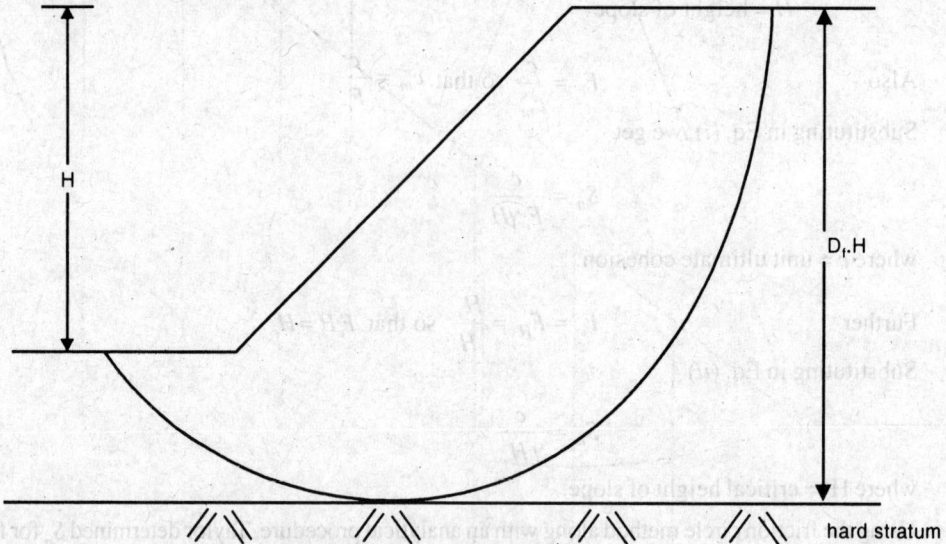

Fig. 13.14 Illustration of depth factor, D_f

Taylor presented table (Table 13.2) and chart (Fig. 13.19) from which one can obtain S_n for different values of i and D_f.

13.9 Bishop's Method of Stability Analysis

In the Bishop's method of stability analysis the following assumptions are made :

1. The slip surface is cylindrical, that is, arc of a circle in section.

2. The sliding soil mass is assumed to consist of a number of vertical slices.

3. The forces of interaction between adjacent slices, which were neglected in the Swedish method, are considered in the Bishop's method.

Analysis. Let AD be a slip circle with radius r and O the centre of rotation. The section of sliding soil mass $ABDA$ is divided into a number of slices. In the Fig. 13.15 is shown the free body diagram of a slice between sections n and $n + 1$. The explanation of notations is as follows.

E_n and E_{n+1} = normal forces on the sections n and $n + 1$, exerted by adjacent slices.

X_n and X_{n+1} = shear forces on sections n and $n + 1$, exerted by adjacent slices.

W = weight of slice

N = normal reaction at the base of slice

S = shear resistance at the base of slice.

z = height of slice

l = length of the base of slice

b = horizontal width of slice

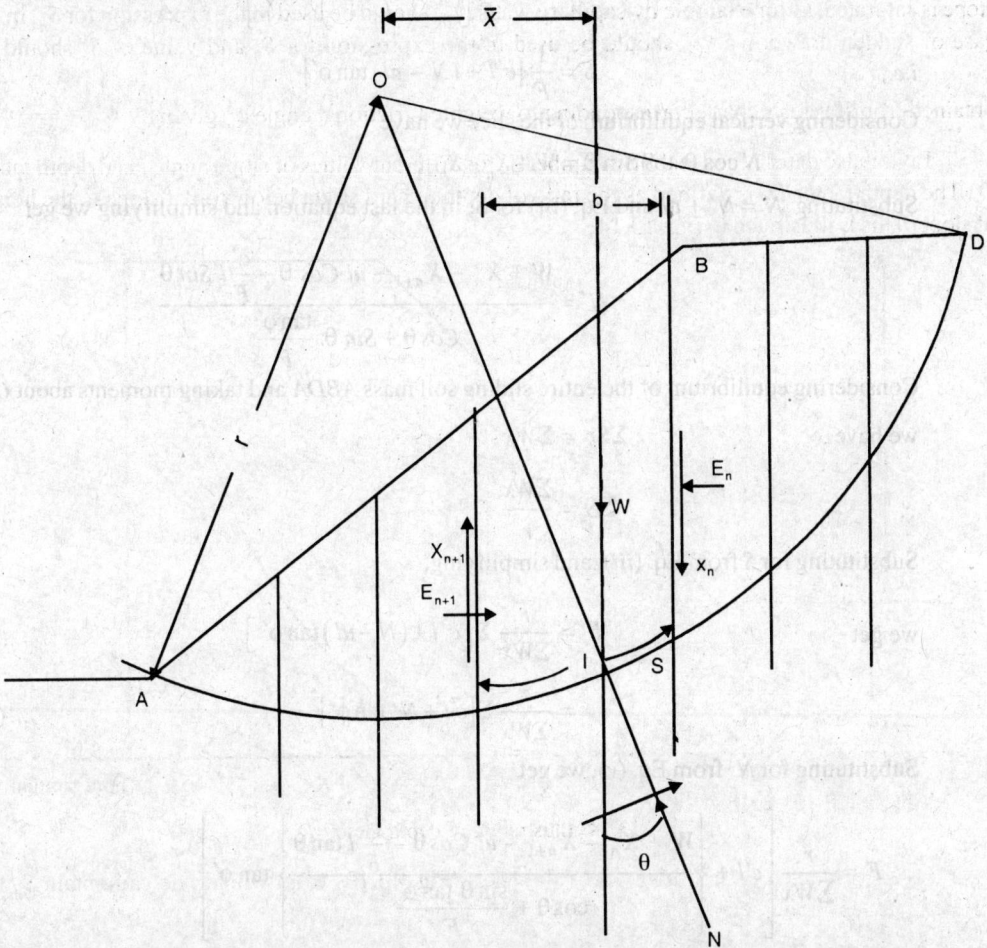

Fig. 13.15 Bishop's method of stability analysis

θ = angle made by base of slice with the horizontlal

x = horizontal distance between mid-ordinate of slice and centre of rotation O.

Considering unit length perpendicular to plane of figure, we have

Normal stress on base of slice, $\sigma = \dfrac{N}{l}$

Effective stress on base of slice, $\sigma' = \sigma - u = \dfrac{N}{l} - u$ ...(i)

where u = pore pressure

Factor of safety against sliding $F = \dfrac{\tau_f}{\tau}$

$$\therefore \qquad\qquad \tau = \frac{\tau_f}{F} = \frac{1}{F}\left[c' + \sigma' \tan \phi'\right] \qquad\qquad ...(ii)$$

Shear force on base of slice, $\quad S = \tau l = \dfrac{1}{F}\left[c' + \left(\dfrac{N}{l} - u\right)\tan \phi'\right]l$

$$i.e., \qquad\qquad S = \frac{1}{F}\left[c'l + (N - ul)\tan \phi'\right] \qquad\qquad ...(iii)$$

Considering vertical equilibrium of the slice we have

$$N \cos \theta + S \sin \theta = W + X_n - X_{n+1} \qquad\qquad ...(iv)$$

Substituting $N = N' + ul$ and Eq. (iii) for S, in the last equation and simplifying we get

$$N' = \frac{W + X_n - X_{n+1} - ul \cos \theta - \dfrac{c'}{F}l \sin \theta}{\cos \theta + \sin \theta.\dfrac{\tan \phi'}{F}} \qquad\qquad ...(v)$$

Considering equilibrium of the entire sliding soil mass *ABDA* and taking moments about *O*,

we have $\qquad\qquad \Sigma Sr = \Sigma Wx$

$$\therefore \qquad\qquad \Sigma S = \frac{\Sigma Wx}{r} \qquad\qquad ...(vi)$$

Substituting for S from Eq. (iii), and simplifying,

we get $\qquad\qquad F = \dfrac{r}{\Sigma Wx}\Sigma\left[c'l + (N - ul)\tan \phi'\right] \qquad\qquad ...(vii)$

$$= \frac{r}{\Sigma Wx}\Sigma\left[c'l + N'\tan \phi'\right] \qquad\qquad ...(viii)$$

Substituting for N' from Eq. (v), we get

$$F = \frac{r}{\Sigma Wx}\left[c'l + \left\{\frac{W + X_n - X_{n+1} - ul \cos \theta - \dfrac{c'}{F}l \tan \theta}{\cos \theta + \dfrac{\sin \theta \tan \phi'}{F}}\right\}\tan \phi'\right] \qquad\qquad ...(ix)$$

Substituting $x = r \sin \theta$, $b = l \cos \theta$

$$\frac{ub}{W} = r_u \qquad \left(\text{where } r_u = \frac{u}{\gamma z}\right)$$

and $X_n - X_{n+1} \simeq 0$, we get

$$F = \frac{1}{\Sigma W \sin \theta}\Sigma\left[\left\{c'b + W(1 - r_u)\tan \phi'\right\}\frac{\sec \theta}{1 + \dfrac{\tan \theta \tan \phi'}{F}}\right] \qquad\qquad ...(x)$$

For partly submerged slope, a similar treatment leads to the following expression :

$$F = \frac{1}{\Sigma(W_1 + W_2)\sin \theta}\Sigma\left[\left\{c'b + (W_1 + W_2 - \bar{u}b)\tan \phi'\right\}\times\frac{\sec \theta}{1 + \dfrac{\tan \theta \tan \phi'}{F}}\right] \qquad\qquad ...(xi)$$

where

W_1 = weight of slice above the free water surface

W_2 = submerged weight of slice below the free water surface.

$\bar{u}$ = pore water pressure expressed as an excess over the hydrostatic pressure corresponding to the water level outside the slope.

Eqs. (x) and (xi) contain factor of safety F on both sides. The solution has to be obtained by trial and error. If one desires to avoid trial and error computations, the following method given by Bishop and Morgenstern (1960) can be used. They gave the following expression for the factor of safety:

$$F = m - n\,r_u \qquad \qquad \qquad ...(xii)$$

where m and n are stability coefficients to be obtained from charts prepared by them. The

charts give m and n for various values of $\dfrac{c}{\gamma H}$, depth factor D_f and r_u

During and immediately after the end of construction, the pore pressure ratio r_u may be written as

$$r_u = \frac{u}{\gamma z} = \frac{u_i}{\gamma z} + \frac{\Delta_u}{\gamma z}$$

$$= \frac{u_i}{\gamma z} + \frac{\bar{B}.\Delta\sigma_1}{\gamma z}$$

Taking $u_i \simeq 0$ at optimum water content, and substituting $\Delta\sigma_1 = \gamma Z$, we get

$$r_u \simeq \bar{B} \qquad \qquad \qquad ...(xiii)$$

where $\bar{B}$ = overall pore pressure coefficient = $\dfrac{\Delta u}{\Delta\sigma_1}$

From Skempton's pore pressure equation

$$\Delta u = B\left[\Delta\sigma_3 + A\left(\Delta\sigma_1 - \Delta\sigma_3\right)\right]$$

$$\bar{B} = \frac{\Delta u}{\Delta\sigma_1} = B\left[\frac{\Delta\sigma_3}{\Delta\sigma_1} + A\left(1 - \frac{\Delta\sigma_3}{\Delta\sigma_1}\right)\right] \qquad \qquad ...(xiv)$$

Putting $\qquad\qquad K = \dfrac{\Delta\sigma_3{}'}{\Delta\sigma_1{}'} = \dfrac{\Delta\sigma_3 - \Delta u}{\Delta\sigma_1 - \Delta u},$

we get $\qquad\qquad \bar{B} = \dfrac{B\left[K + A\left(1 - K\right)\right]}{1 - B\left(1 - A\right)\left(1 - K\right)} = r_u \qquad \qquad ...(xv)$

13.10 Fellenious Method for Locating Centre of Critical Slip Circle

In order to reduce the number of trials to locate the centre of critical slip circle, Fellenious has given this method illustrated in Fig. 13.16 in which the centre of critical slip circle is shown to lie on the line QP. The point Q is located a distance H below the toe and 4.5H away from it. The other point P is located with the help of directional angles α and β obtained from accompanying table (p.250). Trial centres are chosen on line QP and factor of safety corresponding to each centre is computed and plotted normal to line QP as shown in Fig. 13.16 A smooth curve is then drawn and the point corresponding to minimum ordinate represents the centre of critical slip circle. The method is proposed for a homogeneous c - ϕ soil. For a purely cohesive soil ($\phi = 0$) the point P itself represents the centre of critical slip circle.

Slope	Slope angle i (deg)	Directional angles	
		α (deg)	β (deg)
0.58 : 1	60	29	40
1:1	45	28	37
1.5:1	33.8	26	35
2:1	26.6	25	35
3:1	18.4	25	35
5:1	11.3	25	35

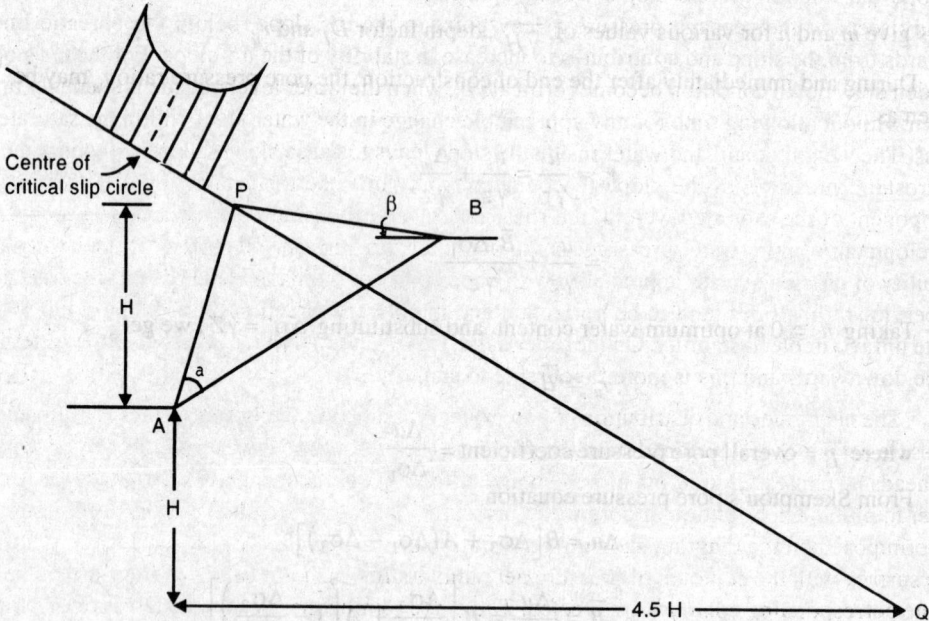

Fig. 13.16 Fellenious method for locating centre of critical slip circle

13.11 Stability of Side Slopes of Earth Dam

The stability of side slopes of an earth dam is checked under the following three conditions:

1. Stability of d/s slope during steady seepage

2. Stability of u/s slope during sudden drawdown

3. Stability of both u/s and d/s slopes during and immediately after construction.

1. Stability of d/s slope during steady seepage :

When the reservoir is full and seepage is at its maximum rate the stability of d/s slope becomes critical. While computing the factor of safety we should consider the effective normal stress at the base of slice. If we determine factor of safety against sliding by Swedish method of slices, then the expression for F should be written as

$$F = \frac{c'\hat{L} + \Sigma(N - U)\tan\phi'}{\Sigma T}$$

The pore pressure at the base of a slice is given by $u = h_w \gamma_w$ where h_w is the vertical distance between point of intersection of an equipotential line with the slip surface at base of slice, and the point at which that equipotential line intersects the phreatic line. The distribution of pore pressure on the slip surface is determined after constructing flow net and ΣU computed.

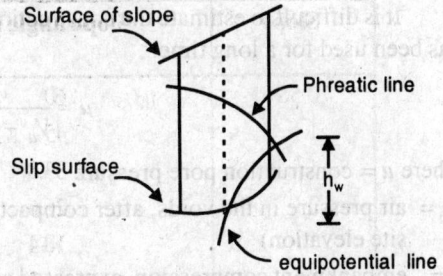

Fig. 13.17 Piezometric head at base of slice

2. Stability of u/s slope during sudden drawdown

The steady seepage condition is critical for the d/s slope but not for the u/s slope. Because during steady seepage the seepage pressure at any point in the u/s slope, below the phreatic line acts inwards from the slope and contributes to increase in stability of the u/s slope. For the u/s slope the sudden draw down condition becomes critical *i.e.*, when the water level outside is suddenly brought down without allowing time for any appreciable change in the water level within the saturated soil mass. The weight of soil and water inside the slope tends to cause slope failure with the removal of hydrostatic pressure on the slope to counteract it. While the disturbing force is the tangential component of the saturated weight, the shear resistance is considerably reduced on account of the development of pore water pressure on the potential slip surface. The effect of drawdown on the stability of u/s slope varies appreciably with opportunity of drainage at the base. If the base is quite impervious, flow lines tend to be horizontal and in outward direction towards the slope, which is quite unfavourable to stability. On the other hand, if the base material is pervious, flow pattern tends to be downwards and this is more favourable to stability.

The magnitude and distribution of pore water pressure on a likely slip surface is estimated from pressure net which is developed from the flow net. The pressure net gives the line of equal piezometric heads h_w expressed in terms of percentage of total hydraulic head when the reservoir is full. In order to plot the distribution of pore water pressure on the assumed slip surface, the pressure net is superimposed on the diagram showing the location of slip surface. At the intersection point of the slip surface with the contours of pressure net radial ordinates of the values of piezometric heads h_w of the corresponding contours are erected. The curve joining these ordinates gives the distribution of pore water pressure. The factor of safety is calculated from the equation

$$F = \frac{c'\hat{L} + \Sigma(N - U)\tan\phi'}{\Sigma T}$$

As it is difficult to obtain flow net for sudden drawdown condition, unlike in the case of steady seepage condition, the following equation can be used in the absence of flow net, to get an approximate value of F.

$$F = \frac{c'\hat{L} + \Sigma N'\tan\phi'}{\Sigma T}$$

in which N' components are computed from submerged weight of slices and the T components are computed from saturated weight of slices. The same procedure can also be used in the absence of flow net for steady seepage condition also.

3. Stability of u/s and d/s slopes during and immediately after construction

During construction of an earth dam soil is compacted in layers. The decrease in volume of soil during compaction and the self weight of soil cause development of excess pore pressure in air and water entrapped in voids. This excess pore pressure is referred to as construction pore pressure.

It reduces the shear strength and affects the stability of both u/s and d/s slopes.

It is difficult to estimate the construction pore pressure. However, Hilf's equation given below has been used for a long time.

$$u = \frac{P_a \Delta}{V_a + h_c V_w - \Delta}$$

where u = construction pore pressure

P_a = air pressure in the voids, after compaction (taken equal to atmospheric pressure, corrected to site elevation)

Δ = embankment compression, expressed as percentage of original total embankment volume

V_a = Volume of free air voids expressed as percentage of original total embankment volume

V_w = Volume of pore water expressed as percentage of original total embankment volume

h_c = Henry's constant of solubility of air in water by volume. It is equal to 0.02 at 68° F.

The use of Hilf's equation in illustrated in the following example. Compression Δ is plotted against effective stress σ' from consolidation test data as shown in Fig. 13.18 (a). The pore pressure u is calculated for various values of Δ using Hilf's equation. Then total stress $\sigma = \sigma' + u$ is calculated for each value of Δ. Pore pressure u is plotted against σ as shown in Fig. 13.18 (b). This plot can be used to find the construction pore pressures at the bottom of various slices in the stability analysis by method of slices. If z denotes the mid-height of a slice then $\sigma = uz$ at the base of slice. Then using Fig. 13.18 (b), corresponding to σ, value of u is obtained. ΣU can then be computed as indicated in the following table -.

Determination of construction pore pressure

Slice No	Width of slice	Mid-height	Total stress	Pore pressure	
(b)	(z)	($\sigma = \gamma z$)	u	U = u × b
					$\Sigma U =$

The factor of safety F is given by $F = \dfrac{c'\hat{L} + \Sigma(N - U)\tan\phi'}{\Sigma T}$

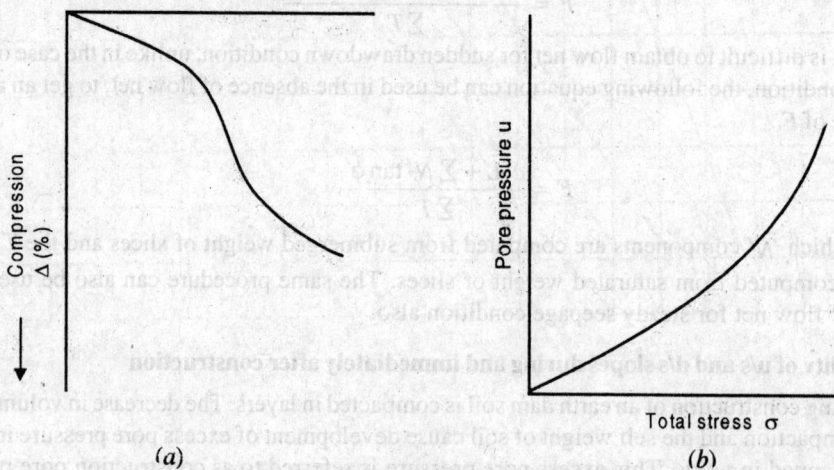

Fig.13.18 Plot of consolidation test data for use in estimating construction pore pressure

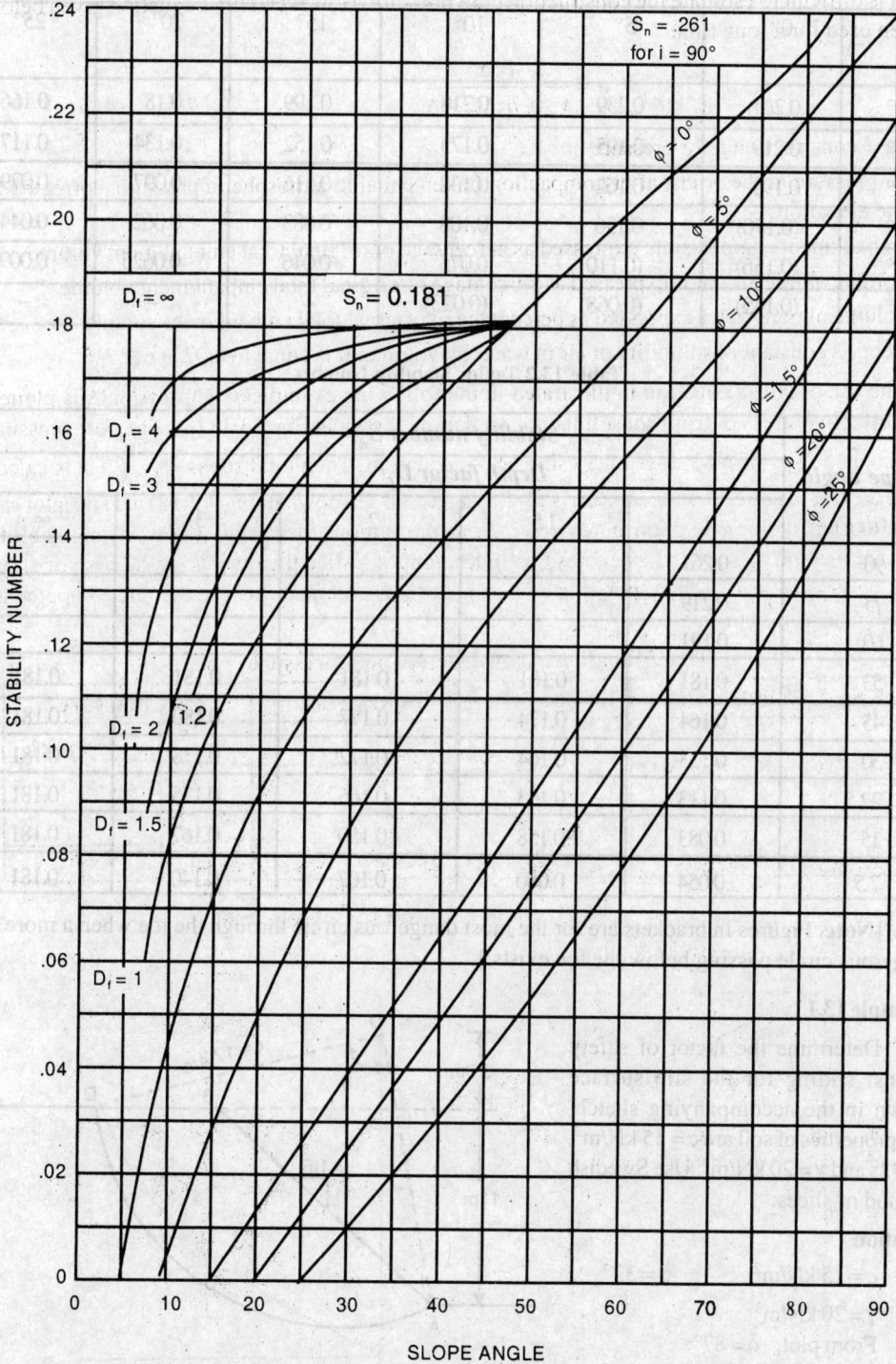

Fig. 13.19 Chart for Taylor Stability Number

Table 13.1 Taylor Stability Number

$\phi \rightarrow$ $i \downarrow$	0°	5°	10°	15°	20°	25°
90°	0.261	0.239	0.218	0.199	0.18	0.166
75°	0.219	0.195	0.173	0.152	0.134	0.117
60°	0.191	0.162	0.138	0.116	0.097	0.079
45°	(0.170)	0.136	0.108	0.083	0.062	0.044
30°	(0.156)	(0.110)	0.075	0.046	0.0625	0.009
15°	(0.145)	(0.068)	(0.023)	–	–	–

Table 13.2 Taylor Stability Number

Slope angle i (deg)	Stability number S_n Depth factor D_f				
	1	1.5	2	3	∞
90	0.261				
75	0.219				
60	0.191				
53	0.181	0.181	0.181	0.181	0.181
45	0.164	0.174	0.177	0.180	0.181
30	0.133	0.164	0.172	0.178	0.181
22.5	0.113	0.153	0.166	0.175	0.181
15	0.083	0.158	0.150	0.167	0.181
7.5	0.054	0.080	0.107	0.140	0.181

[**Note.** Figures in brackets are for the most dangerous cir:le through the toe when a more dangerous circle passing below the toe exists.]

Example 13.1

Determine the factor of safety against sliding for the slip surface shown in the accompanying sketch. The properties of soil are $c = 15$ kN/m², $\phi = 32°$ and $\gamma = 20$ kN/m³. Use Swedish method of slices.

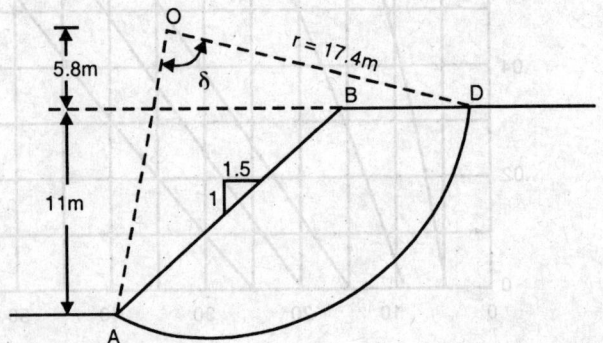

Solution

$c = 15$ kN/m² $\phi = 32°$

$\gamma = 20$ kN/m³

From plot, $\delta = 87°$

Length of are AD,

$$\hat{L} = \frac{\pi r \delta}{180°} = \frac{\pi(17.4)(87)}{180} = 26.42 m$$

Figure for Example 13.1 (c – φ analysis)

Slice No.	Mid-ordinate (m)	Width (m)	Wt. of Slice W (kN)	θ (degrees)	$N = W \cos \theta$ (kN)	$T = W \sin \theta$ (kN)
1	1.35	0.75	20.25	65	8.56	18.35
2	4.35	3.0	261.00	55	149.70	213.80
3	7.65	3.0	459.00	4.	351.61	295.04
4	7.50	3.0	450.00	38	354.60	277.05
5	6.90	3.0	414.00	17	395.91	121.04
6	5.55	3.0	333.00	8	329.76	46.34
7	3.90	3.0	234.00	3	233.68	− 12.25
8	1.35	3.0	81.00	13	78.92	− 18.22
					$\Sigma N = 1902.74$	$\Sigma T = 941.15$

Factor of safety against sliding,

$$F = \frac{c\hat{L} + \sum N \tan \phi}{\sum T}$$

$$= \frac{(15)(26.42) + (1902.74)\tan 32°}{941.15} = 1.68$$

Example 13.2

Calculate the factor of safety for the slip circle shown in Figure on page 257 for a cutting in a purely cohesive soil with $\phi_u = 0$, $c_u = 30$ kN/m^2 and $\gamma = 20.5$ kN/m^3

Solution:

$c_u = 30$ kN/m^2 and $\gamma = 20.5$ kN/m^3

From plot, $\delta = 68°$.

Length of are AD, $\hat{L} = \dfrac{\pi r \delta}{180} = \dfrac{\pi (13.5)(68)}{180} = 16.02 m$

Area of section ABDA = area of sector OAD - area of triangle OAD + area of triangle ABD

$$= \frac{1}{2}(13.5)^2 \left(68 \times \frac{\pi}{180} \right) - \frac{1}{2}(15)(11.1) + \frac{1}{2}(15)(1.2)$$

$$= 33.89 \, m^2$$

Weight of sliding soil mass ABDA = 33.89×20.5 =694.75 kN

Factor of safety against sliding,

$$F = \frac{M_R}{M_D} = \frac{c\hat{L}r}{Wx} = \frac{(30)(16.02)(13.5)}{(694.75)(6)} = 1.56$$

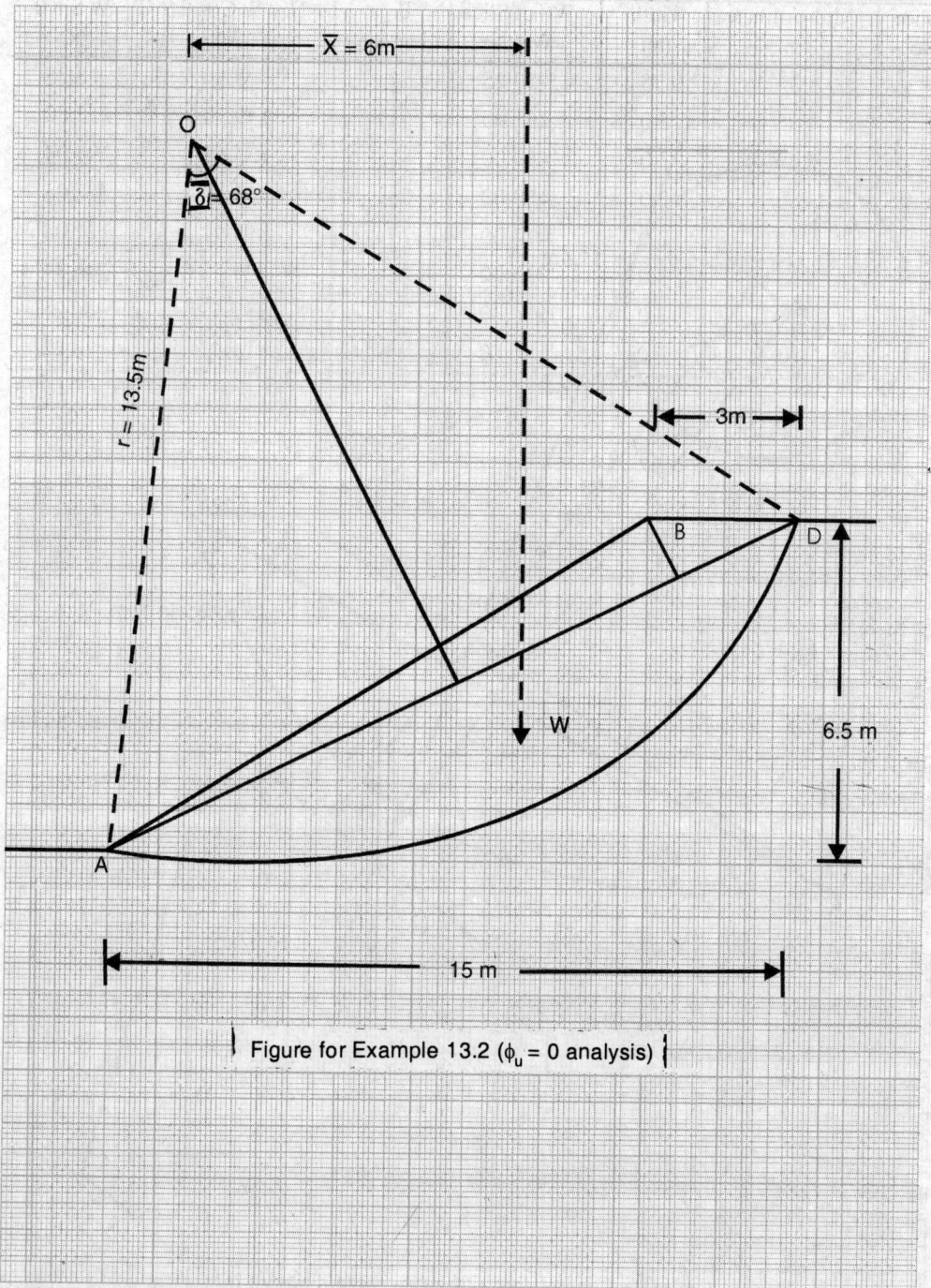

Figure for Example 13.2 ($\phi_u = 0$ analysis)

Example 13.3

Determine the factor of safety against sliding for the slip circle shown in the figure on page 259. The properties of soil are $c = 30 \text{ kN/m}^2$, $\phi = 15°$ and $\gamma = 21 \text{ kN/m}^3$.

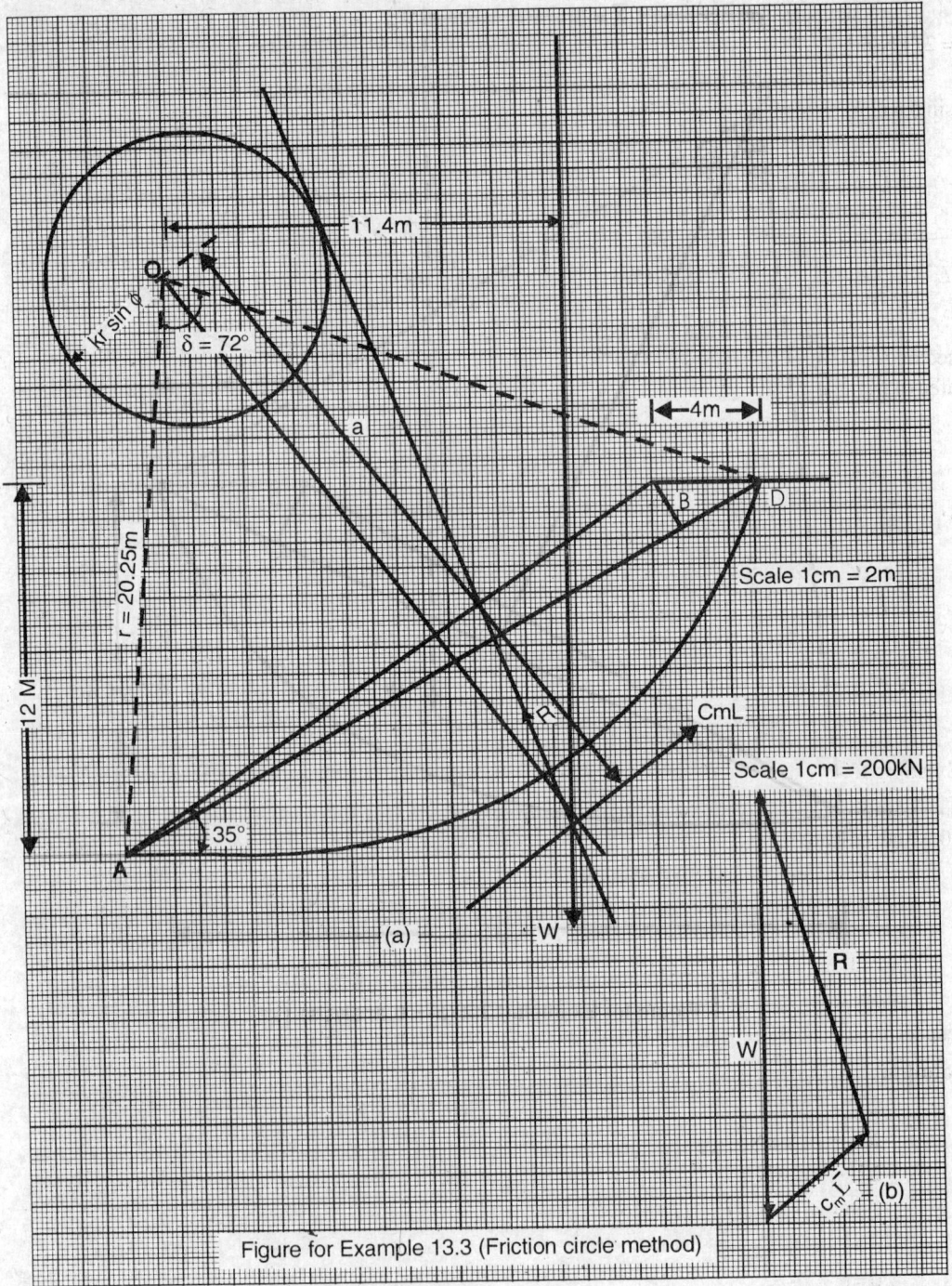

11.4m

δ = 72°

kr sin φ

a

r = 20.25m

12 M

4m

B   D

Scale 1cm = 2m

R   CmL

Scale 1cm = 200kN

35°

A

(a)   W

R

W

CₘL   (b)

Figure for Example 13.3 (Friction circle method)

Solution :

$c = 30$ kN/m^2, $\phi = 15°$, $\gamma = 21$ kN/m^3

Radius of friction circle $= Kr$

$\sin \phi = (1) (20.25)$ Sin $15° = 5.25$m

Length of chord AD,

$\overline{L} = 23.6$ m

Length of arc AD,

$\hat{L} = \dfrac{\pi r \delta}{180} = \dfrac{\pi (20.25)\,(72)}{180} = 25.45$ m

Area of section ABDA = area of sector AOD - area of triangle AOD + area of triangle ABD

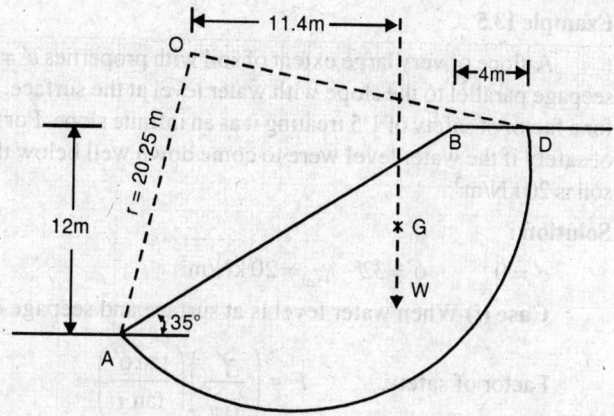

$$= \frac{1}{2} (20.25)^2 \left(\frac{72 \times \pi}{180} \right) - \frac{1}{2} (23.6)(16.6) + \frac{1}{2} (23.6)(2) = 85.36\,\text{m}^2$$

Weight of sliding soil mass ABDA, W $= 85.36 \times 21 = 1792.6$ kN

(Per meter length perpendicular to plane of fig)

From force triangle [Fig (b)], $c_m \, \overline{L} = 2.8 \times 200 = 560$ kN

$\therefore \qquad c_m = \dfrac{c_m \, \overline{L}}{\overline{L}} = \dfrac{560}{23.6} = 23.73$ kN/m^2

$$F_c = \frac{c}{c_m} = \frac{30}{23.73} = 1.26$$

Example 13.4

Stability analysis by Swedish method of slices gave following values per running metre for a 10m high embankment.

  (i)   Total shearing fore = 480 kN
  (ii) Total normal force = 1950 kN
  (iii) Total neutral force = 250 kN
  (iv) Length of arc = 22 m

If the properties of soil arc $c = 24$ kN/m^2 and $\phi = 6°$, calculate the factor of safety with respect to shear strength.

Solution :

$\Sigma T = 480$ kN

$\Sigma N = 1950$ kN

$\Sigma U = 250$ kN

Length of arc, $\hat{L} = 22$ m

$c = 24$ kN/m^2

$\phi = 6°$

$$F = \frac{c\,\hat{L} + \Sigma (N - U) \tan \phi}{\Sigma T}$$

$$= \frac{(24)(22) + (1950 - 250) \tan 6°}{480} = 1.47$$

Example 13.5

A slope of very large extent of soil with properties $c' = 0$ and $\phi' = 32°$ is likely to be subjected to seepage parallel to the slope with water level at the surface. Determine the maximum angle of slope for a factor of safety of 1.5 treating it as an infinite slope. For this angle of slope what will be the factor of safety if the water level were to come down well below the surface? The saturated unit weight of soil is 20 kN/m³.

Solution :

$$c' = 0 \qquad \phi' = 32° \quad \gamma_{sat} = 20 \text{ kN/m}^3$$

Case (*i*) When water level is at surface and seepage occurs parallel to surface,

Factor of safety, $F = \left(\dfrac{\gamma'}{\gamma_{sat}}\right)\left(\dfrac{\tan \phi'}{\tan i}\right)$

$\therefore$ $\tan i = \dfrac{(20 - 9.81)}{20} \cdot \dfrac{\tan 32°}{1.5} = 0.2122$

Slope angle, $i = 12°$

Case (*ii*) When the water table goes well below the surface,

$$F = \frac{\tan \phi'}{\tan i} = \frac{\tan 32°}{\tan 12°} = 2.94$$

Example 13.6

A slope 1 in 2 with a height of 8m has the following soil properties:

$$c = 28 \text{ kN/m}^2, \qquad \phi = 10° \quad \gamma = 18 \text{ kN/m}^3$$

Calculate (*i*) factor of safety with respect to cohesion and (*ii*) critical height of slope.

Solution :

(*i*)   If i is the slope angle, we have

$$\tan i = \frac{1}{2}$$

$\therefore$ $i = \tan^{-1} \dfrac{1}{2} = 26.6°$

From Taylor stablity chart, for $i = 26.6°$ and $\phi = 10°$, $S_n = 0.064$

$$S_n = \frac{c}{F_c \, \gamma H}$$

$\therefore$ $F_c = \dfrac{c}{S_n \, \gamma H} = \dfrac{28}{(0.064)(18)(8)} = 3.04$

(*ii*) $F_c = \dfrac{H_c}{H}$

$\therefore$ $H_c = F_c \cdot H = (3.04)(8) = 24.32 \text{ m}$

Example 13.7

A slope is to be laid at an angle of 30° with the horizontal. Find safe height of slope for a factor of safety of 1.5 if the soil properties are :

$$c = 15 \text{ kN/m}^2 \qquad \phi = 22° \quad \gamma = 18 \text{ kN/m}^3$$

Solution :

Since factor of safety given in the problem is with respect to shear strength, it is applicable to both c and ϕ.

Mobilised frictional angle, ϕ_m is given by

$$F_\phi = \frac{\tan \phi}{\tan \phi_m} \quad \therefore \quad \phi_m = \tan^{-1}\left(\frac{\tan 22°}{1.5}\right) = 15°$$

From Taylor stability chart,

for $i = 30°$ and $\phi_m = 15°, S_n = 0.046$

We have $$S_n = \frac{c}{F \gamma H}$$

$\therefore$ $$H = \frac{c}{S_n \cdot F \cdot \gamma} = \frac{15}{(0.046)(1.5)(18)} = 12.1 \text{ m}$$

Example 13.8

A 5 m deep canal has side slopes of 1:1. The properties of soil are $c_u = 20 \text{ kN/m}^2$, $\phi_u = 10°$, $e = 0.8$ and $G = 2.8$. If Taylor's stability number is 0.108, determine the factor of safety with respect to cohesion, when the canal runs full. Also find the same in case of sudden drawdown, if Taylor's stability number for this condition is 0.137.

Solution :

$$c_u = 20 \text{ kN/m}^2, \qquad \phi_u = 10°, \ G = 2.8$$

$$\gamma_{sat} = \frac{(G + e)\gamma_w}{1 + e} = \frac{(2.8 + 0.8)9.81}{(1 + 0.8)} = 19.62 \text{ kN/m}^3$$

$$\gamma' = \gamma_{sat} - \gamma_w = 19.62 - 9.81 = 9.81 \text{ kN/m}^3$$

Case (*i*) When canal runs full the side slopes are submerged.

$$S_n = \frac{c}{F_c \gamma' H}$$

$$F_c = \frac{c}{S_n \gamma' H} = \frac{20}{(0.108)(9.81)(5)} = 3.8$$

Case (*ii*) Sudden drawdown condition, $S_n = 0.137$

$$F_c = \frac{c}{S_n \gamma_{sat} H} = \frac{20}{(0.137)(19.62)(5)} = 1.5$$

Example 13.9

A canal with a depth of 5m has banks with slope 1:1. The properties of soil are:

$$c = 20 \text{ kN/m}^2 \qquad \phi = 15° \quad e = 0.7 \quad G = 2.6$$

Calculate factor of safety with respect to cohesion (*i*) when canal runs full and (*ii*) it is suddenly and completely emptied.

Solution :

Case (i) When canal runs full, the side slopes are submerged :

$$\gamma_{sat} = \frac{(G+e)\gamma_w}{1+e} = \frac{(2.6 + 0.7)(9.81)}{1+0.7} = 19.04 \, \text{kN/m}^3$$

$$\gamma' = \gamma_{sat} - \gamma_w = 19.04 - 9.81 = 9.23 \, \text{kN/m}^3$$

From Taylor stability chart,

for $i = 45°$ and $\phi = 15°$, $S_n = 0.083$

$$S_n = \frac{c}{F_c \gamma' H}$$

$\therefore$

$$F_c = \frac{c}{S_n \cdot \gamma' \cdot H} = \frac{20}{(0.083)(9.23)(5)} = 5.22$$

Case (ii) When canal is suddenly and completely emptied : For sudden drawdown condition, S_n is to be obtained for slope angle i and weighted frictional angle ϕ_w.

$$\phi_w = \frac{\gamma'}{\gamma_{sat}} \phi = \left(\frac{9.23}{19.04}\right)(15) = 7.3°$$

From Taylor stability chart,

for $i = 45°$ and $\phi_w = 7.3°$, $S_n = 0.122$

$$S_n = \frac{c}{F_c \cdot \gamma_{sat} \cdot H}$$

$\therefore$

$$F_c = \frac{c}{S_n \cdot \gamma_{sat} \cdot H} = \frac{20}{(0.122)(19.04)(5)} = 1.72$$

Example 13.10

The soil at a site has $c = 15 \, \text{kN/m}^2$ and $\gamma = 18 \, \text{kN/m}^3$. A hard stratum exists at a depth of 9 m below ground surface. If a cutting 6 m deep is to be made in the soil, find if a 30° slope is safe. What will be the safe slope angle if a factor of safety of 1.5 is required.

Solution :

Depth factor $D_f = \dfrac{9}{6} = 1.5$

From Taylor stability chart,

for $i = 30°$ and $D_f = 1.5$, $S_n = 0.163$

$$S_n = \frac{c}{F_c \gamma H}$$

$\therefore$

$$F_c = \frac{c}{S_n \gamma H} = \frac{15}{(1.5)(18)(6)} = 0.85$$

From Taylor stability chart,

for $S_n = 0.092$ and $D_f = 1.5$, the slope angle is $i = 9°$

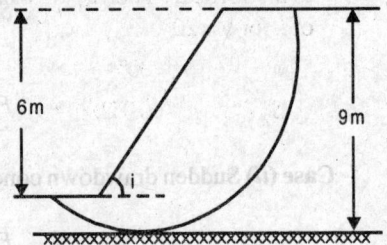

EXERCISE–13

13.1 Derive expressions for factor of safety against sliding for an infinite slope of (i) cohesionless soil and (ii) cohesive soil.

13.2 State and briefly explain the various conditions for which the stability of side slopes of earth dam are to be analysed.

13.3 Define Taylor's stability number and explain how it is used in the stability analysis of slope under (i) moist condition, (ii) submerged condition and (iii) sudden drawdown.

13.4 Explain Fellenious method for locating centre of critical slip circle.

13.5 What are the basic modes of failure of earth slopes? Briefly outline the remedial measures that can be undertaken against failure of slopes.

13.6 A long natural slope in a fissured overconsolidated clay is inclined at $12°$ to the horizontal. The water table is at the surface and the seepage is paralled to the slope. A slip has developed on a plane paralled to the surface at a depth of 5 m. The saturated unit weight of clay is 20 kN/m^3. The peak shear strength parameters are $c' = 10$ kN/m^2 and $\phi^1 = 18°$. Determine the factor of safety for the slip plane in terms of (i) peak shear strength parameters and (ii) residual shear strength parameters.

13.7 An embankment is 9 m high with side slopes 1.5:1. The properties of soil are $c = 25$ kN/m^2, $\phi = 20°$ and $\gamma = 19$ kN/m^3. Determine the factor of safety along a slip circle passing through the toe. Take Fellenious directional angles as $\alpha = 26°$ and $\beta = 35°$. Use Swedish method of slices.

13.8 List the different types of land slides and state the causes for a slide to occur.

13.9 The accompanying sketch shows a slip surface with a radius of 22 m. The slope is 1:1 with a height of 14 m. The soil properties are $c = 30$ kN/m^2, $\phi = 22°$ and $\gamma = 20$ kN/m^3

Determine the factor of safety with respect to cohesion using friction circle method.

13.10 A new canal is excavated to a depth of 5m with banks having slope 1:1. The properties of soil are $c = 14$ kN/m^2, $\phi = 20°$, $e = 0.65$ and $G = 2.70$. Calculate factor of safety with respect to cohesion when the canal is running full. What will be the change in the factor of safety if the slope is changed to be at $40°$ to the vertical? The Taylor stability number is given in the following table for different values of i, for $\phi = 20°$.

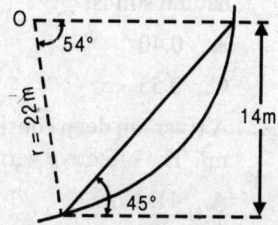

i	$30°$	$45°$	$60°$	$75°$	$90°$
S_n	0.052	0.062	0.097	0.134	0.182

ADDITIONAL QUESTIONS

MULTIPLE CHOICE

1. Which one of the following relations is **not** correct ?

   A. $e = \dfrac{n}{1-n}$ **B.** $n = \dfrac{e}{1-e}$

   C. $e = \dfrac{WG}{S_1}$ **D.** $\gamma_{sat} = \dfrac{(G+e)}{1+e}\gamma_w$

2. If the porosity of a soil sample is 20%, the void ratio is.

   A. 0.20   **B.** 0.80

   C. 1.00   **D.** 0.25

3. A soil sample in its natural state has mass of 2.290 kg and a volume of 1.15×10^{-3} m³. After being oven dried, the mass of the sample is 2.035 kg. G_s for soil is 2.68. The void ratio of the natural soil is

   A. 0.40   **B.** 0.45

   C. 0.55   **D.** 0.51

4. A river 5m deep consists of a sand bed with saturated unit weight of 20 kN/m³ $\gamma_w = 9.81$ kN/m³. The effective vertical stress at 5m below the top of the sand bed is

   A. 41kN/m²   **B.** 51kN/m²

   C. 55kN/m²   **D.** 53kN/m²

5. Principle involved in the relations between submerged unit weight and saturated unit weight of a soil is based on.

   A. Equilibrium of floating bodies   **B.** Archimedes' Principle

   C. Stoke's law   **D.** Darcy's Law

6. The approximate depth at which the effective vertical pressure is equal to 100 kN/m² in a typical deposit of submerged soil is:–

   A. 5m   **B.** 10m

   C. 20m   **D.** 100m

7. The consistency of a saturated cohesive soil is affected by :

   A. water content   **B.** particle size distribution

   C. density index.   **D.** coefficient of permeability.

8. A soil having particles of nearly the same size is said to be :

   A. well graded   **B.** uniformly graded

   C. poorly graded   **D.** gap graded

9. The particle size distribution curves are extremely useful for the classification of

    A. fine grained soils

    B. coarse grained soils

    C. both coarse grained and fine grained soils

    D. silts and clays

10. If soil is dried beyond its shrinkage limit, it will show.

    A. large volume change.   B. moderate volume change

    C. low volume change   D. no volume change

11. Consistency index for a clayey soil is

    A. $\dfrac{w_L - w}{I_\rho}$ B. $\dfrac{w - w_L}{I_\rho}$

    C. $w_L - w_p$ D. $0.5\,w$

12. The values of liquid limit and plasticity index for soils having common geological origin in a restricted locality usually define.

    A. a zone above A-line

    B. a straight line parallel to A-line

    C. a straight line perpendicular to A-line

    D. points may be anywhere in the pasticity chart.

13. The toughness index of clayey soils is given by

    A. plasticity index / flow index   B. liquid limit / plastic limit

    C. liquidity index / plastic limit   D. plastic limit / liquidity index.

14. The values of void ratio of a micaceous sand sample in the densest and the loosest states are 0.4 and 1.2 respectively. The density index of the soil for in place void ratio 0.6 will be :

    A. 60%   B. 75%

    C. 65%   D. 80%

15. Consistency, in general, is that property of soil which is manifested by its resistance to :

    A. impact   B. rolling

    C. flow   D. none of the above.

16. The liquid limit of saturated normally consolidated soil is 50%. The compression index of the soil for virgin compression curve will be :

    A. 0.36   B. 0.505

    C. 0.605   D. 0.705

17. The group index of a soil sub grade is 7. The sub grade soil is rated as.

    A. poor.   B. very poor

    C. good   D. fair.

18. The coefficient of curvature (C_c) is defined by :–

A. $\dfrac{D_{10}^2}{D_{30} \cdot D_{60}}$ **B.** $\dfrac{D_{30}^2}{D_{10} \cdot D_{60}}$

C. $\dfrac{D_{30} \cdot D_{10}}{D_{60}^2}$ **D.** $\dfrac{D_{60}^2}{D_{10} \cdot D_{30}}$

19. The description 'sandy silty clay' signifies that.

   A. The soil contains unequal proportions of the three constituents, in the order, sand > silt > clay.

   B. The soil contains equal proportions of sand, silt and clay.

   C. The soil contains unequal proportion of the three constituents such that clay > silt > sand.

   D. There is no information regarding the relative proportions of the three.

20. The shape of clay particle is usually

   A. angular   **B.** flaky

   C. tubular   **D.** rounded

21. When the products of rock weathering are not transported as sediments but remain in place, the soil is :

   A. alluvial soil.   **B.**   residual soil.

   C. glacial soil.   **D.**   aeolian soil.

22. The structure of a clay mineral as represented in the following figure is of :–

   A. kaolinite

   B. montmorillonite.

   C. halloysite

   D. illite.

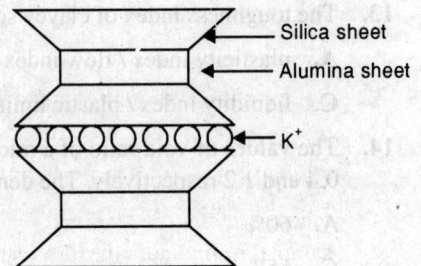

23. Amongst the clay minerals, the one having the maximum swelling tendency is :–

   A. kaolinite

   B. montmorillonite.

   C. halloysite

   D. illite

24. The swelling nature of block cotton soil is primarily due to the presence of

   A. kaolinite   **B.** lllite

   C. montmorillonite   **D.** verimiculite

25. Water chemically combined in the crystal structure of soil particles is called

   A. adsorbed water   **B.** capillary water

   C. structural water   **D.** free water

26. A soil mass has coefficients of horizontal and vertical permeability as 9×10^{-7} cm/s and 4×10^{-7} cm/s, respectively. The transformed coefficient of permeability of an equivalent isotropic soil mass is.

A. 9×10^{-7} cm/s   B. 4×10^{-7} cm/s
C. 13×10^{-7} cm/s   D. 6×10^{-7} cm/s.

27. According to Darcy's law for flow through porous media, the velocity is proportional to :
    A. effective stress   B. hydraulic gradient
    C. cohesion   D. stability number.

28. For anistropic soil, permeabilities in x and y directions are k_c and k_y respectively. In two dimensional flow the effective permeability K_{eq} for the soil is given by
    A. $k_x + k_y$ B. k_x / k_y
    C. $\sqrt{\left(k_x^2 + k_y^2\right)}$ D. $\sqrt{k_x k_y}$

29. The coefficient of permeability of a soil is 5×10^{-5} cm/sec for a certain pore fluid. If the viscosity of the pore fluid is reduced to half, the coefficient of permeability will be
    A. 5×10^{-5} cm/sec.   B. 10×10^{-5} cm/sec.
    C. 2.5×10^{-5} cm/sec.   D. 1.25×10^{-5} cm/sec.

30. The soils most susceptible to liquefaction are
    A. saturated dense sands
    B. saturated fine and medium sands of uniform particle size.
    C. saturated clays of uniform size.
    D. saturated gravels and cobbles.

31. The piezometric head at point C, in the experimental set-up shown in accompanying figures when the flow takes place under a constant head through the soils A ands B is
    A. 0 cm   B. 40 cm
    C. 80 cm   D. 120 cm

32. Seepage force per unit volume (j) can be expressed as.
    A. $i\gamma_w L$ B. iL
    C. $\gamma_w h$ D. $i\gamma_w$
    where i=hydraulic gradient, l = length of soil sample, h=hydraulic head, γ_w=unit weight of water.

33. Flow is taking place through a non-homogeneous soil deposit from zone 1 to zone 2 having the permeabilities as shown in figure. The deflection angle (α_2) of the streamlines as shown in the figure will be.
    A. 66.6°
    B. 14.0°
    C. 8.2°   D. 76.0°

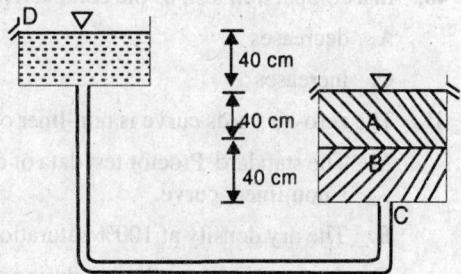

34. Piping in soil occurs when :–
    A. effective stress becomes zero.
    B. sudden change of permeability takes place.
    C. the soil is fissured and cracked.   D. the soil is highly porous.

35. Along a phreatic line in an earth dam.

    A. the total head is constant but not zero. **B.** the total head is everywhere zero.

    C. the pressure head is everywhere zero. **D.** none of the above.

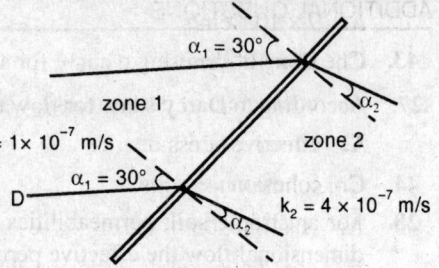

36. The hydraulic gradient needed to make effective stress zero at a point in fine sand will be given by :

    A. $\dfrac{\gamma'}{\gamma_{sat}}$ **B.** $\dfrac{G+1}{1+e}$

    C. $\dfrac{G-1}{1+e}$ **D.** $\dfrac{1+e}{G-1}$

37. A point load of 700 kN is applied on the surface of a thick layer of clay. Using Boussinesq's elastic analysis, the estimated vertical stress (O_v) at a depth of 2m and a radial distance of 1.0m from the point of application of the load is :

    A. 47.5 kPa   **B.** 47.6 kPa

    C. 47.7 kPa   **D.** 47.8 kPa

38. The vertical stress at depth, z directly below the point load Q is (k is a constant)

    A. $k\dfrac{Q}{z}$ **B.** $k\dfrac{Q}{z^3}$

    C. $k\dfrac{Q}{z^2}$ **D.** $k\dfrac{Q}{\sqrt{z}}$

39. The dry unit weight of soil at zero air voids depends on.

    A. specific gravity   **B.** water content

    C. unit weight of water.   **D.** all the three

40. In a compaction test, as the compactive effort is increased, the optimum moisture content.

    A. decreases   **B.** remains same

    C. increases   **D.** increases first and thereafter decreases

41. The zero-air voids curve is non-liner owing to:-

    A. The standard Proctor test data of dry density and corresponding water content plotting as a non-linear curve.

    B. The dry density at 100% saturation being a non-linear function of the void-ratio.

    C. the water content altering during compaction.

    D. The soil being compacted with an odd number of blows.

42. The time for a clay layer to achieve 90% consolidation is 15 years. The time required to achieve 90% consolidation, if the layer were twice as thick, 3 times more permeable and 4 times more compressible would be:

    A. 70 Years   **B.** 75 Years

    C. 80 Years   **D.** 85 Years.

43. The slope of the e-log p curve for a soil mass gives

    A. coefficient of permeability, k   **B.** coefficient of consolidation, C_y

    C. compression index, C_c **D.** coefficient of volume compressibility, M_v.

44. Consolidation in soils.

    A. is a function of the effective stress

    B. does not depend on the present stress

    C. is a function of the pore water pressure.

    D. is a function of the total stress.

45. Terzaghi's one-dimensional consolidation theory assumes that

    A. e Vs. p relation is linear.   **B.** e Vs. $\log_{10} p$ relation is linear

    C. p Vs. $\log_{10}$ e relation is linear.   **D.** e Vs. $\log_{10} p/p_0$ relation is linear.

46. Consolidation time.

    A. increases with increase in compressibility.

    B. increases rapidly with decreasing size of soil mass.

    C. decreases with increase in permeability.

    D. is in dependent of the magnitude of stress.

47. The coefficient of consolidation is used for

    A. establishing the duration of primary consolidation.

    B. estimating the amount of settlement for a load increment

    C. determining the depth to which the soil is stressed when loads are applied on the surface of a soil deposit.

    D. determining the preconsolidation pressure for soil deposit known to be overconsolidated.

48. The appropriate field test to determine the insitu undrained shear strength of soft clay is

    A. plate load test.   **B.** static cone penetration test

    C. standard penetration test   **D.** vane shear test.

49. The unconfined compressive strength of a stiff clay falls in the range.

    A. less than 50 kN/m^2 **B.** 50 to 100 kN/m^2

    C. 100 to 200 kN/m^2 **D.** above 200 kN/m^2

50. The stress-strain behaviour of soils as shown in the following figure corresponds to:

    A. Curve 1 : Loose sand and normally consolidated clay.

        Curve 2 : Loose sand and over consolidated clay.

    B. Curve 1 : Dense sand and normally consolidated clay.

        Curve 2 : Loose sand and over consolidated clay.

    C. Curve 1 : Dense sand and over consolidated clay

        Curve 2 : Loose sand and normally consolidated clay.

    D. Curve 1 : Loose sand and over consolidated clay

        Curve 2 : Dense sand and normally consolidated

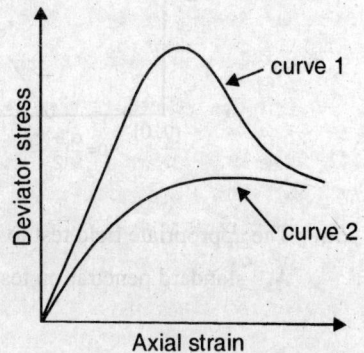

clay.

51. Some of the structural strength of a clayey material that is lost by remoulding is slowly recovered with time. This property of soils to undergo an isothermal gel-to sol-to-gel transformer upon agitation and subsequent rest is termed

   A. isotropy   B. anisotropy

   C. thixotropy   D. allotropy

52. The ratio of unconfined compressive strength of an undisturbed sample of soil to that of a remoulded sample, at the same water content, is known as.

   A. activity   B. damping

   C. plasticity   D. sensitivity.

53. Vane tester is normally used for determining in situ shear strength of

   A. soft clays   B. sand

   C. stiff clays   D. gravel

54. Triaxial compression test of three soil specimens exhibited the patterns of failure as shown in following figure. Failure modes of the samplers respectively are

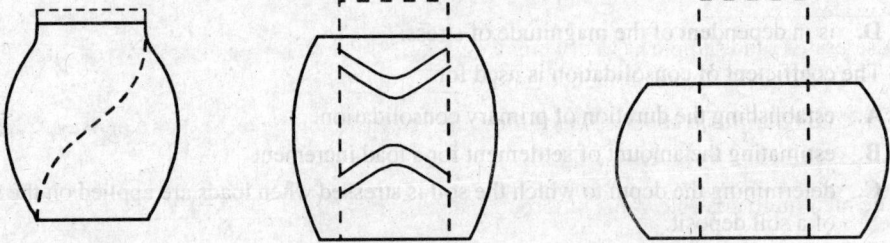

   A. (i) brittle,   (ii) semi-plastic,   (iii) plastic.

   B. (i) semi-plastic,   (ii) brittle,   (iii) plastic.

   C. (i) plastic,   (ii) brittle,   (iii) semi-plastic.

   D. (i) brittle,   (ii) plastic,   (iii) semi-plastic.

55. For a soil specimen the relation between the deviator stress (q) and the mean stress (p) is given below in the figures. Which of the following conditions satisfy ?

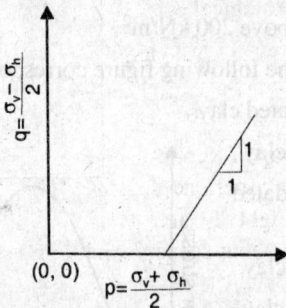

   A. $\Delta\sigma_v = 0; \Delta h < 0$

   B. $\Delta\sigma_n = -\Delta\sigma_v$

   C. $\Delta\sigma_v > 0; \Delta\sigma_n = 0$

   D. None of the above

56. The appropriate field test to determine the insitu undrained shear strength of a soft clay is:

   A. standard penetration test.   B. plate load test.

    C. static cone penetration test.   **D.** vane shear test.

57. For a very heavily over consolidated clay sample the probable value of pore pressure parameter A at failure is likely to be:

    A. 0.85   **B.** 0.35

    C. 0.0   **D.** – 0.20

58. For a saturated normally consolidated soil specimen the pore pressure coefficient B will be:

    A. 1.0   **B.** 0.8

    C. 0.2   **D.** – 0.5

59. In a drained triaxial compression test conducted on dry sand, failure occurred when the deviator stress was 218 kN/m^2 at a confining pressure of 61 kN/m^2. What is the effective angle of shearing resistance and the inclination of failure plane to major principal plane?

    A. 34°, 62°   **B.** 34°, 28°

    C. 40°, 25°   **D.** 40°, 65°

Fill in the blanks:-

60. If the saturation water content of a soil of specific gravity 2.7 is 40%, its void ratio is _____.

61. A saturated sand sample has a dry unit weight of 18kN/m^3 and specific gravity of 2.65. Taking $\gamma_w = 10 \text{kN/m}^3$, the water content of soil is _____.

62. For sand of uniform spherical particles, the void ratio in the loosest and densest states are _____ and_____.

63. The maximum possible value of Group Index for a soil is _____.

64. Soils transported by wind are known as _____.

65. The hydraulic head at a point in soil is the sum of _____ and _____.

66. A 1000 kN load is uniformly distributed on an area 2m × 3m. the approximate average vertical stress at 3_m depth using 2:1 dispersion is _____ kN/m^2.

67. Following curves indicate schematically three standard Proctor compaction curves.
Assuming the three soil types are from the same geological origin, the curves for silty clay, silty sand and sandy silt are _____ , _____ and _____ respectively.

68. An infinite slope with a slope angle of 14°, is made up of a cohesionless soil having $\phi = 30°$ and $\gamma = 20 \text{ kN/m}^3$. It experiences seepage with the water table at surface. If the unit weight of water is 10 kN/m^3, the factors of safety against failure without seepage and with seepage will be _____ and _____ respectively.

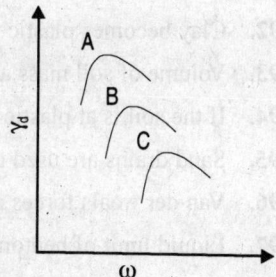

STATE WHETHER THE FOLLOWING STATEMENTS ARE TRUE OR FALSE:-

69. The void ratio of soil can exceed unity.

70. The porosity of soil can be greater than its void ratio.

71. The water content of a soil cannot be greater than one hundred percent.

72. In the IS soil classification system SM stands for sandy silt.

73. A soil having a uniformity coefficient smaller than about 2 is considered 'uniformly graded'.

74. The 'A line in the platicity chart separates organic clays from inorganic clays.

75. The charge on Kaolinite is due to one aluminium substitution for every four hundredth silicon ion.

76. The capillary pressure in a soil may be more than 5m head of water.

77. In some situations effective stress will be greater than the total applied stress.

78. In practically all seepage problems, velocity heads are disregarded.

79. The measure of soil compaction is its wet density.

80. The coefficient of volume compressibility of a soil is always less than its coefficient of compressibility.

81. The total settlement of a soil layer is dependent on the length of drainage path.

82. If the Mohr circle for a given state of stress lies entirely below the Mohr envelope for a soil, then the soil will be unstable for that state of stress.

83. The maximum possible slope angle in a granular soil is equal to the friction angle of the soil.

84. In an earth dam phreatic line is a boundary equipotential line.

85. In an earth dam shell imparts stability and protects the core.

86. In a homogeneous earth dam sudden drawdown causes instability of upstream slope.

87. In Swedish method of slices the forces of interaction between adjacent slices are not considered.

88. In the case of infinite slope the slip surface is parallel to the surface of slope.

89. The porosity of soil can never be greater than 100%.

90. The effect of stratification in soil mass will result in greater horizontal permeability than vertical permeability.

91. When soil deposit is in its densest state, its density index is zero.

92. Clay becomes plastic when mixed with kerosene.

93. Volume of soil mass at shrinkage limit is same as that at its dry state.

94. If the soil is at plastic limit the consistency index is unity.

95. Sand drains are used to decrease the rate of consolidation.

96. Van der waals forces are weaker than hydrogen bond.

97. Liquid limit of bentonite clay is more than 100%.

98. Quick sand is a type of sand.

99. Stability number is a dimensionless quantity.

100. Taylor stability number is a function of slope angle only.

QUESTIONS WITH ANSWERS

1(a). Why is a long stem hydrometer used in laboratory ?

Ans. A long stem hydrometer for which stem is about 30 to 40 cm long is used in laboratory as the spacing between graduations (0.001) is sufficiently large enabling greater accuracy in reading.

(b) Why is the hydrometer marked only upto 1.030 ?

Ans. Hydrometer used in grain size analysis is marked only upto 1.030 as not more than 50 gm of soil is taken for preparing one litre of suspension and the density of such soil suspension will not exceed 1.030 gm/cc.

2. Draw typical particle size distribution curves to indicate uniformly graded, well graded and gap graded soils.

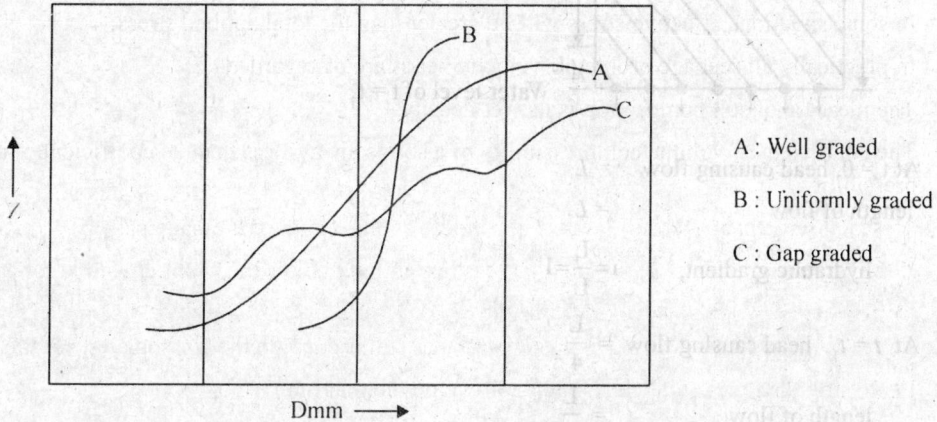

A : Well graded

B : Uniformly graded

C : Gap graded

Dmm ⟶

3. Explain how soils can be identified in the field using simple tests ?

Ans. Gravel and sand can be easily identified since the individual particles are visible to the naked eye. In the case of fine-grained fraction – silts and clays, the following simple tests are used for identification.

(a) **Dry strength test :** A dry lump of clay cannot be easily crushed between thumb and fore finger whereas it is relatively easy to crush a dry lump containing more of silt.

(b) **Dilatancy test :** A wet pat of soil is kept in palm of one hand and struck several times with the other. If water appears quickly on surface and when squeezed it quickly flows back, we say reaction to dilatancy test is quick. Otherwise it is said to be sluggish. The reaction to dilatancy test will be quick to moderately quick in the case of fine sand and silt. It will be sluggish in the case of clay.

(c) **Thread test :** If a small lump of clay can be rolled into a thread of 3 mm dia without crumbling then it is plastic clay. In the case of less or non-plastic soils like silts the soil lump cannot be rolled into 3 mm diameter thread without crumbling.

(d) **Dispersion test.** If a small quantity of soil is dispersed in water taken in a beaker (about 10 cm height) gravel and sand will settle in less than a minute, silts will settle in 15 to 60 minutes and clays will remain in suspension for several hours.

4. Define coefficient of transmissibility.

Ans. Coefficient of transmissibility (T) is defined as the rate of flow of water through a vertical strip of unit width to the full saturation height of an aquifer, under unit hydraulic gradient. In the case of confined aquifer of thickness b we have, $T = bk$, where k is the coefficient of permeability.

5. A tank of height L and cross sectional area A has perforated bottom. It is filled with soil upto the top. The soil is fully saturated with water level initially at the top. The water level falls as the water drains out at the bottom. Determine the time required for water level to drop to a

level $\frac{3}{4}$L from the top. The porosity of soil is n and coefficient of permeability, k.

Ans.

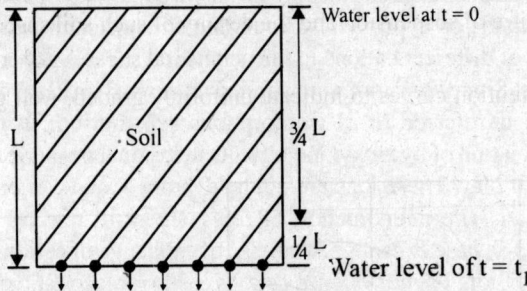

At t = 0, head causing flow   = L

   length of flow   = L

$\therefore$ hydraulic gradient, $i = \dfrac{L}{L} = 1$

At $t = t_1$, head causing flow = $\dfrac{L}{4}$

     length of flow   = $\dfrac{L}{4}$

hydraulic gradient $i = \dfrac{L/4}{L/4} = 1$

Thus hydraulic gradient is always unity as the water drains out.

If water level drops through dh in a small interval of time dt, the discharge is given by

$$dq = \frac{-(nA)\,dh}{dt} \qquad \ldots (i)$$

Negative sign indicates fall in head.

By Darcy's law $q = kiA$

$dq = k\,(1)\,A = kA$ $\ldots (ii)$

From Eq. (i) and (ii), $-n\,A\dfrac{dh}{dt} = kA$

$-n\,dh = k\,dt$

Integrating, $-n \displaystyle\int_{L}^{L/4} dh = k \int_{0}^{t_1} dt$

$n\left(L - \dfrac{L}{4}\right) = kt_1$

$\therefore \;\; t_1 = \dfrac{3}{4}\dfrac{nL}{k}$

6. Write a note on base exchange capacity.

Ans. The two fundamental building blocks of clay minerals are the tetrahedral silica sheet and the octahedral alumina sheet. The particular way in which these sheets are stacked and the type of bonding determine the different clay minerals. The tetrahedral sheet is basically a combination of silica tetrahedral units which consist of four oxygen atoms at the corners of tetrahedron with a silicon atom at the centre. The octahedral sheet is basically a combination of octahedral units consisting of six oxygen atoms or hydroxyls surrounding an aluminium, magnesium, iron, or other atom. Substitution of different cations in the octahedral sheet is rather common leading to different clay minerals. Since the cations substituted are approximately of the same physical size, the phenomenon is referred to as isomorphous substitution. Isomorphous substitution along with the dissociation of hydroxyl ions results in residual negative charge on the surface of mineral particles of clay. These ions are not held strongly and can be replaced by other cations present in water. The phenomenon of replacement of one by the other exchangeable cation is referred to as base exchange. The ease of exchange of cations depends primarily on the valence of each cation. Higher valence cations easily replace cations of lower valence. The cations can be listed in approximate order of their replacement ability. The specific order depends on the type of clay, ion being replaced and concentration of the various ions in the water. In order of increasing replacement power the ions are

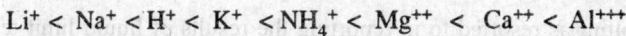

$Li^+ < Na^+ < H^+ < K^+ < NH_4^+ < Mg^{++} < Ca^{++} < Al^{+++}$

Base exchange capacity is expressed in milli-equivalent per 100 gm (meq/100 gm) of soil. It is expressed in terms of the mass of a cation which may be held on the surface of 100 gm of dry mass of mineral. One milli-equivalent is one milligram of hydrogen (atomic weight 1.0) or an equivalent mass of any other ion.

Base exchange capacity is a very useful property with several practical applications. The use of chemicals to stabilize soils is possible because of this property. The swelling of sodium mont-morillonitic clays can be significantly reduced by the addition of lime. Some clays can be hardened by the passage of electric current, as for example from an aluminium anode to a copper cathode.

Problem : If 100 gm of a dry soil is capable of absorbing 50 mg of Calcium, calculate base exchange capacity of soil. Atomic weight of calcium is 40.

Solution. Calcium has two cations

$\therefore$ Base exchange capacity of soil is $\dfrac{50}{40/2} = 2.5$ meq/100 gm of soil

7. Write a short note differentiating between total settlement and rate of settlement.

Total settlement (due to primary consolidation) is given by

$$\rho_f = m_v.H\ \Delta\sigma' = C_c\ \frac{H}{1+e_0}\log\frac{\sigma_0^1 + \Delta\sigma^1}{\sigma_0^1} \qquad \dots(i)$$

$$V(=\%) = f(Tv) \qquad \dots(ii)$$

$$Tv = \frac{c_v t}{d^2} = \frac{k}{m_v y_w}\cdot\frac{t}{d^2} \qquad \dots(iii)$$

From Eq. (i) we see that total settlement or amount of consolidation is dependent on compressibility of soil mass and independent of permeability and drainage path.

From Eq. (iii) we observe that rate of settlement depends on compressibility, permeability and drainage path.

8. Write a short note on sand drains.

Ans. Vertical sand drains are constructed in cohesive soil deposits to accelerate the consolidation process so that the estimated future settlement takes place to a great degree before completion of proposed construction, as for example, in the construction of highway or airport embankment. As the rate of gain of shear strength is also accelerated, it permits more rapid construction.

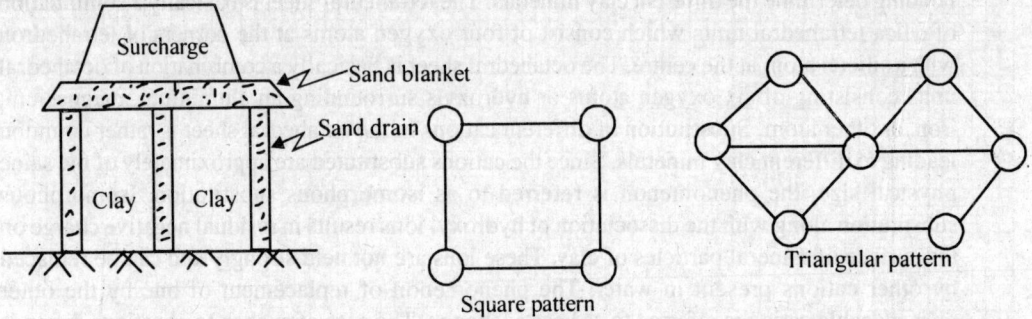

Surcharge
Sand blanket
Sand drain
Clay   Clay

Square pattern

Triangular pattern

Sand drains are constructed by drilling holes at the site in a squar of triangular grid pattern and filling the holes with cohesionless soil. A sand blanket is placed on top of clay layer. It serves as drainage layer and also functions like a graded filter to prevent soft clay from entering into the embarkment material above.

9. Explain why semi-log plotting is resorted to in reporting the results of liquid limit test and grain size analysis.

Ans. In liquid limit test the relation between water content and number of blows is exponential.

$$N \propto 10^w$$

$$Log_{10} N \propto w$$

By plotting w against $log_{10} N$ we get a straight line.

In grain size analysis, the range of particle sizes being too large it necessitates semi-log plotting.

10. Define compression ratios r_0, r_p, and r_s.

Ans. The corrected dial gauge reading, R_C corresponding to zero per cent consolidation does not coincide with initial dial gauge reading, R_i. The compression $(R_c - R_i)$ represents initial compression and is due to factors such as imperfect saturation, elastic compression of soil and lateral compression when not properly mounted. Initial compression is followed by primary compression and secondary compression. The relative magnitudes of the three stages of compression are given by the following ratios.

Initial compression ratio, $r_0 = \dfrac{R_c - Ri}{R_f - R_i}$

Primary compression ratio, $r_p = \left(\dfrac{R_{q0} - R_c}{R_f - R_i} \right) \dfrac{10}{9}$ for $\sqrt{t}$ method

$\qquad\qquad\qquad\qquad\qquad = \dfrac{R_{100} - R_c}{R_f - R_i}$ for log t method

Secondary compression ratio, $r_s = \dfrac{Secondary\, compression}{Total\, compression}$

$\qquad\qquad\qquad\qquad\qquad = 1 - (r_0 + r_p)$

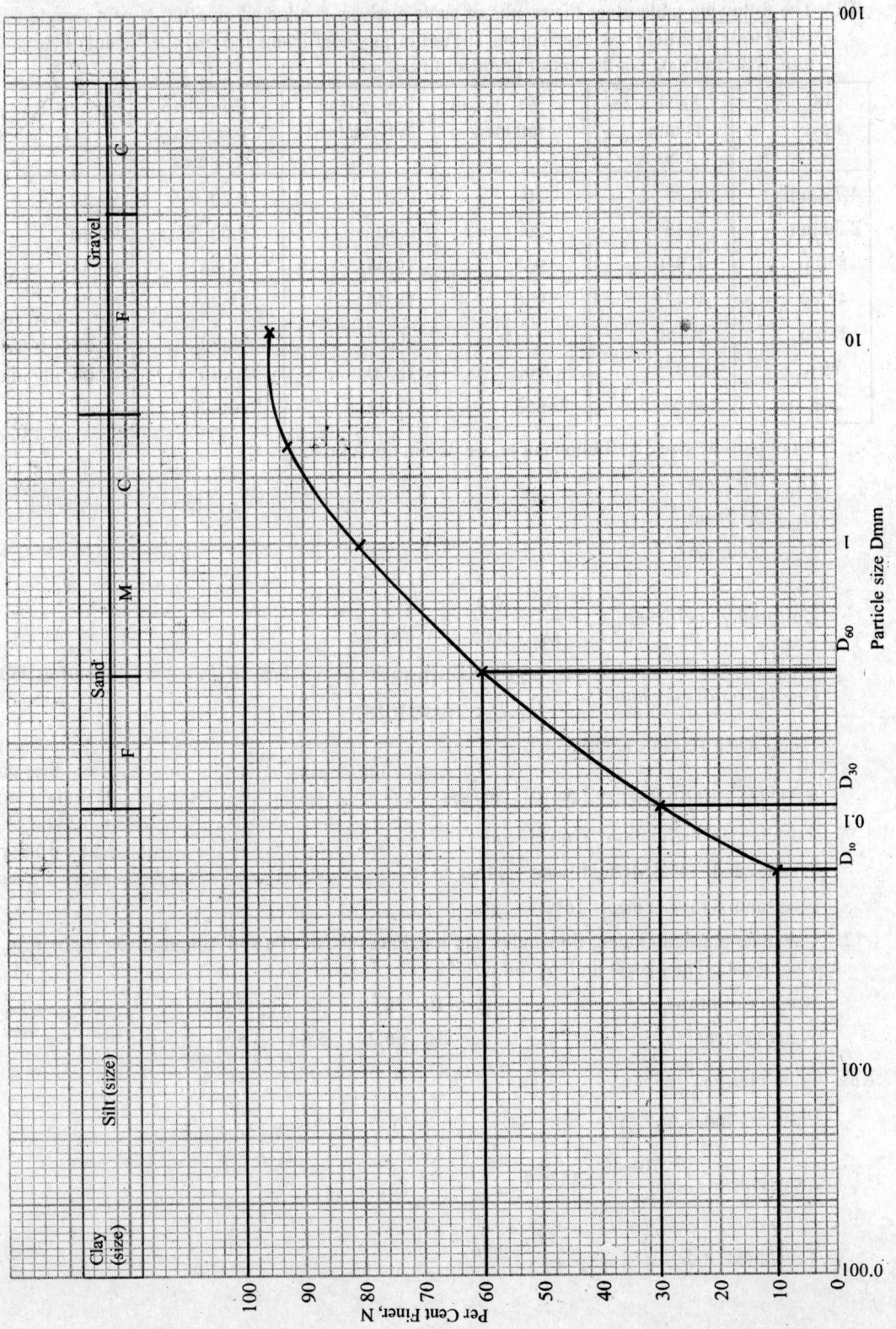

11. The following table gives the results of sieve analysis conducted on 5000 N of a sample of soil. Find (i) particle size distribution, (ii) uniformity coefficient and (iii) coefficient of curvature. Also estimate coefficient of permeability of soil.

IS sieve	Sieve size D mm N	Wt. of soil retained	Percentage Wt. retained	Cumulative percentage retained	Per cent finer N
4.75 mm	4.75	10	0.20	0.20	99.80
2.36 mm	2.36	20	0.40	0.60	99.40
850 μ	0.850	953	19.06	19.66	80.34
425 μ	0.425	930	18.60	38.26	61.74
150 μ	0.150	1335	26.70	64.96	35.04
75 μ	0.075	900	18.00	82.96	17.04
pan	–	852	17.04	100.00	–

From particle size distribution curve, we obtain the following :

$D_{10} = 0.05$ mm

$D_{30} = 0.12$ mm

$D_{60} = 0.40$ mm

Uniformity coefficient, $C_u = \dfrac{D_{60}}{D_{10}} = \dfrac{0.4}{0.05} = 8$

Coefficient of curvature, $C_C = \dfrac{D_{30}^2}{D_{10} \cdot D_{60}} = \dfrac{(0.12)^2}{(0.05)(0.4)} = 0.72$

Using Allen Hazen's equation,

$k = C\, D_{10}^2 = 100(0.005)^2 = 2.5 \times 10^{-3}$ cm / sec

Particle size distribution : Gravel – nil,

Coarse sand – 4%, Medium sand – 34%,

Fine sand – 45%, Fines – 16%

12. Calculate plasticity index, flow index and toughness index of a soil sample given in the following data. The plastic limit of soil is 25%.

Number of blows	10	20	32	40
Water content (%)	70	60.2	55.4	50

Ans. From plot, $\omega_L = 57\%$

Plasticity Index, $I_p = (57 - 25) = 32\%$

Flow Index, $I_F = \dfrac{70 - 60}{\log_{10} 2} = 33.2\%$

Toughness Index, $I_T = \dfrac{I_P}{I_F} = \dfrac{32}{33.2} = 0.96$

13. Write a short note on roofing and piping under weir.

Ans. In the case of flow under weir resting on permeable bed, if the exit gradient close to the toe of weir exceeds critical hydraulic gradient for soil, the soil particles will be carried away by seeping water. In course of time this will lead to creation of a hollow channel just below the base of weir. Since the base of weir will act as roof over this channel, this phenomenon is referred to as roofing.

If the exit gradient exceeds critical hydraulic gradient of soil, at a point away from toe, the soil particles will be carried away by seeping water and after sometime a continuous pipe will be formed well below the base of weir. This phenomenon is referred to as piping.

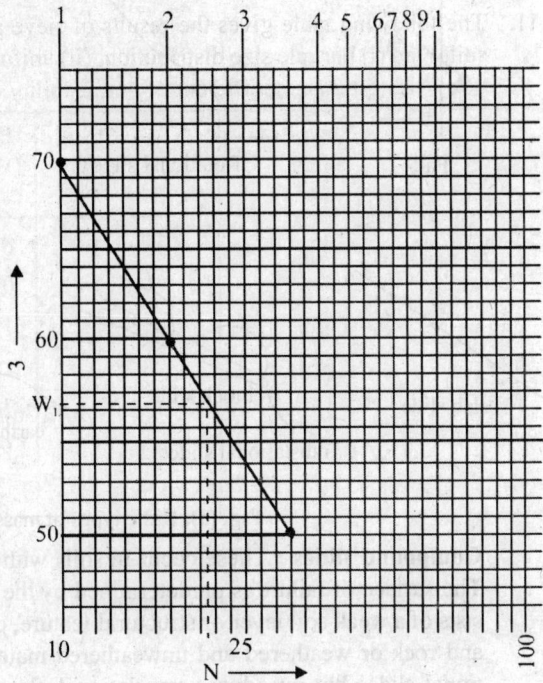

Note: For more explanation on piping, please read Art (8.14) page. 132

14. State the causes of failures of earth slopes.

Ans. The causes of failures of earth slopes may be divided into external and internal ones. External causes are those which produce increase in shearing stresses at unaltered shearing resistance of the material. They include

1. steepening of slope

2. deposition of material along edge of slope

3. vibratory forces

Internal causes are those which lead to slope failure without any change in surface conditions which involve unaltered shearing stresses in the slope material. The causes of such a condition are the decrease in shearing resistance due to development of excess pore pressure, leaching of salts, softening, breakage of cementation bonds and ion exchange.

15. Write a note on classification of landslides.

Ans. A systematic classification of landslides in clays and other mass movements was proposed by Skempton and Hutchison (1969). The five basic types are falls, rotational slides, compound slides, translational slides and flows.

Falls : This type is common in the case of steep slopes of over consolidated fissured clays. Removal of lateral earth support causes bulging at the toe and tension crack at the top. The development of cracks induces additional stresses on the separating mass which ultimately leads to a fall.

Rotational slides : These include slips and slumps and are common in fairly uniform clays or shales. The curved surface of failure being concave upwards imparts a back-tilt to the sliding mass resulting in sinking at the rear and heaving at the toe. Such slides are relatively deep seated with curved failure surfaces.

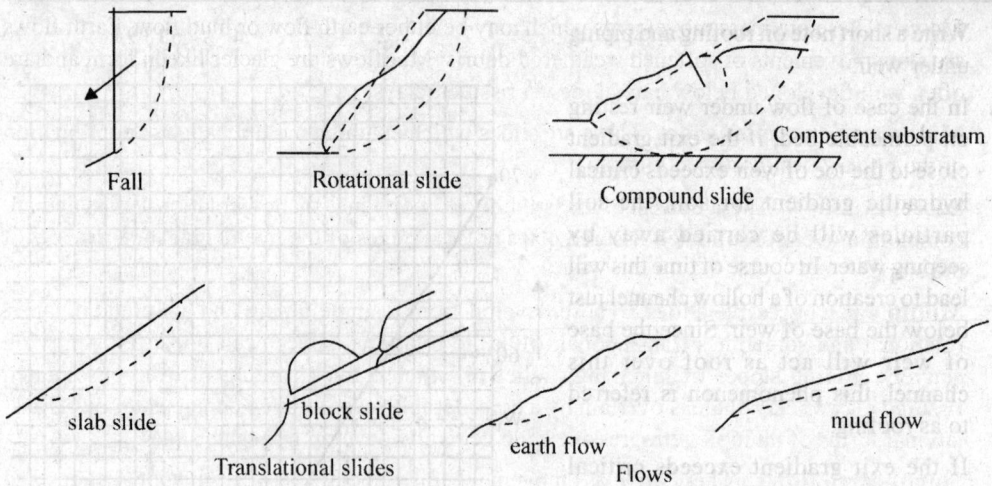

Fig. (a). Basic types of mass movements on clay slopes

Compound slides : These occur in soils with presence of heterogeneity at moderate depth. The surface of failure is predetermined by the presence of heterogencity which usually consists of a weak soil layer or structural feature, or boundry between two materials such as clay and rock or weathered and unweathered material. The heterogeneity prevents simple rotational slides but introduces translational element in the movement in combination with or without rotational slide.

Translational slides : These are planar, block slides or slab slides commonly occuring in a mantle of weathered material with heterogeneity being at shallow depth. Block slides are found in Marls and sandstones. Slab slides are a type translational failure in more weathered clay slopes.

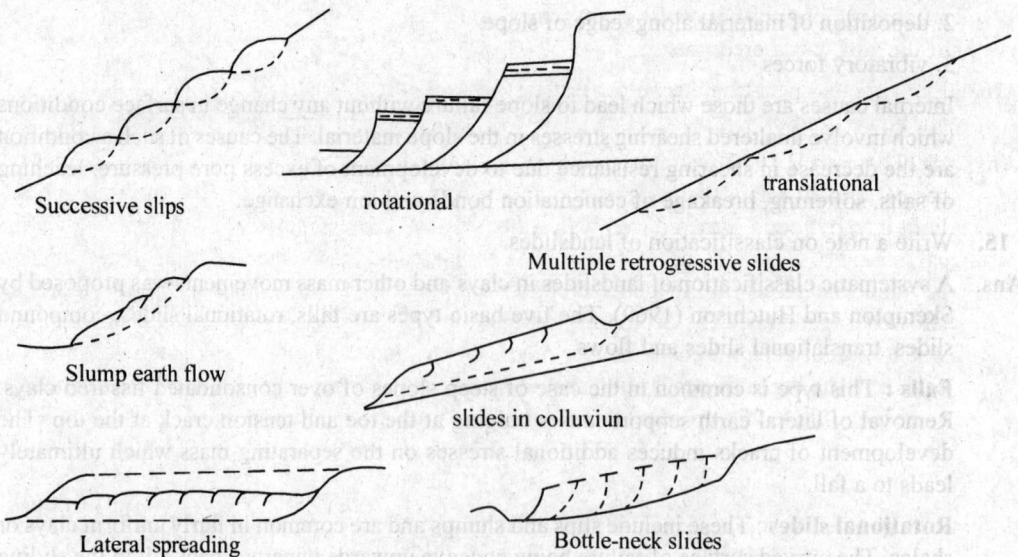

Flows : These are mass movements which may be either earth flow or mud flow. Earth flows are slow movements of softened weathered debris. Mudflows are glacier like in form and are often well developed below bare slopes in fissuied clays.

Multiple and complex slides. These are slides which exhibit a multiplication of combination of the basic types discussed hitherto.

Successive slips : These consist of an assembly of individual shallow rotational slips and are common in over consolidated fissured clays at later stages of the free degradation process of the slopes.

Multiple retrogressive slides : These develop from a single failure and are predominantly rotational and sometimes translational. Multiple rotational slides occur most frequently on actively eroding slopes of fairly high relief in which a thick stratum of overconsolidated fissured clay or clay-shale is overlain by a considerable layer of more competent rock. Translational forms of multiple retrogressive slide generally develop from slab slides.

Slump-earth flows : These are mass movements intermediate between rotational slides and mudflows. They are found to develop in a rotational slide of considerable displacement where the toe of the slipping mass is much broken by overriding which in the presence of water softens and forms into a mudflow.

Slides in colluvium: These are mass movements of debris noticeable in the accumulation zones below freely degrading cliffs.

Spreading failures : These are a particular type of retrogressive translational slide. Because of the gentle slopes involved, initial rapid movement reduces considerably and stops within a few minutes.

Bottleneck slides : These are a peculiar type of landslide associated with quick clays. Such slides generally begin with an initial rotational slip in the bank of a stream incised into quick clay deposits. The slipping mass is in part remoulded to the consistency of a liquid which runs out of the cavity, carrying flakes of the stiff, weathered crust.

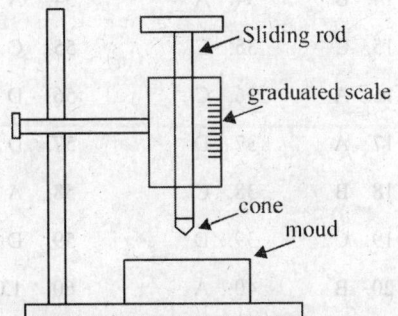

16. Explain cone penetrometer meter test for determining liquid limit of soil.

Ans. IS 2720 - part V, 1970 recommends this method which has certain advantages over the method using Casagrande's apparatus. The soil whose liquid limit is to be found is mixed well with water to a soft consistency and filled into the cylindrical, mould of 50 mm ϕ and 50 mm ht. The cone attached to sliding rod, with central angle of 31° and total mass of 148 gm is brought in contact with the surface of soil in mould and allowed to penctrate. The depth of penetration for 30 seconds is noted and the corresponding water content determined. A number of trials are conducted with varying water content. In each trial the depth of penetration should be between 20 mm and 30 mm. The water content corresponding to 25 mm penetration is taken as the liquid limit of soil.

ANSWER TO ADDITIONAL QUESTIONS

1.	B	21.	B	41.	B	61.	17.8%	81.	F
2.	D	22.	D	42.	C	62.	0.91, 0.35	82.	F
3.	D	23.	B	43.	C	63.	20	83.	F
4.	B	24.	C	44.	A	64.	aeolin soils	84.	F
5.	B	25.	C	45.	B	65.	pressure head and datum head	85.	T
6.	B	26.	D	46.	C	66.	2000 kN/m²	86.	T
7.	A	27.	B	47.	A	67.	C, A, B	87.	T
8.	B	28.	D	48.	D	68.	2.32, 1.16	88.	T
9.	B	29.	B	49.	C	69.	T	89.	T
10.	D	30.	B	50.	B	70.	F	90.	T
11.	A	31.	D	51.	C	71.	F	91.	F
12.	B	32.	D	52.	D	72.	T	92.	F
13.	A	33.	A	53.	C	73.	T	93.	T
14.	B	34.	A	54.	C	74.	T	94.	T
15.	C	35.	C	55.	C	75.	F	95.	F
16.	A	36.	C	56.	D	76.	T	96.	T
17.	A	37.	D	57.	D	77.	T	97.	T
18.	B	38.	C	58.	A	78.	T	98.	F
19.	C	39.	D	59.	D	79.	F	99.	T
20.	B	40.	A	60.	1.08	80.	T	100.	F

APPENDIX—I
[IS SIEVES]

Designation	Size of aperture (mm)	
IS 50 mm	50.0	
40 mm	40.0	
20 mm	20.0	
10 mm	10.0	
4.75 mm	4.75	
* 2.80 mm	2.80	**Note 1:** IS sieves 50 mm, 40 mm, 20
2.00 mm	2.00	mm and 10 mm can be of
* 1.40 mm	1.40	square holes in perforated
1.00 mm	1.00	plates, square holes in wire cloth or round holes in perforated plates
600 μ	0.600	2. Sieves marked with *have
* 500 μ	0.500	been proposed as an International (ISO) standard and are
425 μ	0.425	recommended to be included,
* 355 μ	0.355	if possible, in all sieve analysis
300 μ	0.300	data.
212 μ	0.212	
* 180	0.180	
150 μ	0.150	
* 125 μ	0.125	
* 90 μ	0.090	
75 μ	0.075	
* 63 μ	0.063	
* 45 μ	0.045	

APPENDIX — II

SOLUTION TO TERZAGHI'S ONE-DIMENSIONAL CONSOLIDATION EQUATION

$$\frac{\partial \bar{u}}{\partial t} = C_v \frac{\partial^2 \bar{u}}{\partial z^2} \qquad \qquad \dots (i)$$

With $\bar{u} = \bar{u}(z,t)$, initial condition is $\bar{u}(z,0) = \bar{u}_0$ and boundary conditions are

$\bar{u}(0,t) = 0$ and $\bar{u}(2H,t) = 0$

where H is half the thickness of consolidating stratum.

Denoting $\dfrac{\partial \bar{u}}{\partial t}$ by T' and $\dfrac{\partial^2 \bar{u}}{\partial z^2}$ by Z''

Let $u = Z(z)\, T(t)$ Eq. (i) can then be written as

$$C_v Z'' T = Z T' \qquad \qquad \dots (ii)$$

$\dfrac{Z''}{Z} = \dfrac{T'}{C_v T} =$ separation constant say, K

Eq. (ii) which is a partial differential equation reduces to a coupled system of ordinary differntial equations

$Z'' - KZ = 0$ and $T' - C_v KT = 0$

Using the operator D,

$(D^2 - K).\, Z(z) = 0$

$\Rightarrow D = \pm\sqrt{K} = 0 \pm \sqrt{-K}i$

$(D - CvK).\, T(t) = 0$

$\Rightarrow D = C_v K$

For Fourier Series solution, for K < 0 we can write

$\mu = e^{C_v kt}\,(C_1 \sin\sqrt{-K}\, Z + C_2 \cos\sqrt{-K}\, Z)$

To find C_1, C_2 and K we use the condition

$\bar{u}(0,t) = 0$

$\Rightarrow e^{C_v Kt} C_2 = 0$

For $C_2 = 0$, $\bar{u} = C_1 e^{C_v Kt} \sin\sqrt{-K}z$ $\qquad \dots (iii)$

With Z = 2H, $C_1 e^{C_v Kt} \sin\sqrt{-K}\, 2H = 0$ $\qquad \dots (iv)$

For non-trivial solution, $\sin\sqrt{-K}\, 2H = 0$

$\sqrt{-K}.2H = n\pi; \quad n = 1,2,3,\dots..$

$\therefore \quad \sqrt{-K} = \dfrac{n\pi}{2H}$

Squaring, $\therefore \quad K = \left(\dfrac{n\pi}{2H}\right)^2$

By substitution in Eq. (*iii*)

$$\bar{u} = C_1 e^{-C_v\left(\frac{n\pi}{2H}\right)^2 t} \sin \frac{n\pi Z}{2H} \;;\; n = 1, 2, 3, ..$$

We now write, $\bar{u} = \displaystyle\sum_{n=1}^{\infty} \left\{ C_1 e^{-C_v\left(\frac{n\pi}{2H}\right)^2 t} \sin \frac{n\pi Z}{2H} \right\}$

$\bar{u} = \bar{u}_0$ at $t = 0$. $\bar{u}_0 = \displaystyle\sum_{n=1}^{\infty} C_1 \sin \dfrac{n\pi Z}{2H}$

We can show that $C_1 = \begin{cases} 0; & n \text{ even} \\ \dfrac{4\bar{u}_0}{n\pi}; & n \text{ odd} \end{cases}$

Therefore, $\bar{u} = \displaystyle\sum_{n=1}^{\infty} \left\{ \frac{4\bar{u}_0}{n\pi} e^{-C_v\left(\frac{n\pi}{2H}\right)^2 t} \sin \frac{n\pi Z}{2H} \right\};\; n = 1, 3, 5, 7 \ldots \qquad \ldots (v)$

In general Eq. (v) can be written as

$$\bar{u} = \sum_{m=0}^{\infty} \left[\frac{4\bar{u}_0}{(2m+1)\pi} e^{-C_v\left\{\frac{(2m+1)\pi}{2H}\right\}^2 t} \sin \left\{ \frac{(2m+1)\pi Z}{2H} \right\} \right];\; m = 0, 1, 2, 3 \qquad \ldots (v)$$

REFERENCES

Atterberg, A., 'Uber die Physikalische Bodenunter suchung und uber die Plastizitat der Tone' Int. Mitt fur Bodenkunde, Vol. 1, Berlin, 1911.

Bell, F.G., Fundamentals of Engineering Geology, Butterworths, London, 1983.

Bishop, A.W., 'The use of pore pressure coefficients in practice', Geotechnique, Institution of Civil Engineers, London, U.K., Vol. 4, 1954.

Bishop, A.W., 'The use of the slip circle in the stability Analysis of slopes', Geotechnique, Vol. 5, Institution of Civil Engineers, London, U.K., 1955.

Bishop, A.W., and D.J. Henkel, The Measurement of Soil properties in the Triaxial Test, 2nd ed. Edward Arnold, London, 1962.

Bowles, J.E., Physical and Geotechnical properties of soils, 2nd ed., McGraw-Hill Book Co., New York, 1984.

Casagrande, A., 'Classification and Identification of Soils,' Trans. ASCE, Vol. 113, pp. 901–930, 1948.

Cernica, J.N., Geotechnical Engineering, John Wiley & Sons, New York, 1995.

Darcy, H., Les Fontaines Publiques de la ville de Dijon. Dalmont, Paris, 1856.

Fadum, R.E., 'Influence Values for Estimating Stresses in Elastic Foundations', 2nd ICSMFE, Vol. 3, pp. 77-84, 1948.

Fellenius, W., 'Calculation of the stability of Earth Dams', Trans. 2nd congress on large Dams, Washington, D.V. Vol. 4, 1936.

Grim, R.E., Applied Clay Mineralogy, McGraw-Hill Book Co., New York, 1962.

Harr, M.E., Ground Water and Seepage, McGraw-Hill Book Co., New York, 1962.

IS: 1498 Classification and Identification of soils for General Engineering Purposes, BIS, 1970

IS : 2720, Methods of Test for Soils, BIS

Part 1   Preparation of Dry Soil Samples for Various Tests, BIS, 1983

Part 2,   Determination of Water Content, BIS, 1973

Part 3,   Determination of Specific Gravity, BIS Section 1, 1980; Section 2, 1981.

Part 4,   Grain Size Analysis, BIS, 1975.

Part 5,   Determination of Liquid and Plastic Limits, BIS. 1970.

Part 6,   Determination of Shrinkage factors, BIS, 1972.

Part 7,   Determination of Water content - Dry Density relation using light compaction, BIS, 1974.

Part 8,   Determination of Water content - Dry Density Relation using Heavy compaction, BIS, 1983.

Part 10,   Determination of Unconfined Compressive Strength, BIS, 1973.

Part 11,   Determination of the shear strength parameters of a specimen tested in Unconsolidated Undrained Triaxial Compression without the Measurement of Pore Water Pressure, BIS, 1971.

Part 12.   Determination of Shear strength parameters of soil from consolidated undrained Triaxial Compression Test with Measurement of Pore Water Pressure, BIS, 1981.

Part 13,   Direct shear Test, BIS, 1972.

Part 14,   Determination of Density Index of Cohesionless Soils, BIS, 1983.

Part 15.   Determination of consolidation Properties, BIS, 1986.

Part 17.   Laboratory Determination of Permeability BIS, 1986

Part 28,   Determination of Dry Density of Soils, in place, by the sand-Replacement Method, BIS,. 1974.

Part 29,   Determination of Dry Density of Soils, in places by the core cutter method, BIS, 1975.

Part 30.   Laboratory Vane Shear Test, BIS, 1980.

IS: 5529,   In situ Permeability Tests, BIS.

Lambe,   T.W. and Whitman R.V., Soil Mechanics, Wiley Eastern Ltd., 2nd ed. 1979.

Newmark, N.M. Influence Charts for Computation of Stresses in Elastic Foundations, Univ. of Illinois, Bulletin No. 338, 1942

Singh, A., Soil Engineering in Theory and Practice, Asia Publishing House, New Delhi, Vol I. 1975.

Skempton A.W., The Pore Pressure Coefficients A and B', Geotechnique, Institution of Civil Engineers, Vol. 4, pp. 143-147, 1954.

Skemption, A.W., and J. Hutchinson, Stability of Natural Slopes and Embankment Foundation, 7th INCSMFE, State of Art Volume, pp. 291-340, 1969.

Taylor, D.W., Fundamentals of Soil Mechanics, Asia Publishing House, New Delhi, 1948

Terzaghi, K., Erdbaunechanik, Franz Deuticke, Vienna, 1925.

Terzaghi, K. and R.B. Peck, soil Mechanics in Engineering Practice, 2nd ed. John Wiley & Sons, New York, 1967.

Unified Soil Classification System for Roads, Airfields, Embankments and Foundations, Military Standard; MIL-STD-619A, US. Deptt of Defence, Washington D.C., 1962

Westergaard, H.M., 'A problem of Elasticity Suggested by a Problem in Soil Mechanics: Soft Material Reinforced by Numerous Strong Horizontal Sheets' in contributions to the Mechanics of Solids, Stephen Timoshenko 60th Anniversary Volume, New York, 1938.

Woods, K.B. (ed.), Highway Engineering Handbook, McGraw-Hill Book Co., New York, 1959.

INDEX

A

Air content, 5
Atterberg limits, 29
Atterberg indices, 35
Activity number, 35
AASHTO system, 45
Alluvial soil, 61
Aeolin deposit, 62
Adsorbed water, 72
Allen Hazen's equation, 98
Angle of repose, 222

B

Bulk unit weight, 6
Bulk density, 6
Boulder clay, 62
Boussinesq analysis, 137
Bishop's method of slope stability analysis, 246

C

Coefficient of curvature, 27
Consistency of soils, 28
Consistency index, 35
Core cutter method. 37
Colluvial soil, 62
Coarse grained skeleton structure, 64
Cohesive matrix structure, 64
Clay minerals, 64,65
Colloid, 68, 72
Cumulose soil, 69
Calcareous soil, 69
Cemented soil, 69
Capillary water, 72,73
Contact moisture, 74
Coefficient of permeability, 87
Coefficient of percolation, 87
Compressibility, 164

Compaction, 164
Consolidation, 174
Coefficient of consolidation, 182

D

Degree of saturation, 5
Dry unit weight, 6
Dry density, 6
Density index, 39
Delta, 61
Dune sand, 62
Dispersed structure, 63
Diffuse double layer, 68
Darcy's law, 86
Discharge velocity, 86
Direct shear test, 204
Deviator stress, 209

E

Effective size, 26
Effective stress, 78
Electrical analogy, 130

F

Flow index, 35
Field density, 37
Flood plains,61
Flocculated structure, 63
Flow net, 115
Field compaction methods, 168
Field compaction control, 169
Failure envelope, 202
Friction circle method, 242
Fellenious method,249

G

Geo-technical engineering, 1
Grooving tool, 29
Group index, 45

ANSWER

2.2	16.81 kN/m³, 0.57, 0.36, 90%, 10.62 kN/m³
2.3	15.35 kN/m³, 0.73, 2.71, 18.68 kN/m³
2.4	151095 m³
2.6	0.68, 0.4, 71.4%, 11%
2.7	16.27 kN/m³, 19.98 kN/m³, 10.17 kN/m³, 18.13 kN/m³
2.8	15.86 kN/m³, 19.8 kN/m³, 10 kN/m³, 18.22 kN/m³
2.9	19.26 kN/m³, 0.35, 91.4%, 0.09
3.3	0.36 min., 899.1 min
3.4	47.6 gm, 0.0077 mm, 50.42%
3.5	11.9%
3.8	1.7 gm/cc, 1.52 gm/cc, 42.4%, 1.95 gm/cc
3.9	1.74 gm/cc
3.10	57.6%
6.4	183.8 kN/m²
6.5	No change
7.1	0.0134 cm /sec, 103.5 sec
7.2	4.21 cm/ sec
7.3	27% increase
7.5	30×10^{-4} cm/sec, 26.4×10^{-4} cm/sec
7.6	3.02×10^{-4} m/sec
7.7	0.72 cm
7.8	2.48×10^{-3} m/sec, 7.31×10^{-7} m/sec
8.2	(i) 90 kN/m², 49.05 kN/m², 40.95 kN/m²
	(ii) 90 kN/m², 73.57 kn/M², 16.43 kN/m²
8.3	0.026 m³/sec
8.4	66.6°
8.5	(a) 3m (b) 0.71 (c) 2.16×10^{-2} m/day
9.1	3.76 kN/m²
10.3	1.12m³
11.2	1.63
11.4	0.24
11.5	0.075 mm²/N
11.6	527 mm
11.7	453 days 16.9 hrs, 725 days 22.2 hrs
12.9	52.5 kN/m², 20° (for $\sigma_3 = 50$ kN/m² in triaxial shear test)
12.10	B = 0.64, A = 0.55
13.6	1.27
13.10	4.47, 3.46